6G Communication Network

This book focuses on 6G technology beyond 5G. The objective of next generation 6G wireless communications is to improve the benchmarks while simultaneously delivering additional services. Many widely expected future services, such as life-critical services and wireless brain-computer interactions, will be important to their success. This book presents the evolution of 6G technology, architecture, and implementation. This book provides a comprehensive overview of the theoretical and experimental modelling of 6G communication, providing detailed implementation issues and performance evaluation of emerging technologies along with research results, and networking methods.

This book:

- Contains a comprehensive overview of 6G communication network technology.
- Contains both fundamental theories and cutting-edge technologies.
- Covers implementations, architecture, security, privacy, and reliability in 6G communication and performance analysis of the 6G communication.
- Future trends and applications covered include vehicle-to-everything, massive radio access networks, Massive IoT Access Cybertwin, Sustainable Society 5.0 using 6G communication.
- Covers the challenges and research directions to enable future research to make 6G communication a wireless solution for sustainability.
- Contains a comprehensive overview of 6G communication network technology.
- Contains both fundamental theories and cutting-edge technologies.
- Covers implementations, architecture, security, privacy, and reliability in 6G communication and performance analysis of the 6G communication.
- Future trends and applications covered include vehicle-to-everything, massive radio access networks, Massive IoT Access Cybertwin, Sustainable Society 5.0 using 6G communication.
- Covers the challenges and research directions to enable future research to make 6G communication a wireless solution for sustainability.

This book is primarily written for senior undergraduate students, graduate students, and academic researchers in the fields of electrical engineering, electronics, communications engineering, and computer science and engineering.

Prospects in Networking and Communications – P-NetCom
Series Editor:
Mohammad M. Banat

Low-Power Wide Area Network for Large Scale Internet of Things:
Architectures, Communication Protocols, and Recent Trends
*Mariyam Ouaissa, Mariya Ouaissa, Inam Ullah Khan, Zakaria Boulouard,
and Junaid Rashid*

Internet of Things: A Hardware Development Perspective
Ayoub Khan

6G Communication Network: Architecture, Security and Applications
Ajay Kumar Vyas, Narendra Khatri, and Sunil Kumar Jha

For more information about this series, please visit: https://www.routledge.
com/Prospects-in-Networking-and-Communications/book-series/NETCOM

6G Communication Network

Architecture, Security and Applications

Edited by
Ajay Kumar Vyas, Narendra Khatri, and
Sunil Kumar Jha

CRC Press
Taylor & Francis Group
Boca Raton London New York

CRC Press is an imprint of the
Taylor & Francis Group, an **informa** business

First edition published 2025
by CRC Press
2385 NW Executive Center Drive, Suite 320, Boca Raton FL 33431

and by CRC Press
4 Park Square, Milton Park, Abingdon, Oxon, OX14 4RN

CRC Press is an imprint of Taylor & Francis Group, LLC

© 2025 selection and editorial matter, Ajay Kumar Vyas, Narendra Khatri and Sunil Kumar Jha; individual chapters, the contributors

Library of Congress Cataloging-in-Publication Data
Names: Vyas, Ajay Kumar, author, editor. | Khatri, Narendra, author,
editor. | Jha, Sunil Kumar (Researcher in physics), author, editor.
Title: 6G communication network : architecture, security and applications /
edited by Ajay Kumar Vyas, Narendra Khatri, Sunil Kumar Jha.
Description: First edition. | Boca Raton, FL : CRC Press, 2024. |
Series: Prospects in networking and communications - p-netcom |
Includes bibliographical references and index.
Identifiers: LCCN 2024025792 (print) | LCCN 2024025793 (ebook) |
ISBN 9781032563985 (hardback) | ISBN 9781032862484 (paperback) |
ISBN 9781003522003 (ebook)
Subjects: LCSH: 6G mobile communication systems.
Classification: LCC TK5103.252 .A15 2024 (print) | LCC TK5103.252 (ebook) |
DDC 621.3845—dc23/eng/20240801
LC record available at https://lccn.loc.gov/2024025792
LC ebook record available at https://lccn.loc.gov/2024025793

ISBN: 9781032563985 (hbk)
ISBN: 9781032862484 (pbk)
ISBN: 9781003522003 (ebk)

DOI: 10.1201/9781003522003

Typeset in Sabon
by codeMantra

This book is dedication to the late Shri Chhagan Lal Vyas & Late Smt. Jank Devi Khatri and our dearest family, whose unwavering support has been our guiding light throughout this journey. To our friends, for their encouragement and belief in our creative endeavors. And to all those who find refuge in the pages of this book, may it ignite a spark of joy, wonder, and introspection in your hearts.

Contents

Part I
Architecture

Preface

This book provides an overview of 6G communication and a perspective on the next-generation network. It includes 6G communication opportunities and challenges. It also discusses the opportunities and challenges of the emergence of smart sixth-generation (6G) networks. Implementation and Architecture of 6G Communication: Some chapters discuss various applications of 6G communication in different. The other parts of this book discuss the device-to-device communication system, viz. antenna, band, and gain for a 6G communication network. It also outlines the aim of the sixth-generation communication network, which is to provide revolutionary wireless communication services and applications to fulfill future demands in the 2030s while encouraging environmentally sustainable development. Other chapters provide a systematic overview of security and privacy challenges based on 6G physical, connection, and service layer technologies and lessons learned from current security architectures and state-of-the-art countermeasures. This book provides a good understanding of 6G communication and key enabling technologies, standards, components, terminology, and real-time applications.

This book focuses on three major parts: (i) architecture, (ii) security, and (iii) applications. This book is comprehensive material of 20 chapters.

Chapter 1 presents the background of 6G communication, the primary architecture of the 6G model, and challenges to implantation and briefly describes the perspective in the real-time scenario. This chapter also discusses the present research status of 6G technology, constraints, and solutions.

Chapter 2 focuses on next-generation wireless communication using intelligent reflecting surfaces. The wireless communication technology in its next generation, 6G and beyond, foresees to revolutionize the way we connect and communicate in the future. One of the key features of 6G and beyond communications project is the availability of spectrum in the order of terahertz (THz). However, it comes with several challenges including dominant line-of-sight components and large fading effects for wireless channels. Apparently, intelligent reflecting surface (IRS) is one of the best solution to resolve these challenges. This chapter discusses the theoretical frameworks, mathematical models, and performance metrics for

IRS-assisted wireless communications. It also highlights the current limitations and research directions for interested readers.

Chapter 3 discusses enabler technologies, challenges, and future research directions. The development of mobile communication technologies, particularly wireless communication, is accelerating and will likely continue for some time. This progression is being pushed forward by factors such as an expanding customer base's ever-increasing demand as well as the advent of new applications, increased traffic rates, and more complex data services.

Chapter 4 provides a comprehensive review that delves into the multifaceted aspects of 6G communication networks, covering their architectural foundations, diverse applications, security imperatives, and the transformative integration of antennas with machine learning. This chapter, a comprehensive overview, illuminates the fascinating landscape of 6G communication networks. It showcases the architectural ingenuity, myriad applications, security imperatives, and the transformative integration of antennas with machine learning that collectively propel us toward the next era of connectivity. The possibilities are vast, and as the journey to 6G unfolds, this review serves as a guiding beacon for researchers, engineers, and innovators embracing the frontier of wireless communication.

Chapter 5 explores digital twin architectures tailored for the forthcoming 6G communication network and delves into their prospective advancements in digital twin architectures for the 6G communication network and their future directions. This chapter delineates the imperative role of digital twin technology in orchestrating the intricate design and operation of 6G networks, propelled by real-time interaction between physical systems and their virtual counterparts. Further, this chapter delves into the pivotal role of Digital Twins across various 6G network use cases, encompassing simulation, planning, operation, and management. It elucidates the significance of AI-driven training, inference, and what-if-analysis in optimizing network performance and resilience.

Chapter 6 presents device-to-device communication for 6G communication. Device-to-device (D2D) communication is one of the most exciting developments in wireless cellular networks. It can increase data throughput, reduce connection latency, and improve energy and spectral efficiency. It also aids in network optimization by enhancing data throughput, reducing latency, and permitting more efficient use of resources. It contributes to improving the network as a whole, increasing data throughput, reducing latency, and permitting more efficient use of resources. To fulfill the needs of many new applications, the sixth-generation (6G) mobile network of the future is expected to be highly dynamic, ultra-dense dispersed, and inherently intelligent. It will also transport data at high speeds with very low latency. To develop a workable implementation of intelligent D2D in future 6G, this chapter shows several potential D2D communication trends related to 6G in terms of mobile edge computing, slicers for networks, and non-orthogonal multiple access cognitive networking.

Chapter 7 introduces intelligent reflecting surface (IRS) assisted wireless system design and analysis. IRS has gained significant attention in the field of wireless communications for high energy efficiency next-generation network design. IRS-assisted systems can overcome path loss and shadowing effects by focusing and redirecting the wireless signals toward the desired receivers. Design and analysis of an IRS-assisted wireless system require a deep understanding of wireless propagation, channel modelling, and optimization techniques. By properly designing the IRS, it is possible to improve the performance of wireless systems and achieve higher data rates, longer coverage, and better reliability. Some examples of IRS-assisted systems are also discussed.

Chapter 8 provides a survey on IoT networks enabled via 6G wireless networks. Firstly, the breakthrough IoT technologies are detailed followed by the details of technological drivers foreseen to advance future IoT networks. Next, relevant IoT use cases and applications are presented with details of the specific role of 6G systems. Lastly, many research challenges are highlighted, and a list of the potential research gaps is presented for further research in this area.

Chapter 9 delivers an overview of 6G communications security issues and possible solutions. The developments in new generation Information and Communication Technologies such as AI, VR/AR/XR, IoT, blockchain technology, etc. The 6G has intelligent communication with intelligent deep ubiquitous and holographic connectivity. The challenges in security may be encountered with requirements, key performance indicators, novelty in network architecture, novel applications, enabling technologies, etc. The security solutions provide physical layer security, DLT, distributed AI/ML, quantum communication, etc.

Chapter 10 describes the security challenges for 6G technologies. This chapter delves into the critical aspects of security and quantum computing in the context of 6G communication networks. With the advent of 6G, novel security challenges emerge due to ultra-dense heterogeneous networks and diverse connected devices, necessitating robust solutions across multiple trust domains. Additionally, this chapter discusses the role of deep learning and machine learning in enhancing network intelligence, with a focus on real-time mixed signal processing and the optimization of various network layers for improved data transmission and reception.

Chapter 11 begins with an overview of 6G communication, placing the development of cellular communication from the second to the sixth generation in context. It explores the unique qualities and difficulties presented by this next generation, highlighting the vital role that 6G plays in fulfilling the growing needs for ultra-high data speeds, low latency, and ubiquitous connection. The discussion surrounding 6G highlights security is a top need. To protect user privacy, secure the network environment, reduce cybersecurity threats, and maintain secure key protocols, strong procedures are required, as this chapter outlines many privacy and security problems that come with 6G communication.

Chapter 12 examines the significant consequences and unresolved issues about privacy rights and security protections in the context of emergent high-speed 6G networked environments. This chapter begins by situating the revolutionary enhancements in speeds, capacities, and novel use cases that 6G intends to introduce in comparison to the current 5G infrastructure. Further, the authors explicate how other linchpin 6G capabilities, including pervasive sensing, intelligent coding, and integrated blockchain architectures, contribute to the emergence of new cyberattack surfaces and risk vectors. They describe in detail, using recent instances as illustrations, the various categories of privacy-threatening threats—from sophisticated inference attacks to covert malware.

In Chapter 13, the authors delve into the critical aspect of Cluster Head (CH) selection in 6G IoT networks to optimize communication efficiency and reliability. They propose a novel method, Taylor-PO, integrating particle swarm Optimization (PO) and Taylor series, for CH selection and multi-path routing. The method prioritizes attributes such as energy, delay, and network lifetime, enhancing network longevity and throughput. They present comprehensive experimental analyses across varying network scales, demonstrating the efficacy of Taylor-PO in optimizing delay, energy consumption, and throughput. Their findings underscore the pivotal role of CH selection in shaping the performance and resilience of 6G IoT networks, laying a robust foundation for future advancements in connectivity.

Chapter 14 explores 6G wireless networks for V2X communication. This chapter presents an overview of the challenges and opportunities of using 6G networks for vehicle communication. Firstly, the key features of 6G networks, including ultra-high data rates, ultra-low latency, massive connectivity, and ubiquitous coverage have been discussed. This chapter emphasizes the importance of standardization and interoperability to enable seamless communication across different systems and devices. Emerging technologies such as machine learning, edge computing, and quantum communications find their way in the development of 6G networks making it suitable for the design of autonomous vehicles.

Chapter 15 presents a massive IoT access through 6G that can be used to create smart services in a range of fields, especially transportation, healthcare, and smart cities, which can be made feasible by an increasingly rapidly evolving technology known as the Internet of Things (IoT). This surge in IoT devices also holds implications for the forthcoming sixth-generation (6G) systems. This chapter also discusses securing 6G IoT networks against emerging threats, particularly from inventive attackers, which presents a formidable challenge for infrastructure providers. This example underscores the resilience of artificial intelligence (AI) and key networking factors such as software-driven networking (SDN) and network function virtualization

(NFV) as pivotal elements in devising promising solutions for securing the Internet of Things.

Chapter 16 presents the dynamic test technique of analog-to-digital converter (ADC) for 6G communication. As the speed of ADCs continues to increase, more and more sophisticated applications including medical images, 60 GHz wireless communications, wireless military services, 5G/6G wireless communication networks, and conative radar can benefit from the use of ADCs and digital signal processing. The performance of the system using ADC depends on test conditions for that particular application. Four techniques namely, the histogram technique, Fast Fourier Transform (FFT), sine wave curve fitting technique, and Beat frequency test technique are common for finding out the functional parameters of an ADC. This knowledge will benefit young researchers, designers, and manufacturers in the modern digital world.

Chapter 17 describes intelligent traffic management using VANET simulations with 6G for smart traffic management. This chapter aims to underscore the importance of integrating VANETs and 6G communication technology within the framework of intelligent traffic management. Special emphasis is placed on exploring the practical application of Python programming, machine learning methodologies, and computer vision techniques to enhance traffic flow management. Through a comprehensive examination of these technologies, this chapter seeks to demonstrate their significant potential to revolutionize the transportation industry and elevate the efficacy of traffic control measures.

Chapter 18 proposes a gallium nitride (GaN)-based design approach of a class F −1 power amplifier (PA) suitable for wireless power transfer (WPT) applications in 6G communication. To validate the efficacy of the proposed PA, the optimized output is used for simulations, fabrication, and measurements of a proposed PA. The achieved measurement results of PA for DE of 58.4%, PAE of 55.1%, Pout of 38.21 dBm at 28 dBm Pin, and Gain of 10.2 dB, IMD3 of −32.69 dBc, IMD5 of −64.01 dBc respectively. The measured output of GaN-based PA gives the prominent efficiency and output power required for WPT applications. This proposed work ensures the highest possible WPT system efficacy for advancing efficiency and reliability in the field.

Chapter 19 specifies the 6G wireless communication as prospective technological advances. In the telecom sector today, network evolution to serve the 6G vision, use cases, and requirements within the 2030-time frame is a topic of great interest. Both in the telecom sector and academia, 6G research is well advanced, with a planned commercial deployment date of 2030. This provides a comprehensive analysis of 6G systems with the way that changes in society and lifestyle are what are driving the demand for next-generation networks. Subsequently, it delves into the technical prerequisites for the implementation. This chapter presents the analysis of the principal obstacles and opportunities for workable system solutions in every tier of the Open Systems Interconnection hierarchy, spanning from applications to the physical layer. The use of frequencies between 100 GHz and 1 THz becomes crucial since many 6G applications will require access to an order of magnitude more spectrum. The details of fundamental modifications that will be needed in the future core networks, including the large reduction or redesign of the transport architecture, which is a major source of latency for applications that need to respond quickly.

Chapter 20 delves into the exciting realm of 6G technology and its potential applications and use cases. 6G, or the sixth generation of wireless communication technology, is expected to bring about revolutionary developments that will influence many facets of life. This chapter examines the many implications of 6G and how they affect many industries, from ultra-fast networking to holographic communication and sophisticated transportation systems. It also looks at the possible advantages in public safety, sustainability, healthcare, and the Internet of Things (IoT). Readers will thoroughly grasp the enormous potential of 6G and its social ramifications by the end of this chapter.

Ajay Kumar Vyas, Ph.D.

Narendra Khatri, Ph.D., PostDoc.

Sunil Jha, Ph.D.

Endorsements

I commend editors' effort in compiling topical research in the area of 6G communication network. The articles are thoughtfully placed in three broad sections: Architecture, Security, and Applications. In my view, this book is a comprehensive, yet accessible resource to the researchers and professionals exploring this fast-changing field.

Pancham Shukla, Imperial College, London, UK

This book will provide the experience of the dawn of a new era in connectivity with '6G Communication Network Architecture'. Dive deep into its resilient security strategies and visionary applications, heralding a transformative future. The chapters are witnessing the revolution unfold with '6G Communication Network Architecture'. Unravel its intricate security framework and groundbreaking applications, redefining the way we connect and communicate.

Celestine Iwendi, University of Bolton, UK

The book provides the bridge to stay ahead of the curve with our indispensable guide to 6G communication. This will also explore the nuances of Network Architecture, Security, and Applications and gain a competitive edge in the rapidly evolving tech landscape.

Valentina Emilia Balas, Aurel Vlaicu University of Arad Romania

Embark on an exhilarating journey into the future with '6G Communication Network: Architecture, Security, and Applications,' a visionary exploration of the indispensable role 6G communication plays in our intricately interconnected world. This book delves deep into the intricacies of next-generation networks, shedding light on their architecture, formidable security protocols, and game-changing applications. As 6G revolutionizes connectivity, its impact resonates across the global village, shaping industries, communities, and daily lives. Whether you're a tech enthusiast, business leader, or curious mind, this book serves as an indispensable guide to understanding and embracing the transformative power of 6G communication for generations to come.

Vincent Olema, Kyambogo University, Uganda

This book will discover the next phase of connectivity evolution with '6G Communication Network Architecture'. Explore its robust security mechanisms and transformative applications, defining the future of communication.

Patrick Acheampong, Ghana Communication Technology University, Ghana

It is anticipated that 6G networks will operate at a considerably higher speed than their predecessors, support a wider range of applications, and facilitate utilization scenarios that extend beyond immediate messaging, intelligent systems, and the Internet of Things (IoT).

6G internet is expected to launch commercially in 2030. It will provide all individuals with a transformative experience by facilitating hyperconnectivity between all entities and individuals. Furthermore, it is anticipated to expand the scope of mobile communication in areas where previous generations were unable to advance. It is anticipated that a few prospective technologies will form the bedrock of 6G networks. Post-quantum cryptography, artificial intelligence (AI), machine learning (ML), enhanced edge computing, molecular communication, THz, visible light communication (VLC), and distributed ledger (DL) technologies such as blockchain are among these emerging and existing technologies. From the standpoint of security and privacy, these advancements necessitate a reevaluation of previous security protocols. Anti-malicious activity detection, access control, authentication, and encryption must all be updated to meet the increasingly stringent requirements of future networks. Furthermore, novel security methodologies are required to guarantee confidentiality and reliability.

Against this backdrop, the present edited volume entitled, "6G Communication Network: Architecture, Security, Applications", by Dr. Ajay Kumar Vyas, Dr. Narendra Khatri and Dr. Sunil Jha, focuses on three sections, Architecture of IoT enabled 6G Technology involving digital twins, Security involved in 6G communication, technologies and its Applications. The book explores the cutting-edge advancements in 6G communication networks. it provides the understanding of the intricate architecture and protocols shaping the future of connectivity. The chapters are also focus on dive into security and real time applications. This material discover how 6G is poised to revolutionize industries and redefine global connectivity standards. The 6G network should be a deep integration of currently emerging AI tools and networking functions. Moreover, as issues on the security and privacy. It is believed that the present edited volume will bring out new dimensions for different sectors of IoT to become sustainable in the way ahead.

Sandeep Poddar, Deputy Vice Chancellor (Research & Innovation), Lincoln University College, Malaysia

Ignite your curiosity and expand your knowledge with our illuminating book on 6G Network Architecture, Security, and Applications. This book is compressive studies from theoretical foundations to practical insights, this is the ultimate resource for aspiring innovators.

Muthupandian Cheralathan Warsaw University of Technology, Poland

6G is a futuristic technology in which India aims to be a leader. Ajay's book lays the foundation of technical know-how for 6G implementation which will go a long way in preparing youth and professionals with requisite knowledge and skills for its roll out in India and across the world. Dr. Vyas is an avid researcher and this being his seventh book will add feather to his glowing career. All the best!

Ravi Prakash Singh, Provost Adani University, India

About the editors

Sunil Kumar Jha is an Associate Professor and Associate Dean of Research at Adani University, Ahmedabad, India. He received a B.Sc. and M.Sc. in Physics from the VBS Purvanchal University, India, in 2003 and 2005, respectively, and a Ph.D. in Physics from Banaras Hindu University, India, in 2012. From 2012 to 2014, he was a postdoctoral researcher at Kyushu University, Japan, and Hanyang University, Korea. From February 2015 to September 2017, he worked as an Assistant Professor at the University of Information Science and Technology, Macedonia, and the University of Information Technology and Management, Poland. Currently, he is working as an Associate Professor at the Nanjing University of Information Science and Technology. He is the author of two book chapters and more than 80+ SCI articles in the journal of repute. He is actively working as a reviewer of several journals IEEE, Elsevier, and Springer. His research interests include chemical sensors, machine learning, data mining, pattern recognition applications, and health informatics. He worked as publication chair and program committee member of several international conferences.

Narendra Khatri is an Assistant Professor at the Department of Mechatronics, Manipal Institute of Technology, Manipal, India. He worked as a postdoctoral research associate (Agri-drones) at the World Bank and the Indian Council of Agricultural Research, a New Delhi-sponsored project under NAHEP-CAAST-DFSRDA, Centre of Excellence for Digital Farming Solution for Enhancing Productivity Using Robots, Drones, and AGVs at VNMKV, Parbhani, India. He was awarded a research fellowship on a project titled "Smart Embedded Control System for Energy Independent Wastewater Treatment Plant". Dr. Khatri is a Senior Member of IEEE, LM-ISTE, and Faculty Advisor IEEE Robotics and Automation Society Student Branch Manipal. He has published various International Journal articles and conference papers. He has been granted with an Indian patent, nine Indian design registrations, and two Indian patents published. He is a peer reviewer of IEEE Systems, Elsevier's Water Research, Environmental Pollution and Bioavailability, Journal of Water Process Engineering, MDPI's

Sustainability, Agriculture, etc. Dr. Khatri was awarded the best paper presentation award at IEEE conferences viz. ICCIKE-2023 and ICATCS-2023. He has also worked as publication chair (RCAAI-2022, MIT Manipal) and publicity chair (ICIBT-2022, BIT Mesra, Ranchi) for international conference.

Ajay Kumar Vyas is an Associate Professor & Head of Department, at the Department of Information and Communication Technology, Adani University, Ahmedabad, India. He is a fellow member of IETE, a senior member of IEEE, a life member of ISTE, a senior member of IACSIT (Singapore) and the Internet Society (USA), and a senior member of the Institute of Research Engineers and Doctors, USA. He is a certified peer reviewer with the Elsevier Researcher Academy and the Publons Academy. He is a reviewer and editorial board member of several peer-reviewed journals. He earned his Ph.D. from Maharana Pratap University of Agriculture and Technology, Udaipur, India. He is the author of several research papers in peer-reviewed international journals and conferences, book texts on Digital Electronics published by Book India Publication, and many book chapters. Dr. Vyas is a distinguished expert serving as a Topical Advisory Panel Member of Photonics Journal. Dr. Vyas's multifaceted contributions across various prestigious journals underscore their profound expertise and commitment to advancing knowledge and innovation in their respective domains. Dr. Vyas has also published five edited books on ML, blockchain for suitable development, AI of renewable energy systems, RF Circuits For 5G Applications: Contemporary Trends & Opportunities, and Precision Agriculture for Sustainability Use of Smart Sensors, Actuators, and Decision Support Systems relevance with international Publishers.

Contributor lists

Mohd Mohsin Ali
Bennett University
Greater Noida, Uttar Pradesh,
    India

Ajay Sudhir Bale
New Horizon College of
    Engineering
Bengaluru, KA, India

Gaurav Bhargava
National Institute of Technology
    (NIT) Meghalaya
Shillong, Meghalaya, India

Punitkumar Bhavsar
Visvesvaraya National Institute of
    Technology (VNIT)
Nagpur, Maharashtra, India

Sheeja Pon Chakravarthy
Coimbatore Institute of Technology
Tamil Nadu, India-641028

Nitesh Chouhan
MLV Textile & Engineering
    College
Bhilwara, Rajasthan, India

Subrata Chowdhury
Sreenivasa Institute of Technology
    and Management Studies
Chittoor, Andhra Pradesh, India

Saumya Das
Sikkim Manipal University
Gangtok, Sikkim, India

Sreejith L. Das
Sri Venkateswara College of
    Engineering and Technology
Chittoor, Andhra Pradesh, India

Dhivya Priya E. L.
Erode Sengunthar Engineering
    College
Erode, Tamil Nadu, India

Dharmendra Dixit
Motilal Nehru Nation Institute of
    Technology
Allahabad, Uttar Pradesh, India

Radheshyam Gamad
Shri G. S. Institute of Technology
    & Science
Indore, M.P., India

Anish Kumar Gupta
Visvesvaraya National Institute of
    Technology (VNIT)
Nagpur, Maharashtra, India

Shilpi Gupta
Sardar Vallabhbhai National
    Institute of Technology
Surat, Gujrat, India

Sridhar Iyer
Department of CSE(AI),
    KLEMSSCET
Belagavi, KA, India-590008

Manish Jain
Mandsaur University
Mandsuar, M.P., India

Rachit Jain
Madhav Institute of Technology &
    Science
Gwalior, M.P., India

Sunil Jha
Adani University
Ahmedabad, Gujarat, India

Rakhee Kallimani
Department of EE, KLEMSSCET
Belagavi, KA, India-590008

Manivel Kandasamy
Karnavati University
Ahmedabad, Gujarat, India

K. Kavitha
Kumaraguru College of Technology
Coimbatore, Tamil Nadu, India

C. Kavya
Sri Padmavathi Mahila Viswa
    Vidyalayam
Tirupati, Andhra Pradesh, India

Hameed Khan
Jabalpur Engineering College
Jabalpur, M.P., India

Rajashri Khanai
Department of CSE, KLEMSSCET
Belagavi, KA, India-590008

Narendra Khatri
Manipal Institute of Technology
Manipal Academy of Higher
    Education
Manipal, India

Awanit Kumar
Sangam University
Bhilwara, Rajasthan, India

Naresh Kumar
Kurukshetra University
Kurukshetra, Haryana, India

Kamal Kumar Kushwah
Jabalpur Engineering College
Jabalpur, M.P., India

Shubhankar Majumdar
National Institute of Technology
    (NIT) Meghalaya
Shillong, Meghalaya, India

Swati Mavinkattimath
Department of ECE, KLEMSSCET
Belagavi, KA, India-590008

D. Nageswari
Nehru Institute of Engineering and
    Technology
Coimbatore, Tamil Nadu, India

Nageswari P.
Mahendra Institute of Technology
Namakkal, Tamilnadu, India

V. Jayachandra Naidu
Sri Venkateswara College of
    Engineering and Technology
Chittoor, Andhra Pradesh, India

Neelima K.
Mohan Babu University
Tirupati, Andhra Pradesh, India

Krishna Pai
Department of ECE, KLEMSSCET
Belagavi, KA, India-590008

Digvijay Pandey
Department of Technical
    Education
Kanpur, Uttar Pradesh, India

Rahul J. Pandya
Department of EE, Indian Institute
    of Technology, Dharwad,
    WALMI Campus
Dharwad, KA, India-580011

Neelash Patel
Guru Ramdas Khalsa Engineering
    College
Jabalpur, M.P., India

Rucha Patel
Adani University
Ahmedabad, Gujarat, India

Om Prakash
Sri Venkateswara College of
    Engineering and Technology
Chittoor, Andhra Pradesh, India

Ayushman Pranav
Bennett University
Greater Noida, Uttar Pradesh,
    India

Krupa Purohit
Nirma University
Ahmedabad, Gujarat, India

Ramya Radhakrishnan
Nirma University
Ahmedabad, Gujarat, India

Manish Raj
Bennett University
Greater Noida, Uttar Pradesh,
    India

Monika Sahu
Guru Ramdas Khalsa Engineering
    College
Jabalpur, M.P., India

Raju Shanmugam
Karnavati University
Ahmedabad, Gujarat, India

Sanjeev Sharma
Indian Institute of Technology
    (BHU)
Varanasi, Uttar Pradesh, India

Sharmila A.
Bannari Amman Institute of
    Technology
Sathyamangalam, Tamil Nadu,
    India

Pramod Kumar Singhal
Madhav Institute of Technology &
    Science
Gwalior, M.P., India

Palakshi Sinha
Karnavati University
Ahmedabad, Gujarat, India

Rekha Garg Solanki
Dr. Hari Singh Gour University
Sagar, India

T. Somassoundaram
Sri Venkateswara College of
    Engineering and Technology
Chittoor, Andhra Pradesh, India

Suresh Chinnathampy M.
Francis Xavier Engineering College
Tirunelveli, Tamil Nadu, India

Vandana Vikas Thakare
Madhav Institute of Technology &
    Science
Gwalior, M.P., India

Dattaprasad A. Torse
Department of ECE, KLEMSSCET
Belagavi, KA, India-590008

Ajay Kumar Vyas
Adani University
Ahmedabad, Gujarat, India

Acknowledgements

We would like to express our gratitude to all those who contributed to the creation of this edited volume. Firstly, we extend our heartfelt appreciation to the contributors who generously shared their expertise and insights.

We are especially thankful to all the contributors for their invaluable contributions to individual chapters, which have greatly enriched this collection.

We are indebted to all reviewer for their guidance and support throughout the editing process. Their expertise and encouragement have been instrumental in shaping this book.

We would also like to thank the CRC Press production team for their assistance in proofreading and editing the manuscript.

Furthermore, we are grateful to Adani University, Ahmedabad, and Manipal Institute of Technology, Manipal Academy of Higher Education, Manipal for their support, which made this project possible.

Lastly, we express our sincere appreciation to our families for their patience, understanding, and unwavering support during this endeavor.

Thank you all for your contributions and support.

Ajay Kumar Vyas, Ph.D.

Narendra Khatri, Ph.D.

Sunil Jha, Ph.D.

Part I

Architecture

An overview prospective of 6G communication

Ajay Kumar Vyas, Narendra Khatri, and Sunil Jha

1.1 INTRODUCTION

According to a report of Cybersecurity Ventures, projected Internet users will be 6 billion by 2022 and it is 75% of the world population and this figure will be increased up to 7.5 billion by 2030, which will be 90% of the projected world population. The report also shows that presently in a developed country like the USA, this numeral is 84% of the population which is connected on the Internet [1]. The communication network is a new wave of the technological ocean, which will accelerate the growth of economies and society through rapid digitalization. The telecom sector is looking for continuous improvement in the communication network in the form of different generation xG (i.e. 1G, 2G, 2.5G, 3G, 4G, 5G, and 6G). xG is the vision of the imminent wireless communication systems that empower a truthfully wireless Internet. The technologies developed under the xG vision will enable continued scaling of network capacity, low latency, radio flexibility, and high reliability [2]. The development stages of xG network are shown in Figure 1.1.

The release of 15 of 3GPP, which pertains to 5G mobile communications technology, signifies its potential to significantly enhance energy efficiency and serve as a crucial facilitator for advanced, integrated, and autonomous applications across various industries. It is anticipated that the next generation, known as "6G," will emerge by 2030 [3]. The expert predicted that the 6G technology market will facilitate substantive improvements in the areas of telecom, data science, image processing, speech processing, e-health, e-Commerce, e-Governs, sensing, and many solutions in human safety, education, agriculture, etc. 6G will incorporate cutting-edge technologies to meet communication needs that surpass the advancements of 5G. Increased frequencies will allow for faster sampling rates and improved accuracy, reaching measurements as precise as centimeters [4].

It is currently an opportune moment to analyze the forthcoming communication requirements, performance expectations, system and radio obstacles, and significant technical possibilities for the development of 6G. This will help set the research objectives for the next decade, leading up to the 2030s [5]. This study aims to explain the implementation and prediction of

DOI: 10.1201/9781003522003-2

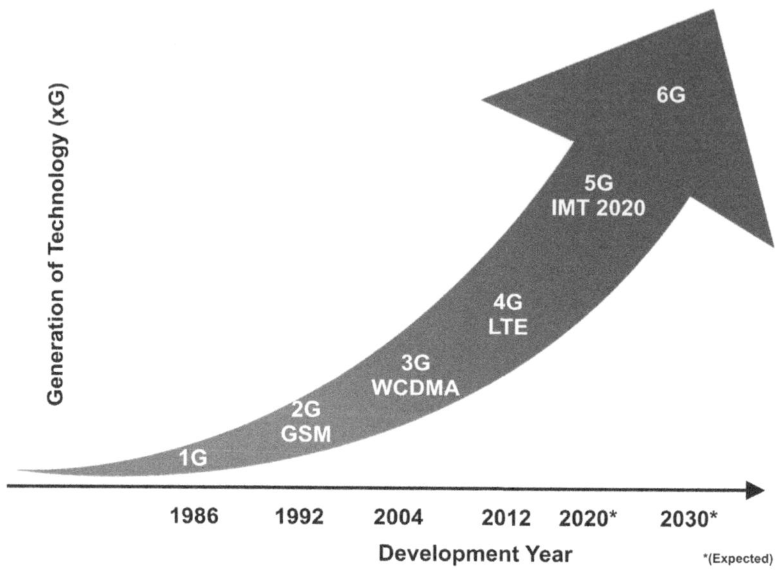

Figure 1.1 Evolution of *x*G communication networks.

the 6G communication network, which will propel the wireless community into the next generation (*x*G), beyond that of the 5G wireless.

1.1.1 Terminology

3GPP	Third-generation partnership project
AI	Artificial intelligence
AR/VR	Augmented reality/virtual reality
BCI	Brain computer interface
D2D	Device to device
DL	Deep learning
IMT 2020	International mobile telecommunications
IoE	Internet of everything
IoT	Internet of things
MIMO	Multiple input multiple output
M2M	Machine to machine
ML	Machine learning
NR	New radio
THz	Terahertz
VLC	Visible light communication
UMMIMO	Ultra massive MIMO
D-OMA	Delta-orthogonal multiple access
NOMA	Non-orthogonal multiple access
B5G	Beyond 5G

1.2 TECHNOLOGICAL BACKGROUND

We are expecting the implementation of 6G communication will take place within the next decade. This advanced form of communication will offer incredibly fast speed, increased capacity, and extremely low latency. It has the potential to support a wide range of new applications, including advanced medical treatments, intelligent disaster prediction, and immersive virtual reality experiences. The foundation of 6G builds upon the current 5G architecture, and it adheres to the previous guidelines for mobile network evolution in order to carry over the advantages of 5G [6].

The specification of 6G network expecting is mentioned in Table 1.1.

6G is considered to be a significant advancement beyond 5G, which will effectively address the severe limitations of the current access scheme in use. In the OFDMA scheme, the number of users who can simultaneously use the spectrum resources is limited by the number of available subchannels or resource blocks (RBs). Lately, there has been a lot of interest in a new technology that shows promise in addressing the challenges of OMA-based schemes.

Two multi-access techniques have been proposed for the 6G network: Delta-orthogonal multiple access (D-OMA) and nonorthogonal multiple access (NOMA) are shown in Figure 1.2. The NOMA is already used for 5G network, and the DOMA is relatively similar to NOMA, but in DOMA, it allows the overlapping with adjacent frequency bands with δ percentage overlapping (maximum value of allowed subband). NOMA has been viewed as a key enabler for catering to the inconveniences of OFDMA scheme in 5G networks and $\delta = 0$ corresponds to NOMA [7].

The transmission intensity for m-th user inside the cluster using D-OMA method [8] is as given in Equation (1.1)

$$R_m = B \log_2 \left(1 + \frac{\sum_{k=1}^{K} P_{m,k} |h_{m,k}|^2}{\sum_{k=1}^{K} \Lambda_k |h_{m,k}|^2 + \delta I_{\mathrm{ICI}} + N_m} \right)$$

Table 1.1 Expected specification for 6G

Parameter	Range
Data rate	>1,000 Gpbs
Switching	Packet switching
Modulation	UMMIMO
Network	Internet (CARN)
Standard	B5G
Frequency	THz range
Bandwidth	20 MHz

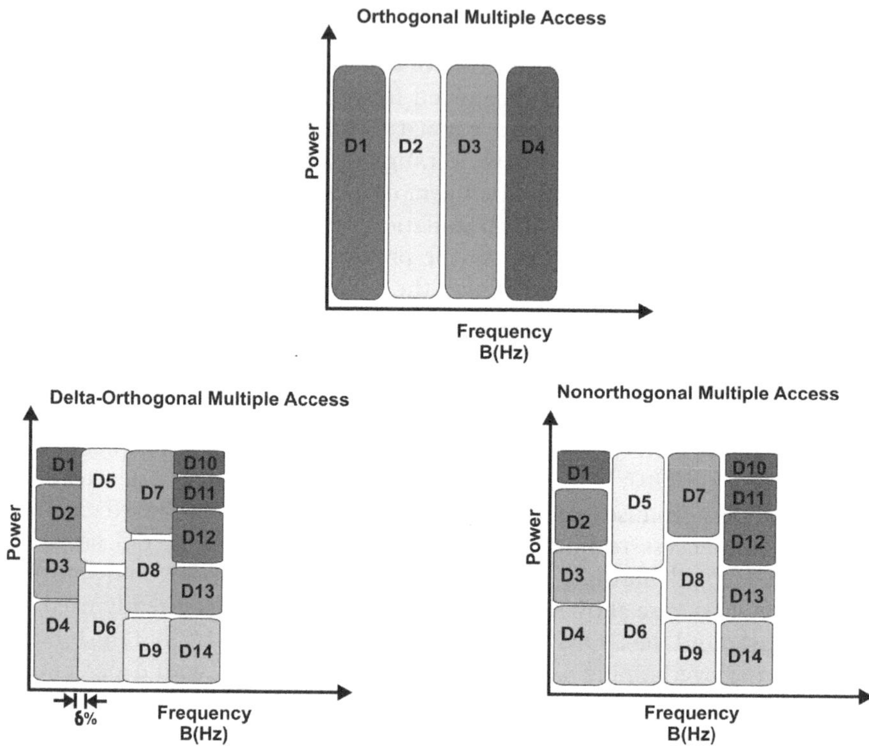

Figure 1.2 Two version of orthogonal multiple access: DOMA and NOMA.

And

$$\Lambda_k = \sum_{j=\beta+1}^{M} P_{j,k}, \, M \le M$$

Notation

K: Number of APs,
β: Degree of ordering rank of the mth device with respect to all APs,
M: Number of all Ads within the cluster,
B: Bandwidth of the channel transmission,
$P_{m,k}$: Transmission power of kth AP for mth of the device,
I_{ICI}: Interstation interference gives the value of internal interference in the D-OMA method,
N_m: Power of AWGN on the input mth device AD,
$h_{m,k}$: Channel gain between AP and mth device.

The transmission intensity for m-th user inside the cluster using N-OMA method [9] is as given in Equation (1.2)

$$R_m = B \log_2 \left(1 + \frac{\displaystyle\sum_{k=1}^{K} P_{m,k} \left| h_{m,k} \right|^2}{\displaystyle\sum_{k=1}^{K} \Lambda_k \left| h_{m,k} \right|^2 + \delta I_{\mathrm{ICI}} + N_m} \right) \tag{1.2}$$

$\delta = 0$ corresponds to traditional power-domain (e.g., the expanded approved frequency bands and the streamlined decentralized network architecture) and prodigiously changing the way we work and play.

For the 6G network architecture, the design of cluster is critical due to several factors to be considered such as data rate, interference, inter-cluster interference, error propagation channel capacity, etc.

1.3 CHALLENGES

The commercialization of 5G communication is still in its early stages, as it will require a comprehensive system approach that includes precise data analytics, artificial intelligence, and advanced computational capabilities through high-performance computing (HPC) and photonic computing. Definitely, the need beyond the 5G is unquestionable, but it is not easy to meet technical requirements and standards for the next generation of 6G. The 6G will provide the means of communications and data gathering necessary to accumulate information, but it cannot completely match the quickly rising technical requirements [10].

1.3.1 Infrastructure

When the sketch of new technology comes into the real picture, from then there are numerous issues open regarding the hardware infrastructure. The hardware constraints will play critical roles when designing an extremely dense network. As discussed earlier, the road map of 6G network includes connection extended reality, sensors and distributed ledger technologies, IoT devices, autonomous driving, smart city and smart grid applications, industrial control and smart manufacturing, surveillance and safety, and health-monitoring services. The sensors, antennas, and discrete components should be compatible with millimeter-wave bands, THz frequencies, nano-level and ultra-dense integration. Those devices and components should be operated at ultra-low power because the power consumption of hardware components will expressively affect the transceiver design and algorithm.

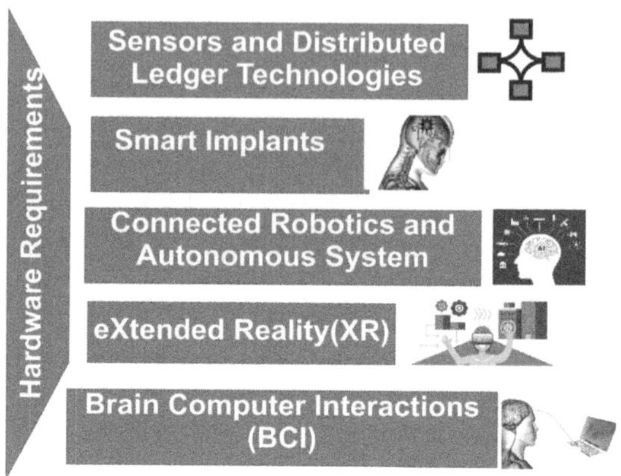

Figure 1.3 Info graph for hardware requirement for 6G network.

Such network needs an energy-efficient system for power management optimization. The hardware requirements of the 6G network are shown in Figure 1.3. One of the most important issues for such hardware infrastructure is the power supply.

Now, most smartphones can only be powered to last for one day, which narrows the bottleneck of telecom business providers. The new high-potential power supply system will be required based on hybrid and energy harvesting like the power over fiber (PoF) wireless electricity transmission [11].

1.3.2 Public safety issue

Over the past two years, more than 250 scientists from over 50 countries have published reports on the widespread and growing exposure to electromagnetic fields (EMF) emitted by electric and wireless devices, even prior to the introduction of 5G. It is very sure that it is "serious concerns," and when we have dealt with B5G (i.e. 6G), this issue will have unintended harmful consequences. The study of health effects on humans, plants, and animals of high-frequency radio waves is determined, and possible solutions to avoid these damages should be mentioned before technology implementation [12].

Several scientific journals have indicated that electromagnetic fields have an impact on living organisms even at levels that are lower than the regulatory standards set by many countries. There are various effects that can occur as a result of exposure to certain factors. These effects may include heightened cellular stress, an elevated risk of cancer, an increase in harmful free radicals, genetic damage, alterations in the structure and function of the reproductive system, impairments in learning and memory, neurological disorders, and overall negative impacts on human well-being. The detrimental impact extends far beyond the human population, as mounting evidence suggests that plants and animals are also experiencing harmful consequences [13].

The advent of THz band communications gives rise to substantial health and safety considerations owing to its potential effects on human tissues and biological systems. THz radiation is classified as non-ionizing within the electromagnetic spectrum. However, its ability to pass through biological tissues has led to inquiries regarding potential thermal effects and biological interactions. Continual research is being conducted to gain a deeper understanding of the possible hazards linked to extended exposure to THz radiation and to establish safety protocols for its utilization in communication technologies. It is of utmost importance to consider these concerns in order to guarantee the secure implementation and widespread use of THz band communications in a variety of applications.

1.3.3 Data security and privacy

The organizations around big data and cloud computing deal enormous amount of data, and information and such data and its security which includes intentional or accidental destruction, modification, or disclosure of data is the biggest challenge for 6G networks.

Security is of utmost importance for 6G wireless networks, particularly when utilizing the STIN technique. In the context of 6G, it is important to not only focus on traditional physical layer security but also take into account other forms of security, such as integrated network security. Payment apps, eWallet, health, food, and online shopping will collect extremely personal data about you. Vehicle maintenance services will track our mobility patterns, while intelligent urban applications will gather data on our daily routines. Social media, networking sites, and job portals are all internet sites getting user information and credentials. These devices are interconnected with other devices through the Internet of Things (IoT). Given the vast number of connected UEs and IoT devices, each with varying capabilities, the development of 6G will require a comprehensive strategy to ensure the security of the immense amount of mobile data across a wide range of platforms. Additionally, it will be crucial to meet stringent privacy and security standards. Blockchain technology is anticipated to have a significant impact on ensuring and verifying future mobile communication requirements. The modification or disclosure of data poses the greatest challenge for 6G networks [14].

6G networks will facilitate the development of novel applications and expand the capacity to connect a greater number of devices to the network, fostering the collection and exchange of an increasing amount of personal data. Some of which may not have been previously recorded in a digital format. As a result, it is important to conduct more in-depth research on new security approaches that prioritize simplicity and offer strong security measures.

1.3.4 Material

In Section 3.2, we discussed the required hardware infrastructure must be compatible with high frequencies. Antenna technology is witnessing a lot of activity in wireless products. In the 6G network, we are considering unlicensed wireless mmWave band of 60 GHz, and there is 7 GHz of bandwidth, thus the antenna design approaches necessitate accuracy to the given spectrum. The selection of material for such designs is done by considering the electrical and mechanical properties, and it should also be with high reliability and low energy consumption.

Conventional components are typically based on semiconductor materials like Silicon (Si) and Gallium Arsenide (GaAs). The heat dissipation in such material-based devices is more, and energy efficiency is less as compared CMOS-based devices. Due to limitations in the manufacturing process after 65 nm, the maximum operating frequency of CMOS transistors has not improved beyond approximately 300 GHz. As a result, it is extremely challenging to develop CMOS-integrated circuits for 300 GHz-band amplifiers. Consequently, their capabilities fall short when it comes to handling computationally demanding and lightning-fast applications for 6G. Efforts must be made to explore novel materials and design concepts in order to surpass the existing boundaries of UE and achieve operation across a broader range of frequencies. These advancements should also aim to enhance diversity, multiplexing gains, and intelligence [15].

1.3.5 Bandwidth

With the advancement of technology, there is a growing need for enhanced efficiency in various parameters. Technologists have expressed concerns about the bandwidth and have suggested enhancing the integration of space communications to increase bandwidth and improve the spatial acuity of the 6G system.

Looking ahead, it will be crucial to assess success based on the capacity to transmit bits per second per m^3, rather than just bits per second.

MIMO technique has a limitation of spatial multiplexing angle, and it is removed by the massive MIMO technique enabled in 5G. The high accuracy is expected beyond the 5G by using the ultra-massive MIMO as state of the art with 6G to individual users as shown in Figure 1.4. The large network is accomplished by numerous devices, integration of space, and terrestrial networks, and users interfacing which need to design novel sharing mechanisms and large bandwidth is required. The large bandwidth is compensated with the spectrum occupancy, which is associated with financial issues. The network with higher bandwidth suffers from traffic congestion, network attack, cross talk, energy consumption, and propagation losses due to high data rates [16].

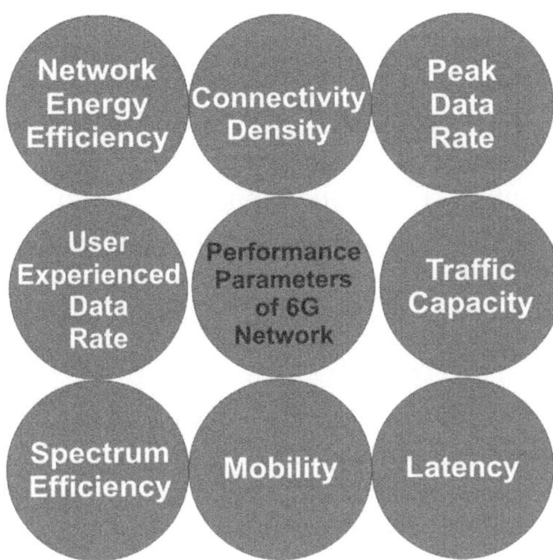

Figure 1.4 Performance parameters of 6G network.

1.4 PROSPECTS

1.4.1 Healthcare applications

The smart healthcare systems are integrated with 6G network. Remote access facilities will be helpful for patient monitoring and remote surgery. For the better diagnosis of syndromes, a large volume of data are required with low latency, which will be an offer by the ultra-reliable 6G network. The 6G network will be enhanced by cutting-edge technologies such as AR/VR, holographic telepresence, mobile edge computing, and AI devices. These advancements have the potential to greatly enhance access to healthcare and improve the overall quality of life [17]. As a result of this integration, patients will benefit from lower costs and remote surgeries due to remote access of skilled physicians. Additionally, by enabling seamless connectivity, technology integration improves the quality of the procedure or operation. In addition to ensuring that patients receive their care in the appropriate city, this also lessens turmoil for the specialty physicians.

1.4.2 Manufacturing

Every manufacturing unit desires for full automation and intelligence mechanism to increase financial health and improve the quality of the outcomes and has achieved it to a great extent with the help of AI and IoT-based systems. In the 6G network, IoT-based networks will be modified as IoE-based network, and this technique is to allow the control and monitors every device/system from remote stations with error-free data transfer without any data loss. The uses, operations, and troubleshooting of systems will be optimized, and it will increase the production ratio and decrease the failure.

The advanced network is driving the shift toward automation and data exchange in the industry. This is known as Industry 4.0 and encompasses various technologies such as the IoT, Industrial Internet of Things (IIoT), cyber-physical systems (CPS), cloud computing, cognitive computing, and AI. These technologies are enhancing manufacturing technologies and processes to add value [18].

1.4.3 Society 5.0

With the advent of Industry 4.0, we are now witnessing the emergence of Society 5.0, a society that surpasses the Information Society (Society 4.0) by leveraging cloud computing through the IoT. "A society that prioritizes the well-being of individuals while also addressing social issues through a highly integrated system that combines cyberspace and physical space" [19]. Society 5.0 is the integration of cyberspace and physical space. Social security, health care, food, pollution, cybercrime, industrialization, redistribution of wealth, and correction of regional inequalities are the social problems that can be reduced in Society 5.0 by using sensors-based network because it works on the big data explored by AI, and the investigation outcomes are useful to redesign society in a new form. The theme notion is a sustainable society, which entails safeguarding public safety and security as well as integrating sustainability into daily life by integrating diverse and sustainable technology. Although it is always a challenge for scientists and engineers to educate the public, there are still a number of defects and low awareness that are leading to financial crimes on the innocent public. However, in order to lessen it, the government is actively creating and putting into practice laws and technological solutions. Moreover, several cybercrime rules are being developed in order to regulate the frauds.

1.4.4 Transport

One more tremendous application of 6G network is the Internet of Vehicles (IoV). IoV is based on the distributed system and worked on the basis of the data of the connected vehicle and vehicular ad hoc networks (VANETs). There are some constraints of the exiting the vehicle tracking system which is based on GPS suffers from the deal zone in tunnels or underground.

A crucial goal of the Internet of Vehicles (IoV) is to enable seamless communication between vehicles, pedestrians, human drivers, other vehicles, roadside infrastructure, and fleet management systems in real time [20]. Improved speed and connectivity guarantee the deployment of several autonomous on-road and off-road vehicles. With the help of 6G connectivity, it will be possible to integrate more sophisticated AI algorithms into the automotive environment to facilitate communication between different sensors and actuators and to improve the control of autonomous vehicles.

1.4.5 Higher education

There is a need for the development of communication networks, with universities taking the lead. The implementation of advanced communication systems in higher education, particularly in universities, will bring about significant changes in the way universities operate and will greatly enhance student learning across various disciplines and at all levels [21]. Aside from the possibility of ensuring widespread access to experimental facilities and information distribution, the integration of 6G technology into education technology will result in diverse learning experiences. Even now, there is access to the distant server and instruments, but there is always concern about the quality and speed of such connectivity. Hence, integration of it may result in the establishment of different platforms that give researchers and students access to a variety of remotely accessible labs and instruments.

1.4.6 Smart agriculture

The applications of the sensor node network are not limited to monitoring the process of agronomy but also provide the widespread solutions for the problem accompanying with Agro-businesses by big data analytics. Different tools like IoT, machine learning, and edge computing are useful for data scientists and agro-engineers to study extreme weather conditions, environmental impact fertilizer, cedes, climate change soil, and water to improve the intensive farming practices. Such agriculture based on state-of-the-art technologies will support farmers and shareholders to moderate waste and enrich productivity [22]. Additionally, it brings in a new era of remote agricultural machinery monitoring and operation. Many intelligent precision agricultural robots and equipment are being created these days. To guarantee that, all farm operations are carried out effectively, and these can all be remotely managed and operated. By using a high-speed communication system to operate the machine efficiently, less manpower will be needed in the field, maximizing profit. Unfavorable weather conditions, peak harvesting times, etc. can be handled with ease, preventing significant losses from occurring, as in the past.

1.5 CONCLUSION

The concept of the connectivity of anywhere and anytime can be successfully implemented and optimized by the integration of the latest trends in the communication. It provides not only the progress in business but also we can enrich our culture, humanity, sociality, and lifestyle. The adoption of real-time applications in the 6G network presents a range of opportunities and challenges.

REFERENCES

[1] Morgan, Steve. "Humans on the Internet Will Triple from 2015 to 2022 and Hit 6 Billion." *Cybercrime Magazine*, July 2019. https://cybersecurityventures.com.

[2] Zhang, Lin, Ying-Chang Liang, and Dusit Niyato. "6G Visions: Mobile Ultra-Broadband, Super Internet-of-Things, and Artificial Intelligence." *China Communications* 16, no. 8 (2019): 1–14.

[3] Ghosh, Amitabha, Andreas Maeder, Matthew Baker, and Devaki Chandramouli. "5G Evolution: A View on 5G Cellular Technology Beyond 3GPP Release 15." *IEEE Access* 7 (2019): 127639–127651.

[4] Kalbande, Dhananjay, Sana Haji, and Rukhsar Haji. "6G-Next Gen Mobile Wireless Communication Approach." In *2019 3rd International Conference on Electronics, Communication and Aerospace Technology (ICECA)*, pp. 1–6. IEEE, India, 2019.

[5] Zhang, Zhengquan, Yue Xiao, Zheng Ma, Ming Xiao, Zhiguo Ding, Xianfu Lei, George K. Karagiannidis, and Pingzhi Fan. "6G Wireless Networks: Vision, Requirements, Architecture, and Key Technologies." *IEEE Vehicular Technology Magazine* 14, no. 3 (2019): 28–41.

[6] Rappaport, Theodore S., Yunchou Xing, Ojas Kanhere, Shihao Ju, Arjuna Madanayake, Soumyajit Mandal, Ahmed Alkhateeb, and Georgios C. Trichopoulos. "Wireless Communications and Applications above 100 GHz: Opportunities and Challenges for 6G and Beyond." *IEEE Access* 7 (2019): 78729–78757.

[7] Chen, Y., Bayesteh, A., Wu, Y., Ren, B., Kang, S., Sun, S., Xiong, Q., Qian, C., Yu, B., Ding, Z. and Wang, S. "Toward the Standardization of Non-orthogonal Multiple Access for Next Generation Wireless Networks." *IEEE Communications Magazine* 56, no. 3 (2018): 19–27.

[8] Al-Eryani, Yasser, and Ekram Hossain. "The D-OMA Method for Massive Multiple Access in 6G: Performance, Security, and Challenges." *IEEE Vehicular Technology Magazine* 14, no. 3 (2019): 92–99.

[9] Mohammadi, Mohammadali, Xiaoyan Shi, Batu K. Chalise, Zhiguo Ding, Himal A. Suraweera, Caijun Zhong, and John S. Thompson. "Full-Duplex Non-Orthogonal Multiple Access for Next Generation Wireless Systems." *IEEE Communications Magazine* 57, no. 5 (2019): 110–116.

[10] Yang, Ping, Yue Xiao, Ming Xiao, and Shaoqian Li. "6G Wireless Communications: Vision and Potential Techniques." *IEEE Network* 33, no. 4 (2019): 70–75.

[11] Vyas, Ajay Kumar. "Modified Power over Fiber Link Architecture for High Power Applications and Its Implementation Challenges." In *2018 International Conference on Advanced Computation and Telecommunication (ICACAT)*, pp. 1–5. IEEE, India, 2018.

[12] Zong, Baiqing, Chen Fan, Xiyu Wang, Xiangyang Duan, Baojie Wang, and Jianwei Wang. "6G Technologies: Key Drivers, Core Requirements, System Architectures, and Enabling Technologies." *IEEE Vehicular Technology Magazine* 14, no. 3 (2019): 18–27.

[13] Chowdhury, Mostafa Zaman, Md Shahjalal, Moh Hasan, and Yeong Min Jang. "The Role of Optical Wireless Communication Technologies in 5G/6G and IoT Solutions: Prospects, Directions, and Challenges." *Applied Sciences* 9, no. 20 (2019): 4367.

[14] Nawaz, Syed Junaid, Shree Krishna Sharma, Shurjeel Wyne, Mohammad N. Patwary, and Md Asaduzzaman. "Quantum Machine Learning for 6G Communication Networks: State-of-the-Art and Vision for the Future." *IEEE Access* 7 (2019): 46317–46350.

[15] Wu, Yanghui, Junjie Wang, Senfeng Lai, Xiaobo Zhu, and Wenhua Gu. "Transparent and Flexible Broadband Absorber for the Sub-6G Band of 5G Mobile Communication." *Optical Materials Express* 8, no. 11 (2018): 3351–3358.

[16] Xia, Qing, and Josep Miquel Jornet. "Expedited Neighbor Discovery in Directional Terahertz Communication Networks Enhanced by Antenna Side-Lobe Information." *IEEE Transactions on Vehicular Technology* 68, no. 8 (2019): 7804–7814.

[17] Tariq, Faisal, Muhammad Khandaker, Kai-Kit Wong, Muhammad Imran, Mehdi Bennis, and Merouane Debbah. "A Speculative Study on 6G." *arXiv preprint arXiv:1902.06700* (2019).

[18] Mahmood, Nurul Huda, Hirley Alves, Onel Alcaraz López, Mohammad Shehab, Diana P. Moya Osorio, and Matti Latva-aho. "Six Key Enablers for Machine Type Communication in 6G." *arXiv preprint arXiv:1903.05406* (2019).

[19] Sato, Y. "Japan pushing ahead with Society 5.0 to overcome chronic social challenges." (2019). https://www.gov-online.go.jp.

[20] Birgisson, Bjorn. "The Role of Modern Research Universities in Advancing Innovative Transportation Infrastructure Renewal." In Elizabeth Deakin (ed.), *Transportation, Land Use, and Environmental Planning*, pp. 555–568. Elsevier, 2020.

[21] Fernández-Caramés, Tiago M., and Paula Fraga-Lamas. "Towards Next Generation Teaching, Learning, and Context-Aware Applications for Higher Education: A Review on Blockchain, IoT, Fog and Edge Computing Enabled Smart Campuses and Universities." *Applied Sciences* 9, no. 21 (2019): 4479.

[22] Lovén, Lauri, Teemu Leppänen, Ella Peltonen, Juha Partala, Erkki Harjula, Pawani Porambage, Mika Ylianttila, and Jukka Riekki. "EdgeAI: A Vision for Distributed, Edge-Native Artificial Intelligence in Future 6G Networks." *The 1st 6G Wireless Summit* 1 (2019): 1–2.

Chapter 2

Next generation wireless communication using intelligent reflecting surfaces

Anish Kumar Gupta, Ajay Kumar Vyas, and Punitkumar Bhavsar

2.1 INTRODUCTION

Consumer demand for higher data rates and quality of service is increasingly transforming wireless communication networks. In response, 5G wireless communication technology is being deployed across the globe for its specific use cases. Thus, scientists are currently concentrating on 6G wireless communication. Higher data rates with reliability, low latency, worldwide coverage, and energy efficiency are anticipated to be the primary characteristics of 6G wireless networks [1,2]. To achieve these, sophisticated network equipment and effective wireless communication mechanisms are necessary [3]. Recent research has identified THz communication, artificial intelligence, and intelligent reflecting surfaces (IRSs) or programmable reflecting surfaces as the key concepts for 6G wireless communication. NTT DoCoMo, a Japanese mobile enterprise, and the smart radar company MetaWave have shown the use of IRS to improve data transfer in the 28 GHz band in November 2018 [4].

2.1.1 Specular or anomalous reflection

Specular reflection refers to the reflection of waves from the surface of the IRS in a single direction according to the law of reflection – "the angle of incidence is equal to the angle of reflection". Specular reflection is what occurs when the surface is smooth and flat, and the waves reflect off the surface in a well-defined manner.

On the other hand, anomalous reflection occurs when the wavefronts are reflected at different angles that deviate from the law of reflection due to the phase shift caused by the irregularities on the surface as shown in Figure 2.1. This phenomenon can be used to create multiple beams with different angles and polarizations to increase the capacity of the wireless channel.

DOI: 10.1201/9781003522003-3

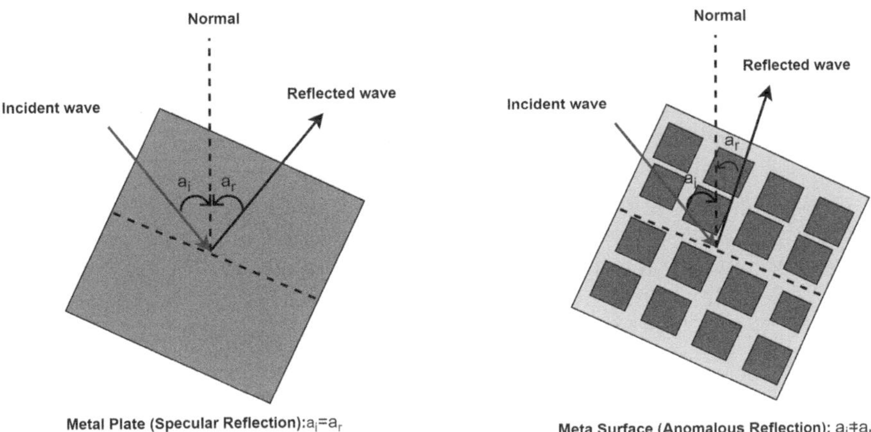

Figure 2.1 Illustration of specular and anomalous reflection.

2.1.2 Intelligent reflecting surface: principles and architecture

The IRS is a digitally controllable device that utilizes a two-dimensional meta-material, also known as meta-surface [5]. To function within the desired subwavelength frequency range, the surface is constructed using a two-dimensional arrangement of multiple meta-atoms, each carefully chosen to possess a specific electrical thickness [6]. Through careful engineering of the individual components, including their geometric characteristics (e.g. dimensions, orientation and spatial arrangement), it becomes possible to manipulate the signal properties of the IRS. This manipulation alters the reflection behavior by controlling the amplitudes and phase shifts of the incident wavefronts, thereby enabling control over the signal response. It is important to adjust the reflection coefficient of each element to accommodate the dynamic wireless channels caused by the mobility of user equipment (UE) in the wireless communication system. This allows the IRS to respond to changing channel conditions and optimize the signal transmission. The IRS efficiently addresses variations in the wireless channel resulting from the movement of UE through the dynamic manipulation of the reflection coefficients associated with each element. This ensures reliable and efficient communication between the transmitter and receiver.

Existing literature highlights three main approaches for controlling the reflection properties of IRS – (i) mechanical attenuation (ii) functional materials, and (iii) electronic devices. Mechanical attenuation involves physical manipulation of the IRS elements to modify the reflection behavior.

Techniques that involve mechanical translation and rotation of the individual elements can be employed to adjust the reflected signal phase. The overall reflection properties of the IRS can be controlled by mechanically altering the position or orientation of the elements. The use of specialized materials, such as liquid crystals and graphene, also offers another avenue for controlling the reflection of the IRS. These materials have unique properties that can be electrically or optically manipulated to change their refractive index or conductivity. By leveraging these material properties, the reflection characteristics of the IRS can be dynamically adjusted which allows flexible control of the signal propagation. Electronic devices, such as PIN diodes, MEMS switches, and FETs, provide electronic means of controlling the IRS reflection. These devices enable active control over the reflection coefficient of individual elements. By adjusting the biasing voltage, the reflection properties of the IRS can be modified to meet the requirements which offers precise control over signal reflection and propagation. Among these approaches, electronic devices have gained significant attention and adoption in controlling the reflection properties of IRS. This preference can be attributed to several advantageous characteristics including their quick response, minimum losses, and lower energy consumption. To achieve real-time adaptive reflection, it is essential to adjust reflection coefficients of the IRS elements to the network and gather information about the external communication environment [7].

Figure 2.2 illustrates a typical architecture of an IRS device having three layers. The initial layer of the setup consists of metal patches that are fixed on a substrate made of dielectric substance. These patches play a crucial role in manipulating the incident waves. A copper plate is positioned behind this layer to prevent any signal energy leakage during the reflection process.

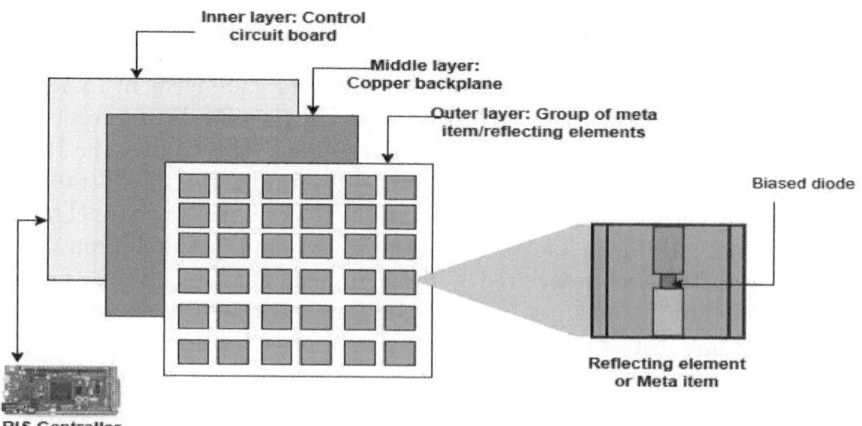

Figure 2.2 Structure of the reconfigurable reflecting surface and element that utilizes a PIN diode.

It ensures that the reflected signals remain contained within the desired propagation path. A control circuit is employed to control the reflection properties of each element. This board adjusts the reflection amplitude and phase shift of the individual elements by suitably selecting the electrical current and voltage levels. By modulating these parameters, the control circuit board enables precise control over the reflected signals. The smart controller is an integral component of the IRS. It is connected to the IRS and functions as the central control unit. The smart controller triggers the adoption of reflection by managing the "ON" or "OFF" condition of the IRS elements. By selectively activating or deactivating the individual elements, the smart controller effectively determines which signals are reflected and which are not.

Figure 2.2 presents a schematic of an IRS element structure, wherein each element contains a PIN diode placed at its midpoint. By manipulating the PIN diode voltage through a feed line, it is possible to control the diode's operational states (i) ON state (ii) OFF state. In turn, it affects the phase shift of the incoming signal. For instance, ON state allows current flow and results in 0 radians phase shift of the incident signal. Conversely, OFF state blocks the current flow and results in phase shift of π radians (180 degrees) of the incident signal. The smart controller regulates the biasing voltages, allowing for individual phase shift realization for each IRS element.

As mentioned in [8], the PIN diode can function at 5 megahertz (MHz) which corresponds to a time variation of 0.2 microseconds (μs), making it particularly suitable for mobile applications that involve time-varying wireless channels. In addition to phase shift control, amplitude control of every IRS element can also be used to enhance the strength of communication link. Furthermore, amplitude control typically offers a more cost-effective solution compared to phase control, and various methods, such as adjusting the load resistance in each IRS element [9], can achieve it in the IRS network. To achieve an effective reflection amplitude scale between 0 and 1, each element of IRS can vary its resistance to convert the incident energy of the signal into heat at a particular location. This is analogous to how passive radio frequency identification tags modulate the reflected signal power by changing their load impedance for data transmission. However, achieving an optimal reflection design in IRS systems often requires independent control of both the amplitude and phase shift of each element. Implementing this requires more sophisticated hardware compared to only having individual control. Ideally, each element can be tuned to adjust the phase and amplitude requirement to optimize the IRS reflection requirement.

2.1.3 Analogy with relay

IRS and relays are used to enhance wireless communication performance; however, both operate in different ways and are used for different applications. A relay retransmits the received signal from transmitter to a receiver, typically to extend the coverage area or to overcome obstacles that may attenuate the signal. Relays are active devices that require a power source to operate and perform signal-processing tasks such as amplification and filtering [10]. For instance, Amplify-and-Forward (AF) relay increases the amplitude of the received signal and sends it towards receiver. AF relays are usually employed in long-range wireless communication systems, where the signal is weak and needs intermediate amplification to retain the required signal-to-noise ratio (SNR) at the receiver. Decode-and-Forward type of relay decodes and again encodes the received signal before sending it to the receiver. This kind of relay is useful in wireless communication where error correction is important in the presence of noisy channels and fading effects. Several other types of relays such as frequency selective relays [11] are also important depending on the requirements of wireless networks.

Unlike relay, an IRS is a passive reflecting surface, typically with more number of small elements that can be configured to strengthen the reflected wave in a particular direction. Generally, phase shift from each element is achieved without the need for an external power source. The primary function of IRS is to strengthen the signal by improving SNR, reducing the interference, and increasing coverage area.

In addition, IRS does not require signal processing tasks such as amplification and filtering, as it reflects the incident signal using passive elements. Since the IRS does not generate or amplify signals, and it operates by modifying the properties of the electromagnetic waves that pass through it, it is classified as a passive element [12]. In contrast, a relay typically amplifies and filters the received signal before retransmitting it to the receiver.

2.2 WIRELESS CHANNEL MODEL WITH IRS

Mathematical model of wireless channels with IRS is useful to understand the impact of different parameters on the system performance and evaluate the effectiveness of different IRS designs and configurations. These models are used to develop efficient algorithms for beamforming, power allocation, and resource management that optimize the performance of the wireless communication system. However, the IRS channels are complex to model due to their dependencies on various factors, such as relative location of IRS, size, and configuration of the IRS, the frequency and bandwidth of the signal, and the mobility of the user. The IRS channel can affect the signal quality and reliability, the data rate, the latency, and the energy consumption of the wireless communication system [13,14].

Usually, wireless channel models consist of deterministic and statistical components. The deterministic part is described as a set of equations and is usually derived from the governing laws of electromagnetic wave propagation in addition to the knowledge of the physical world such as geometry of the reflecting surfaces. However, the statistical component represents the stochastic behavior of the wireless channel, it models the random components that arise due to fluctuations due to the multipath effects and scattering. The statistical component is derived from statistical methods and represented by the probability distribution of the received signal strength, delay, and phase shift [14].

2.2.1 Rayleigh versus Rician distributed channel

Any communication system model generally comprises a transmitter, a receiver and a communication medium, channel. Channel may be wired or wireless depending upon the type of communication system. Wireless channels are random in nature, which may change with time, frequency and geographical position. Various statistical channel models are used to illustrate the random behavior of the transmission medium on the signal which is benign passed through it. Rayleigh distribution, Rician or Rice distribution and Nakagami or Nakagami-m distribution are most common in use. Propagation environment demands a specific distribution, or in other words, which distribution is to be used depends primarily on the obstacles in between the transmitter and receiver end. In this chapter, we restrict our discussion on Rayleigh and Rice distributions.

The term 'fading' refers to the attenuation of the transmitted signal due to multipath propagation, weather conditions or shadowing. In wireless communication, transmitted signals get scattered when a number of objects are present in the propagation environment, thus multiple copies of the same signal are received at the destination. In general multiple copies encounter different attenuation and phase shifts while traveling. Signal gets attenuated, diffracted, refracted, reflected, and attenuated by many other objects.

Rayleigh distribution approximation is most appropriate when there are many scatterers between transmitter and receiver and the gain of multipath signals is uniformly distributed, i.e. none of the dominant paths is present. This distribution arises when the real and imaginary parts of a complex Gaussian random number are independently distributed with zero mean assumption. The magnitude of that complex number represents the Rayleigh distribution. Rayleigh fading is to be best fit in urban areas where due to heavily built-up cities there is no line of sight (LOS) between the transmitting and receiving ends.

Stochastic model is used for radio propagation when one of the received signals at the receiver reaches from different paths is dominant; typically the line of sight signal is a Rician distributed. The absolute value of non-zero mean (non-central) complex random variable is given by Rician distribution.

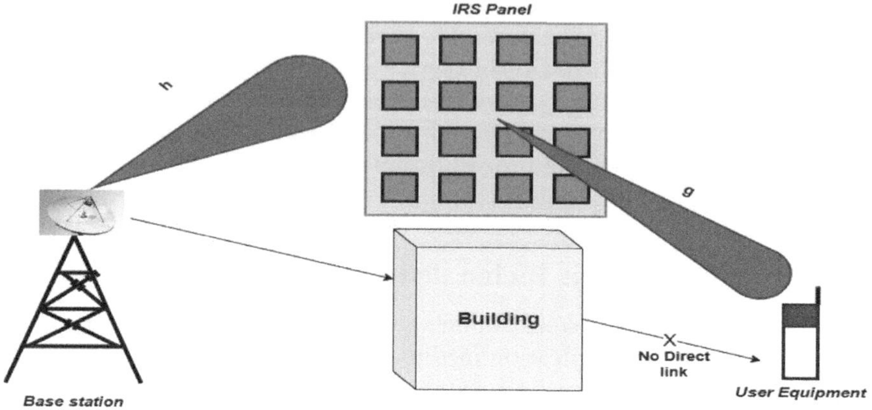

Figure 2.3 An IRS-assisted wireless communication system.

2.2.2 IRS channel model

As shown in Figure 2.3, a wireless communication system consists of a base station (BS), UE and an IRS which assist the transmission between BS and UE in the absence of direct path. Line of Sight (LOS) component is assumed to be present from BS to IRS and from IRS to UE. For simplicity, we consider the base station and user equipment are equipped with single transmitting and receiving antennas respectively, for more complex systems multiple antennas can be used at either transmitter or receiver or at both ends. Let us consider a squared shape IRS having $N = N_\mathrm{H} \times N_\mathrm{V}$ reflecting elements, where N_H denotes the number of elements in horizontal direction and N_V denotes the number of elements in vertical direction. IRS is a passive device, which can modify the phase of the impinging wave to direct it towards the destination.

The reflection coefficient of nth element is denoted by $\eta_n e^{j\theta_n}$, where η_n and θ_n denotes the amplitude and phase of nth reflecting elements respectively. Let $\Theta = \mathrm{diag}\left(\left[\eta_1 e^{j\theta_1}, \eta_2 e^{j\theta_2}, \ldots, \eta_N e^{j\theta_N}\right]\right)$ be the diagonal matrix of reflection coefficients. Consider, The complex channel vector between BS and IRS is $\bar{h} = [h_1, h_2, \ldots, h_N]$, and that between IRS and UE is $\bar{g} = [g_1, g_2, \ldots, g_N]$.

$h_i = |h_i| e^{j\alpha_i}$ and $g_i = |g_i| e^{j\beta_i}$ is the polar form representation of ith h and g channels as shown in Figure 2.3.

We can assume planar wavefront striking to the surface of IRS when the source is far away from the IRS. Anomalous reflection is required for IRS operation at higher frequencies, so the separation between elements should be the order of wavelength.

For the considered model let p_s be the transmitted power and then received signal at the receiver can be expressed as:

$$y_n = \sqrt{p_s}\,h_1\phi_1 g_1 s_n + \sqrt{p_s}\,h_2\phi_2 g_2 s_n + \cdots + \sqrt{p_s}\,h_N\phi_N g_N s_n + n$$

Or

$$y_n = \sum_{i=1}^{N} \sqrt{p_s}\,h_i\phi_i g_i s_n + n \tag{2.1}$$

The same signal can also be expressed in vector form as follows:

$$y_n = \sqrt{p_s}\,\bar{h}\Theta\bar{g}'s_n + n \tag{2.2}$$

where s_n be the nth transmitted symbol with $E\left[|s_n|^2\right]=1$, $E[.]$ denotes the expectation operation, $\phi_i=\eta_i e^{j\theta_i}$ be the reflection coefficient of ith IRS element, and $n \sim CN\left(0, \sigma_n^2\right)$ be the additive white Gaussian noise (AWGN) with zero mean and variance σ_n^2.

For simplicity, Equation (2.1) can be rewritten for $\eta_i = 1$, $i = 1, 2, \ldots, N$

$$y_n = \sum_{i=1}^{N} \sqrt{p_s}\,|h_i||g_i|\,e^{j(\theta_i+\alpha_i+\beta_i)} s_n + n \tag{2.3}$$

Equation (2.3) clearly indicates that the phases of different copies of message signal received at the receiver can be controlled by IRS, for optimal reception $\theta_i = -\left(\alpha_i + \beta_i\right)$.

In the optimal reception, the y_n is given by

$$y_n = \sum_{i=1}^{N} \sqrt{p_s}\,|h_i||g_i|\,s_n + n \tag{2.4}$$

According to Equation (2.4), the SNR at the receiver for decoding the transmitted symbol s_n is given by

$$\gamma_R = \frac{p_s \left(\sum_{i=1}^{N}|h_i||g_i|\right)^2}{\sigma_n^2} \tag{2.5}$$

where $\left(\sum_{i=1}^{N}|h_i||g_i|\right)^2$ is the channel gain depends on the number of IRS elements. It increases with the number of IRS elements. Hence IRS can be deployed in future 5G and 6G wireless communication for spectral and energy efficiency at very high frequencies. However, there are certain limitations of IRS that we will discuss later on in this chapter.

2.3 PERFORMANCE COMPARISON

IRS has emerged as a promising technology to strengthen the link of wireless communication systems, the individual channel links are usually characterized as independent and zero-mean Gaussian distributed. This assumption captures the stochastic nature of wireless channels, where the amplitudes experience multipath fading due to reflections and scattering in the environment. Next we show the system performance when utilized in conjunction with IRS by evaluating the outage probability and bit error rate (BER).

2.3.1 Outage probability

The outage probability represents the probability that the received signal power falls below a certain predefined threshold, indicating a failure to meet the desired quality of service. In IRS-assisted channels, the outage probability is affected by both the direct link and the reflected link. The combined effect of these links can be analyzed by considering the channel gains, path loss, and power allocation at the IRS. Analytical expressions can be derived to characterize the outage probability in different scenarios, such as varying the number of IRS elements, transmit power, and channel conditions.

Let's consider a communication system where an IRS is placed between transmitter and receiver. The channel between the transmitter and the IRS is denoted as h_i and the channel between the IRS and the receiver is denoted as g_i both h_i and g_i follow zero-mean Rayleigh fading, and they are i.i.d. across elements, i.e., $h_i \sim \mathrm{CN}\left(0, 2\sigma_{h_i}{}^2\right)$ and $g_i \sim \mathrm{CN}\left(0, 2\sigma_{g_i}{}^2\right)$.

Hence, Equation (2.4) can be rewritten as:

The signal received at the receiver in optimal condition is given as

$$y_n = \sum_{i=1}^{N} \sqrt{p_s} \, |h_i| |g_i| s_n + n \tag{2.6}$$

Or

$$y_n = \sqrt{p_s} X_{hg} s_n + n \tag{2.7}$$

where $X_{hg} = \sum_{i=1}^{N} |h_i| |g_i|$ is an effective channel between the transmitter and the receiver, considering the combined effects of the IRS elements.

The SNR in terms of channel gain is defined as

$$\gamma_R = \frac{p_s X_{hg}{}^2}{\sigma_n{}^2} \tag{2.8}$$

or

$$\gamma_R = \gamma_T X_{hg}{}^2 \tag{2.9}$$

where $\gamma_T = \dfrac{P_s}{\sigma_n{}^2}$ is transmit SNR and $X_{hg}{}^2 = \left(\sum_{i=1}^{N} |h_i||g_i|\right)^2$ is effective channel gain.

The outage probability is defined as the probability that the received signal power falls below a certain threshold and is expressed as:

$$P_{out} = P_r\left(\gamma_R \leq \gamma_{th}\right)$$

Or

$$P_{out} = P_r\left(\gamma_T X_{hg}{}^2 \leq \gamma_{th}\right)$$

Or

$$P_{out} = P_r\left(X_{hg} \leq \sqrt{\frac{\gamma_{th}}{\gamma_T}}\right)$$

P_{out} is the CDF of random variable, X_{hg}, i.e.

$$P_{out} = P_r\left(X_{hg} \leq \sqrt{\frac{\gamma_{th}}{\gamma_T}}\right) = F_{X_{hg}}\left(\sqrt{\frac{\gamma_{th}}{\gamma_T}}\right) \tag{2.10}$$

where γ_{th} is the threshold level, the minimum acceptable SNR for proper signal reception.

In IRS-assisted wireless communication systems, obtaining a closed-form expression for outage probability can be challenging due to the intricate interaction between the channels and the IRS. Numerical methods or simulations are often used to evaluate the outage probability performance in practical scenarios. However, the distribution of X_{hg} is made available using the moment matching method and approximated as gamma distribution with the following parameters:

$$k_{X_{hg}} = \frac{\mu_{X_{hg}}{}^2}{\sigma_{X_{hl}}{}^2} \text{ and } \omega_{X_{gh}} = \frac{\sigma_{X_{hg}}{}^2}{\mu_{X_{hg}}} \tag{2.11}$$

$k_{X_{hg}}$ is shape and $\omega_{X_{gh}}$ is the scale parameter.

Therefore, CDF of X_{hg} is obtained as

$$F_{X_{hg}}(x) = \frac{\gamma\left(k_{X_{hg}}, \dfrac{x}{\omega_{X_{hg}}}\right)}{\Gamma\left(k_{X_{hg}}\right)} \tag{2.12}$$

Or

$$F_{X_{hg}}\left(\sqrt{\frac{\gamma_{th}}{\gamma_T}}\right) = \frac{\gamma\left(k_{X_{hg}}, \dfrac{\sqrt{\dfrac{\gamma_{th}}{\gamma_T}}}{\omega_{X_{hg}}}\right)}{\Gamma\left(k_{X_{hg}}\right)} \tag{2.13}$$

Or

$$P_{\text{out}} = \frac{\gamma\left(k_{X_{hg}}, \dfrac{\sqrt{\dfrac{\gamma_{th}}{\gamma_T}}}{\omega_{X_{hg}}}\right)}{\Gamma\left(k_{X_{hg}}\right)} \tag{2.14}$$

where $\Gamma(.)$ is Gamma function and $\gamma(,.,)$ is lower incomplete Gamma function. The first moment $\mu_{X_{hg}}$ and variance $\sigma_{X_{hl}}{}^2$ of X_{hg} are evaluated as:

Mean of X_{hg}:

$$\mu_{X_{hg}} = E\{X_{hg}\} = E\left\{\sum_{i=1}^{N}|h_i||g_i|\right\} = \sum_{i=1}^{N}E\{|h_i||g_i|\}$$

$$= \sum_{i=1}^{N}E\{|h_i|\}E\{|g_i|\} \tag{2.15}$$

$$= \frac{\pi}{2}\sum_{i=1}^{N}\sigma_{h_i}\sigma_{g_i}$$

Here $|h_i|$ and $|g_i|$ follow the Rayleigh distribution with mean $\sigma\sqrt{\dfrac{\pi}{2}}$.

Variance of X_{hg}:

$$\sigma_{X_{hl}}{}^2 = E\{X_{hg}{}^2\} - \mu_{X_{hg}}{}^2 \tag{2.16}$$

where the second moment is

$$E\left\{X_{hg}^{2}\right\} = \sum_{i=1}^{N} E\left\{|h_{i}|^{2}\right\} E\left\{|g_{i}|^{2}\right\} + \sum_{i=1}^{N} \sum_{j=1, j \neq i}^{N} E\left\{|h_{i}||h_{j}|\right\} E\left\{|g_{i}||g_{j}|\right\}$$

$$= \sum_{i=1}^{N} E\left\{|h_{i}|^{2}\right\} E\left\{|g_{i}|^{2}\right\} + \sum_{i=1}^{N} \sum_{j=1, j \neq i}^{N} E\left\{|h_{i}|\right\} E\left\{|h_{j}|\right\} E\left\{|g_{i}|\right\} E\left\{|g_{j}|\right\}$$

$$(2.17)$$

where $E\left\{|h_{i}|^{2}\right\} E\left\{|g_{i}|^{2}\right\} = 2\sigma_{h_{i}}^{2} \cdot 2\sigma_{g_{i}}^{2} = 4\sigma_{h_{i}}^{2} \cdot \sigma_{g_{i}}^{2}$.

Substituting Equation (2.17) and Equation (2.15) in Equation (2.16), the variance is obtained.

2.3.2 Bit error rate

Bit error rate (BER) indicates the reliability and performance of wireless communication systems. It is a metric that quantifies the number of bits received in error compared to the total number of bits transmitted. It is the ratio of incorrectly received bits to the total number of bits sent in a communication system. BER quantifies the probability of erroneous bit transmission, considering the impact of channel fading and noise at the receiver. In IRS-assisted wireless channels, the BER can be analyzed by considering the joint effect of the direct and reflected paths, channel fading statistics, and the modulation scheme employed. By studying the BER, insights can be gained into the achievable data rates and impact of system parameters, such as transmit power, channel conditions and coding schemes.

Let us consider the BPSK modulation scheme for signal transmission. Rewrite Equation (2.9), the SNR is defined as

$$\gamma_{R} = \gamma_{T} X_{hg}^{2}$$

Or

$$\gamma_{R} = \gamma_{T} Y_{hg} \qquad (2.18)$$

where $Y_{hg} = X_{hg}^{2} = \left(\sum_{i=1}^{N} |h_{i}||g_{i}| \right)^{2}$ is a random variable, so γ_{R} is also a random variable.

The instantaneous Bit Error Rate (BER) is defined as:

$$P_{e} = Q\left(\sqrt{2\mathrm{SNR}_{R}}\right) = Q\left(\sqrt{2\gamma_{R}}\right) \qquad (2.19)$$

where $Q(.)$ is a Q function and can be defined as:

$$Q(x) = (1/2) \times \operatorname{erfc}\left(x/\sqrt{2}\right) \tag{2.20}$$

where $\operatorname{erfc}(x)$ is the complementary error function.

Now the average BER is evaluated to know the expected error performance of a communication system over various channel conditions it allows us to assess the system's reliability and determine the overall quality of transmitted signal.

The average BER is defined as:

$$\overline{P_e} = \int_0^\infty Q\left(\sqrt{2\gamma_R}\right) f_{\gamma_R}(\gamma_R) \cdot d(\gamma_R) \tag{2.21}$$

where $f_{\gamma_R}(\gamma_R)$ is the distribution or PDF of γ_R.

Like closed-form solutions for outage probability, the closed-form solution for average BER, integral may not always be possible, and numerical methods or approximations might be necessary for evaluation.

Since Y_{hg} is positive random variable hence $\gamma_R = \gamma_T Y_{hg}$ is also a positive random variable and their distributions can be approximated with gamma distributions by using the moment matching method.

2.4 SIMULATION RESULTS AND DISCUSSION

In this section, Monte-Carlo (MC) simulation results for average BER and outage probability (Pout) are presented to demonstrate its performance for different number of IRS elements (N) when a direct path is obstructed between base station and user equipment. This section also presents the analytical and MC simulation results for system outage probability for validating the derived expression.

We consider all Rayleigh fading channels for all links, carrier and system bandwidth BW = 100 MHz. We consider the free space path loss at a distance, $d = 100$ m, carrier frequency $f = 3$ GHz and length and width of each IRS element is $\frac{\lambda}{2}$, where λ is the wavelength of the transmitted signal. We also consider $N = 15, 20, 25$, respectively and noise power No = -93.8 dBm. We set Path loss exponent $\alpha = 2$ and for average value, 10^6 MC simulation points in MATLAB environment.

Figure 2.4 depicts the outage probability performance with varying transmit power levels and different number of IRS elements (i.e. $N = 15, 20,$ 25). The MC simulation results, represented by circle markers on the graph, provide system's performance under different configurations. The analytical calculations, represented by solid lines, are based on mathematical models and approximations. They offer theoretical estimations of the system's probability, accounting for transmit power, number of IRS elements, and channel characteristics. The graph shows a close alignment between the simulation results and the approximation curves validating the accuracy of the analytical model in predicting the system's performance. The plot also reveals the impact of transmit power on the system's performance. The outage probability decreases as transmit power increases from 0 to 10 dBm, indicating improved system performance and reliability. Results in the graph also show the system experiences notable improvements in outage performance with the increase in the number of IRS elements. A large number of IRS elements enables more efficient channel manipulation and signal enhancement, leading to a reduction in outage probability.

Figure 2.5 illustrates the average BER performance of an IRS-assisted wireless communication system under varying transmit power levels and different numbers of IRS elements ($N=15, 20, 25$). The x-axis represents the transmit power in dBm, while the y-axis denotes the average BER. The MC simulation results demonstrate consistent trends across the different cases. Increasing the transmit power leads to a significant reduction in BER for all values of N. Higher power levels result in improved system performance and lower error rates, indicating the positive impact of increased transmit power on the system's reliability.

As the transmit power increases from 0 to 30 dBm, there is a consistent decrease in BER across all cases of N. However, the graph reveals that the reduction in BER is more pronounced at lower transmit power levels, with diminishing returns observed as the power levels increase, i.e. the rate of reduction diminishes at higher power levels. The differences in BER performance can also be observed among the different values of N, where a large number of IRS elements tends to yield lower BER values for a given transmit power. This indicates that the number of IRS elements has an impact on the system's BER performance along with the transmit power.

In conclusion, Figures 2.4 and 2.5 demonstrate the relationship between transmit power and system outage probability and average BER performance respectively for different values of number of IRS elements. They provide valuable insights into the role of the number of IRS elements. By incorporating a greater number of IRS elements, the system can effectively combat fading and interference, resulting in reliability and lower BER and outage probability suggesting the system's performance benefits from a greater level of IRS assistance.

2.4.1 Challenges associated with IRS based communication system

It is evident to observe the benefits of IRS-based communication systems in terms of enhanced performance such as reduced outage and bit error rate. However, it poses several challenges to achieve performance gain which we describe next.

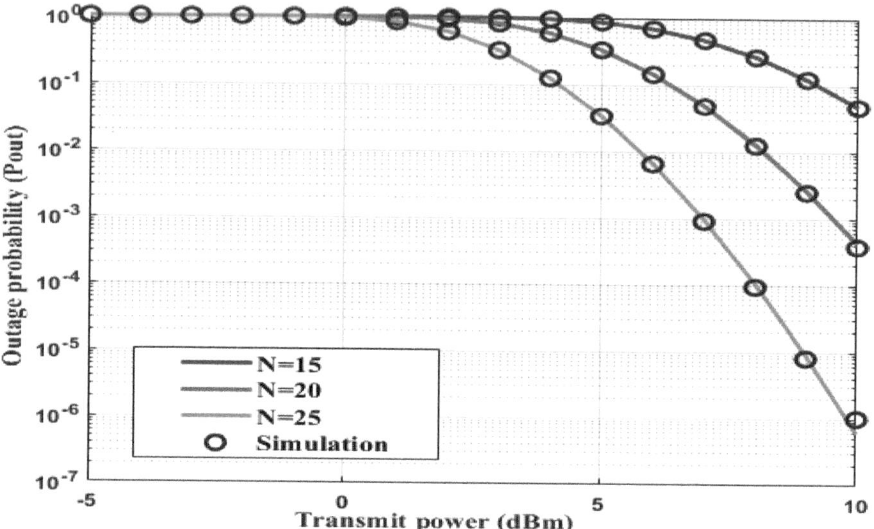

Figure 2.4 Outage probability versus transmit power for different values of *N*.

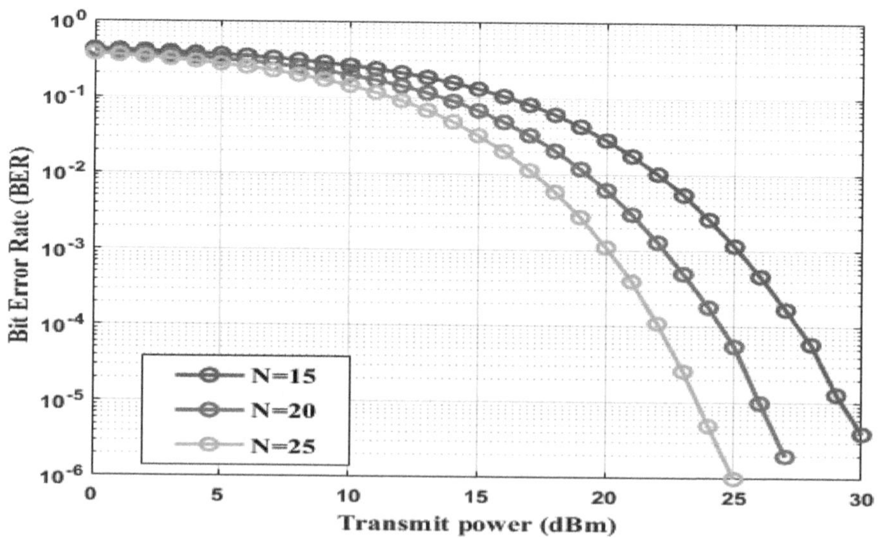

Figure 2.5 Average BER versus transmit power for different values of *N*.

1. *Channel Estimation*: Accurate channel estimation is required to calculate the phase contribution of each IRS element. Inaccuracy in channel estimation results in inaccurate phase modification which deviate the performance of the system from the optimality. Hence, developing efficient channel models and optimization algorithms that account for these factors is crucial [15].

2. *Dynamic Environments and Mobility*: Users and devices are allowed to move freely in wireless communication environments which require the real-time adoption and adaptation to the changing channel conditions. Developing mechanisms and algorithms to ensure IRS full beamforming gain of the system poses challenges related to estimation, hardware configuration, and control [16].

3. *Interference Management*: Multiple IRSs and other coexisting wireless systems lead to interference. Mitigation of disturbance from interference requires efficient interference cancellation and resource allocation techniques to improve overall system performance [17,18].

4. *Cost and Power Efficiency*: The implementation of IRSs involves deploying a large number of passive reflecting elements. Achieving cost and power efficient IRS designs is essential for its widespread adoption. These challenges require advancements in materials, fabrication techniques, and energy harvesting for IRS elements [19].

5. *Regulatory and Policy Considerations*: Regulatory and policy frameworks are essential for widespread deployment and operation of IRS-based systems. Addressing legal, privacy, and spectrum management aspects are important to ensure compliance and foster the adoption of IRS technology [20].

2.5 FUTURE DIRECTION

It is anticipated that similar to the impact of massive MIMO of 5G networks, 6G wireless communication networks may require large-scale deployment of IRS and other support infrastructures. To harness the benefits of IRS technologies, further research is necessary in the following directions.

1. *Validation on Real World Scenario*: Majority of the current literature on IRSs are based on theoretical analyses and simulations, it is crucial to validate these theoretical findings through real-world system implementations and experiments.

2. *Incorporation of Sophisticated Models*: Current models describing the behavior of IRSs in altering incoming waves are simplistic. However, an IRS's performance is influenced by its material properties and fabrication procedures. Developing models that account for these factors will enable more accurate optimization of IRSs to enhance wireless communications.

3. *Development of Scaling Laws*: The performance limits of IRS-assisted communication systems should be established on the basis of scaling laws and more sophisticated information theoretic frameworks. Addressing this question necessitates a comprehensive knowledge of IRSs' impact on conventional information theoretic frameworks of communication systems.

Further work in the above direction will advance the understanding and practical implementation of IRS technology which will unlock its full potential in enhancing wireless communication systems.

REFERENCES

1. Wild, T.; Braun, V.; Viswanathan, H. Joint design of communication and sensing for beyond 5G and 6G systems. *IEEE Access*, 2021, 9, 30845–30857, doi: 10.1109/ACCESS.2021.3059488.
2. Rajatheva, N.; Atzeni, I.; Bicais, S.; Bjornson, E.; Bourdoux, A.; Buzzi, S.; D'Andrea, C.; Dore, J.B.; Erkucuk, S.; Fuentes, M.; et al. Scoring the terabit/s goal: Broadband connectivity in 6G. *arXiv* 2020, arXiv:2008.07220.
3. Dang, S.; Amin, O.; Shihada, B.; Alouini, M.S. What should 6G be? *Nature Electronics*, 2020, 3, 20–29.
4. NTT DoCoMo and Metawave announce successful demonstration of 28GHz-band 5G using world's first meta-structure technology, https://www.businesswire.com/news/home/20181204005253/en/NTT-DOCOMO-and-Metawave-Announce-Successful-Demonstration-of-28GHz-Band-5G-Using-Worlds-First-Meta-Structure-Technology: Accessed 2 April, 2023.
5. Cui, T.J.; Qi, M.Q.; Wan, X.; Zhao, J.; Cheng, Q. Coding metamaterials, digital metamaterials and programmable metamaterials. *Light: Science & Applications*, 2014, 3, e218.
6. Liaskos, C.; Nie, S.; Tsioliaridou, A.; Pitsillides, A.; Ioannidis, S.; Akyildiz, I. A new wireless communication paradigm through software-controlled metasurfaces. *IEEE Communications Magazine*, 2018, 56, 162–169.
7. Wu, Q.; Zhang, S.; Zheng, B.; You, C.; Zhang, R. Intelligent reflecting surface-aided wireless communications: A tutorial. *IEEE Transactions on Communications*, 2021, 69, 3313–3351.
8. Zhang, L.; Chen, X.Q.; Liu, S.; Zhang, Q.; Zhao, J.; Dai, J.Y.; Bai, G.D.; Wan, X.; Cheng, Q.; Castaldi, G.; et al. Space-time-coding digital metasurfaces. *Nature Communications*, 2018, 9, 4334.
9. Yang, H.; Chen, X.; Yang, F.; Xu, S.; Cao, X.; Li, M.; Gao, J. Design of resistor-loaded reflectarray elements for both amplitude and phase control. *IEEE Antennas and Wireless Propagation Letters*, 2016, 16, 1159–1162.
10. Ni, W.; Collings, I.B.; Liu, R.P.; Chen, Z. Relay-assisted wireless communication systems in mining vehicle safety applications. *IEEE Transactions on Industrial Informatics*, 2013, 10(1), 615–627.
11. Deng, Q.; Klein, A.G. Relay selection in cooperative networks with frequency selective fading. *EURASIP Journal on Wireless Communications and Networking*, 2011, 2011, 1–16.
12. Wu, Q.; Zhang, R. Intelligent reflecting surface enhanced wireless network via joint active and passive beamforming. *IEEE Transactions on Wireless Communications*, 2019, 18(11), 5394–5409.
13. Zhao, J. A survey of intelligent reflecting surfaces (IRSs): Towards 6G wireless communication networks. *arXiv* 2019, arXiv:1907.04789.

14. Basar, E.; Di Renzo, M.; De Rosny, J.; Debbah, M.; Alouini, M.S.; Zhang, R. Wireless communications through reconfigurable intelligent surfaces. *IEEE Access*, 2019, 7, 116753–116773.
15. Zheng, B.; You, C.; Mei, W.; Zhang, R. A survey on channel estimation and practical passive beamforming design for intelligent reflecting surface aided wireless communications. *IEEE Communications Surveys & Tutorials*, 2022, 24(2), 1035–1071.
16. Huang, Z.; Zheng, B.; Zhang, R. Transforming fading channel from fast to slow: IRS-assisted high-mobility communication. In *ICC 2021-IEEE International Conference on Communications*,Montreal , pp. 1–6. IEEE, 2021.
17. Sur, S.N.; Bera, R. Intelligent reflecting surface assisted MIMO communication system: A review. *Physical Communication*, 2021, 47, 101386.
18. Xu, S., Liu, J.; Zhang, J. Resisting undesired signal through the IRS-based backscatter communication system. *IEEE Communications Letters*, 2021, 25(8), 2743–2747.
19. Imran, M.A.; Mohjazi, L.; Bariah, L.; Muhaidat, S., Cui, T.J.; Abbasi, Q.H., eds. *Intelligent Reconfigurable Surfaces (IRS) for Prospective 6G Wireless Networks*. John Wiley & Sons, 2023.
20. Rappaport, T.S.; Xing, Y.; Kanhere, O.; Ju, S.; Madanayake, A.; Mandal, S.; Alkhateeb, A.; Trichopoulos, G.C. Wireless communications and applications above 100 GHz: Opportunities and challenges for 6G and beyond. *IEEE Access*, 2019, 7, 78729–78757.

Chapter 3

Revaluation towards 6G

A survey of enabler technologies, challenges, and future research directions

Krupa Purohit and Rucha Patel

3.1 INTRODUCTION

The development of mobile communication technologies, particularly wireless communication, is accelerating and will likely continue for some time. This progression is being pushed forward by factors such as an expanding customer base's ever-increasing demand as well as the advent of new applications, increased traffic rates, and more complex data services. Interactions between individual nodes are only one example of the wide variety of uses that are needed for effective wireless technology. Some examples of these uses include the smart grid, smart homes, smart cities, and healthcare (Bethala, 2022).

For a unified wireless technology to be useful, it must be able to satisfy the many communication requirements posed by different applications. Only then can it be considered successful. The tremendous developments that are taking place in mobile communication technology are further highlighted by the growing popularity of mobile devices and the growth of the industries that are directly tied to mobile technology (Bhairanatti & Mohan Kumar, 2023). The continuous advancement of this subject is very necessary in order to satisfy the ever-changing needs of contemporary communication systems and to make it possible for diverse nodes and applications to connect in a smooth manner. It is anticipated that networks of the sixth generation, or 6G, would make immersive communications feasible and bridge the gap between the real and virtual worlds (Kaur et al., 2021). People's ways of working, having fun, and communicating will be completely transformed as a result of the advent of immersive communications, which will make realistic interactions possible and integrate augmented reality, holography, and haptics (Shen et al., 2023).

It is projected that the next communication technologies that will come after 5G and 6G will give services with improved system capacity, decreased latency, better dependability, increased spectrum efficiency, and seamless interaction with the Internet of Things (IoT). Network automation is going to be very necessary in order to accomplish these lofty ambitions. The administration and management of networks will become much more automated, which will pave the path for the development of intelligent networks (Figure 3.1).

DOI: 10.1201/9781003522003-4

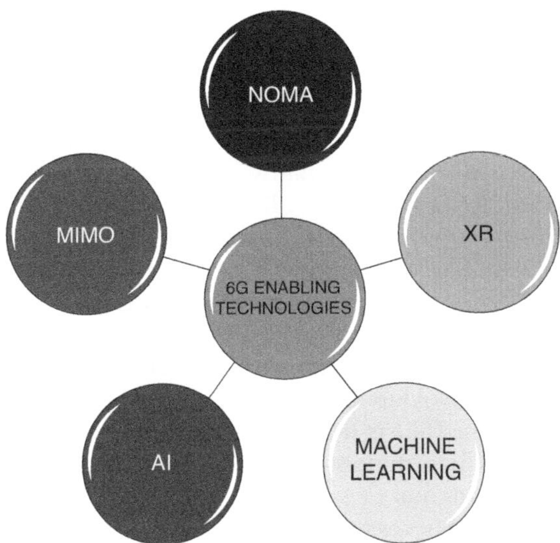

Figure 3.1 Key enabling technologies of 6G.

As a result of this, it is very necessary to include Artificial Intelligence (AI) and Machine Learning (ML) into the design of the infrastructure of future wireless networks. If AI and ML are included in the architecture of the network from the beginning, it will be able to support a wide range of activities, including spectrum sensing and sharing, slicing, radio resource management, and mobility management. These capabilities will be made possible by the network's design (Akyildiz & Guo 2022). This transition towards cognitive networks will result in network operations that are both more efficient and more adaptable.

It is anticipated that ML will radically alter the capabilities of 6G, with ML-based services eventually becoming an intrinsic part of the air interface. Existing roadmaps for the development of 6G technology, on the other hand, have a tendency to ignore the standardization timetable and the accompanying problems connected to the use of ML in 6G networks (Nawaz et al., 2019). While certain research initiatives do give insights into ML and standardization, the majority of these efforts fail to address the technological complexity underlying ML applications and its consequences for both the industry and the standardization procedures (Alsabah et al., 2021).

It is anticipated that forthcoming developments in sixth-generation (6G) networks will offer very high levels of connection, along with enhanced customer performance and service quality. Nevertheless, in order to achieve these capabilities (Shahjalal et al., 2022), it will be necessary to overcome considerable technical hurdles, in particular those connected to cutting-edge technologies using numerous antennas.

This study presents a thorough and understandable review of important Multiple-Input Multiple-Output (MIMO) technologies for 6G networks, such as massive MIMO (mMIMO), exceptionally large MIMO (XL-MIMO), Intelligent Reflecting Surfaces (IRS), and Cell-Free mMIMO (CF-mMIMO) (De Figueiredo, 2022). These technologies include massive MIMO (mMIMO), exceptionally large MIMO (XL-MIMO), and Intelligent Reflecting Surfaces (IRS). Because of the crucial part that these technologies play in conforming to the demanding criteria of 6G networks (Shehzad et al., 2022), it is imperative that they be implemented.

In the realm of 5G mobile technology, non-orthogonal multiple access (NOMA) stands as a vital component. NOMA enhances spectrum performance and facilitates a higher number of connections while reducing signalling time and associated costs. This is accomplished by allocating resources in a manner that is not orthogonal. Power NOMA adds a new power dimension, which makes it possible to multiplex inside the access domain. This is analogous to superimposing many signals onto a single orthogonal resource. Successive interference cancellation (SIC) becomes a crucial technique for extracting information from these superposed signals, eliminating duplicate signals based on related Channel State Information (CSI).

3.2 KEY ENABLING TECHNOLOGIES IN 6G

In contrast to 5G, which only offers a few information services concerning UE positioning and network information provision, 6G is anticipated to offer complete wireless sensing and positioning services and improve network information provision. Furthermore, 6G may gather industry-specific public data, including sensor data and Geographic Information System (GIS) data, to empower people from every aspect of life. This can prevent the information from being gathered repeatedly by several business apps. Wireless sensing, enhanced network information delivery, and industry-specific public information are all part of 6G's core service offering (Nawaz et al., 2019).

The enabling technologies employed in 6G are summarized in this section:

3.2.1 MIMO evolution

Massive MIMO is an essential part of 5G networks because it provides a major advantage over more conventional MIMO techniques. This benefit comes in the form of antenna arrays that are theoretically multiplied by an order of magnitude. This technique makes considerable improvements in both the efficiency of using spectrum and energy in communication networks (Shehzad et al., 2022).

The concept of very large XL-MIMO, which is based on huge MIMO, is introduced when even larger arrays of antennas are deployed at the base station. This is done in order to increase the capacity of the network. This might be a promising area of study for the multi-antenna technology that is used in 6G networks (De Figueiredo, 2022). It is possible for XL-MIMO to considerably improve the spectrum efficiency as well as the energy efficiency of some kinds of 6G network situations (Zhao et al., 2021). It is possible to use it in a wide variety of settings, such as motorways, shopping centres, runways, and building surfaces, and it may be set up in either a centralized or decentralized configuration over large regions. XL-MIMO offers non-stationary spatial characteristics, which, in comparison to conventional mMIMO systems, results in greatly higher throughput and spectrum efficiency. The increased coverage area made possible by the additional antennas installed at the base station makes it possible to provide service to a greater number of customers.

The Cell-Free CF-massive MIMO paradigm is yet another alternative technique for 6G networks. This paradigm does away with the need to have well-defined cell boundaries. In spite of this, it is still unknown whether or not CF-mMIMO is capable of meeting the stringent standards for energy economy and spatial accuracy, particularly in light of mMIMO and XL-MIMO. In addition, the viability of adopting mMIMO, XL-MIMO, and CF-mMIMO in sparsely inhabited regions is dubious owing to the increased deployment costs associated with the large number of antennas. It is possible that these charges may provide difficulties for operators in such places.

3.2.2 NOMA and 6G

The non-orthogonal multiple access, or NOMA, strategy is very important to the development of mobile technology. NOMA improves spectrum performance by using a non-orthogonal resource allocation, which permits bigger connections with less transmission latency and signalling costs (De Alwis et al., 2021). Within the current access domain, multiplexing may be achieved via the use of Power NOMA thanks to the addition of a new power dimension. This is akin to the process of superimposing many signals onto a single orthogonal resource. As a consequence of this, an essential method known as successive interference cancellation (SIC) is used in order to extract information from these superposed signals. This method involves the elimination of duplicate signals based on related Channel State Information (CSI).

NOMA is a promising new contender in the contest to create numerous access mechanisms for sixth-generation (6G) networks (Shah, 2022). The "One Basic Principle Plus Four New Ideas" proposal, which seeks to incorporate NOMA in 6G networks, is where the emphasis of this discussion is at the moment. When you begin with the basic notion of NOMA, the relevance of SIC becomes clear. SICs that are based on quality-of-service and channel-state information also have a large amount of relevance. The implementation of NOMA is now being researched by researchers as a potential means of meeting the increased performance criteria for 6G, in particular those pertaining to handling a high number of connections. In addition, they are researching the possibilities of using NOMA's capabilities in 6G by merging it with new physical layer technologies (Wang & Rahman, 2022).

In addition, the incorporation of ML into NOMA networks ushers in the age of "NGMA," which is characterized by the fact that ML gives NOMA networks additional capabilities (Liu et al., 2022). This combination of non-orthogonal multiple access (NOMA) and ML has a great deal of promise to significantly advance the capabilities of future wireless networks. NOMA networks have the potential to attain even higher levels of efficiency and flexibility via the use of ML, therefore opening the way for a revolutionary new age in wireless communication.

3.2.3 Sixth generation networks and the importance of AI and machine learning

When it comes to 6G wireless systems, dealing with complex optimization restrictions becomes very necessary. This is particularly true in cases where the network topology and application conditions are constantly changing. Two completely separate approaches are often used in order to overcome these difficulties.

The first method is known as analytical optimization, and it often entails either simplifying the complicated problem at hand by dividing it up into a number of more manageable issues or loosening the constraints so that the issue may be recast as a convex one. On the other hand, such simplifications can cause the final solution to deviate from the ideal one, which would then result in the conclusion not being what was expected (Shahjalal et al., 2022).

The use of AI/ML (that is, AI and ML) is the focus of the second approach. The majority of the challenges that arise in 6G networks due to their complex topology and nonlinear components may be efficiently addressed by using AI and ML technologies. Despite this, it is still necessary to carefully construct the AI/ML model, pick the proper instruction dataset, and apply strategies to assure convergence and accelerate the algorithm's rate of convergence (Shehzad et al., 2022).

6G wireless systems are able to overcome complex optimization issues and adapt to the dynamic nature of network topology and application situations because they use both analytical optimization and AI/ML approaches. This results in wireless communication solutions that are more efficient and effective as shown in Figure 3.2.

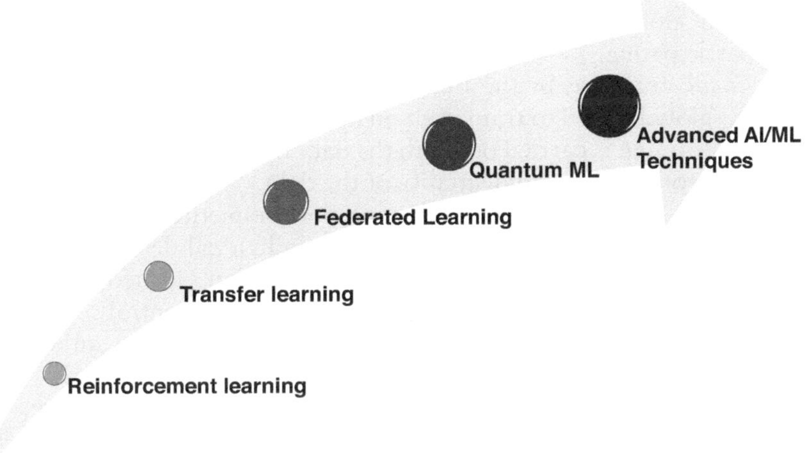

Figure 3.2 AI/ML techniques used in 6G.

1. *Reinforcement Learning (RL)*: RL, which stands for "reinforcement learning," is a ML approach that helps an agent learn to make choices by seeing how those decisions are received in context. This technique is often commonly referred to as "RL." This methodology is also recognized by its abbreviation. The purpose of RL is to teach the agent the optimal actions or strategies that will allow them to accumulate the most rewards possible over a period of time.

   In RL, the agent performs actions in the environment and earns rewards or penalties depending on how the environment responds to those actions. The goal of the agent is to determine which actions lead to the greatest benefits, and after it has accomplished this, it will modify its decision-making process so that it reflects this new knowledge (Kaur et al., 2021).

2. *Transfer Learning (TL)*: TL stands for the transfer of learned information from seasoned/source activities to fresh/target jobs. The key challenge in many real-world situations, especially ML-based techniques, is data efficiency. By using the gained information, TL may significantly enhance data efficiency (Shen et al., 2023).

3. *Federated Learning (FL)*: Federated Learning, often known as FL, is a technique to ML that is decentralized and enables several devices or edge nodes to train a single shared ML model jointly without exchanging raw data. It is designed to protect users' data privacy and security while also allowing a vast number of devices to work together to form a more intelligent whole.

The following activities are commonly included in the FL procedure: The initialization process begins when a central server starts up a global model and sends it out to all of the devices that are participating.

Local Training: In the local training process, each device uses the global model to train itself on its own local data. Because the local training is carried out with the data from the device in isolation, it protects the confidentiality of the data.

Aggregation of Models: Following the completion of the local training, the revised models from each individual device are uploaded to the centralized server. The information gleaned from each of the participating devices is combined by the central server, which then combines the resulting models to produce an enhanced global model.

Model Update: After that, the newly revised global model is redistributed to all of the devices, and the process of recursively repeating both the local training and model aggregation steps is carried out.

The following are some of the advantages of federated learning: FL guarantees that data privacy and secrecy are maintained since the raw data does not leave the individual devices where it was originally stored and is not transferred to the central server or any other devices.

Reduced Communication Overhead Because FL eliminates the need that vast volumes of data to be sent to a centralized server, both costs and delays associated with communication are significantly reduced.

FL supports distributed learning, which makes it suited for edge devices and instances where a centralized method would be difficult or wasteful. Decentralization is one of the key features of FL.

Better Model Generalization The global model tends to generalize better across different data distributions as a result of its training on a variety of data gleaned from many devices because of this (Kaur et al., 2021).

4. *Quantum Machine Learning (QML)*: Quantum Machine Learning, or QML for short, is a relatively new area of research that investigates the interaction of quantum computers with various ML strategies. It intends to do this by capitalizing on the one-of-a-kind qualities of quantum systems in order to improve ML algorithms and more effectively tackle difficult issues.

The following is a list of fundamental ideas and procedures involved in quantum machine learning:

Quantum Bits, Which Are Sometimes Referred to as Qubits: The fundamental constituents of quantum information are referred to as "quantum bits." In contrast to conventional bits, known as bits, which can only stand for either 0 or 1, qubits are capable of being in a superposition of states, which means that they may stand for both 0 and 1 at the same time.

The phenomenon known as quantum entanglement occurs when two or more qubits get correlated with one another in such a manner that the state of one qubit is reliant on the state of the other qubit or qubits. Due to the presence of this attribute, quantum systems are able to carry out certain calculations at a rate that is exponentially higher than that of classical systems.

Circuits Quantum: Quantum algorithms are performed using quantum circuits, which consist of a sequence of quantum gates that control qubits. Quantum circuits are used to construct quantum computing. Quantum processes such as quantum Fourier transformations and quantum phase estimation are also within the capabilities of these circuits.

The term "quantum supremacy" refers to the point at which a quantum computer is able to complete a particular job at a rate that is superior to that of the most powerful classical supercomputers.

Quantum Variation Algorithms are a kind of quantum algorithm that use traditional methods of optimization in conjunction with quantum circuits in order to tackle issues associated with ML. Examples of quantum variation algorithms include the vibrational Quantum Eigen solver (VQE) and the Quantum Neural Networks (QNN).

Quantum Kernel Techniques: Quantum kernel techniques make use of quantum interference to calculate quantum kernel matrices, which, when applied to ML, may be an improvement on traditional kernel approaches.

The following are some of the potential benefits of quantum machine learning:

Increase in Speed: When compared to traditional ML methods, QML has the potential to give an increase in speed that is exponential for certain computing jobs.

Memory Capacity Improvements: In comparison to classical systems, quantum systems are able to store and process a greater amount of data, which results in an improvement in memory capacity for learning activities.

Using QML to Address Quantum Issues: QML may be used to address issues that are intrinsically quantum in nature, such as the preparation of quantum states and the modeling of quantum systems (Nawaz et al., 2019).

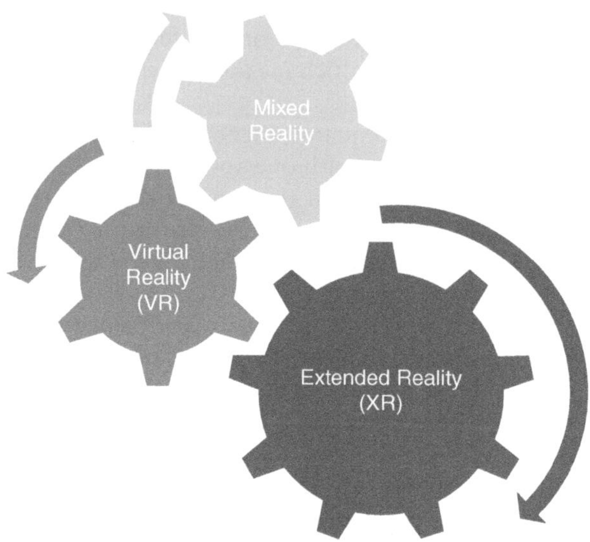

Figure 3.3 Extended reality (XR), mixed reality (MR), and virtual reality (VR).

3.2.4 Wireless extended reality (XR)

The next generation of mobile computing platforms, called Extended Reality (XR), will include Virtual Reality (VR), Mixed Reality (MR), and Augmented Reality (AR) may quickly and significantly impact our lives, just as how computers and smartphones have altered ours. However, the majority of XR gadgets on the market now are cable-tethered, which restricts their mobility and usefulness as shown in Figure 3.3. We present wireless XR systems in this article and go over their use cases, wireless data speeds, and latency requirements. There are presented research issues and suggested remedies to realize the environment-related interior and outdoor applications. Users of XR will be able to move around freely thanks to 6G wireless systems and next-generation Wi-Fi, which can also handle multiple users at once (Akyildiz & Guo, 2022).

3.3 RESEARCH CHALLENGES AND FUTURE DIRECTIONS (DE FIGUEIREDO, 2022)

The investigation into the existence of the 6G network is currently in the beginning stages. As a result, many issues and challenges need to be resolved. 6G networks are considered intelligent and versatile, with various potential approaches and use cases. A few issues associated with 6G communication networks are addressed in this section (Table 3.1).

Table 3.1 Research challenges and future directions

Enabling technologies	Research challenges	Future directions
MIMO evolution	• Maximize sum rate for massive MIMO • Issues related to beam forming • Energy efficiency • Precoding	• Analog beamforming, hybrid beam forming and full digital beam forming • Multiple-input, multiple-output (MIMO) technology, full-duplex transmission, and network assistance are all used by network-assisted full-duplex distributed (NAFD) huge MIMO systems to improve the network's performance and efficiency
NOMA	• Allocation of resources and interference mitigation for NOMA – traditional approaches employ omnidirectional antennas at the base station, however this model takes D2D-NOMA into account • Challenges with power allocation, QOS and complexity of computation • Connectivity problems • Users' cooperation, overload, and conflict	• Interplay between NOMA and other technologies • Reconsideration of SIC • Network interaction between NOMA and intelligent reflecting surfaces (IRS) • Artificial intelligence-enabled NOMA for B5G communication
Artificial intelligence & machine learning	• Allocation of resources due to dynamic variations in channel and traffic, as well as multiple requirements for quality of service (QoS) • ML-based manifoldness analysis, because Multidimensional data is growing in quantity due to significant developments in IoT networks • Employing ML 6G approaches that may shift power dynamically in response to network characteristics and varying physical channels is a big deal • Data management at application level • Wireless channel modelling	• Artificial neural network • Deep transfer learning • Deep neural network • Deep unfolding • Deep learning for cognitive communication • Quantum machine learning • Converged architecture enabling communication and information processing

(Continued)

Table 3.1 *(Continued)* Research challenges and future directions

Enabling technologies	Research challenges	Future directions
Extended reality	• Centralize online management and learning demands for QoE maximization • Need to distinguish user behaviours and manage data to overcome delay • Require more storage space while rendering systems smarter • Data rate requirement in 6G • Multiple-tier computing for cooperation • Wireless connectivity and encoded videos conflict • Security, privacy, sensing • Operating system conflicts	• Immersive communications driven by AI • Predictable architecture that keeps into consideration • Multi-tiered computing for collaboration • Collaborative XR

REFERENCES

Akyildiz, I. F., & Guo, H. (2022). Wireless extended reality (xr): Challenges and new research directions. *ITU Journal on Future and Evolving Technologies*, 3, 1–15.

Alsabah, M., Naser, M. A., Mahmmod, B. M., Abdul hussain, S. H., Eissa, M. R., Al-Baidhani, A.,… & Hashim, F. (2021). 6G wireless communications networks: A comprehensive survey. *IEEE Access*, 9, 148191–148243.

Bethala, C. (2022). *Research Track towards Spectral Efficiency and Energy Efficiency in Massive MIMO Communication System: 5G to 6G Scope*. ACS Publications.

Bhairanatti, S., & Mohan Kumar, S. (2023) Evolution of 6G era: A brief survey of massive MIMO, mm wave, NOMA-based 5G and 6G communication protocols, role of deep learning and inherent challenges. *International Journal of Electronics Engineering*, 10 (1), 2348–8379.

DeAlwis, C., Kalla, A., Pham, Q. V., Kumar, P., Dev, K., Hwang, W. J., & Liyanage, M. (2021). Survey on 6G frontiers: Trends, applications, requirements, technologies and future research. *IEEE Open Journal of the Communications Society*, 2, 836–886.

deFigueiredo, F. A. P. (2022). An overview of massive MIMO for 5G and 6G. *IEEE Latin America Transactions*, 20 (6), 931–940.

Kaur, J., Khan, M. A., Iftikhar, M., Imran, M., & Haq, Q. E. U. (2021). Machine learning techniques for 5G and beyond. *IEEE Access*, 9, 23472–23488.

Liu, Y., Yi, W., Ding, Z., Liu, X., Dobre, O. A., & Al-Dhahir, N. (2022). Developing NOMA to next generation multiple access: Future vision and research opportunities. *IEEE Wireless Communications*, 29 (6), 120–127.

Nawaz, S. J., Sharma, S. K., Wyne, S., Patwary, M. N., & Asaduzzaman, M. (2019). Quantum machine learning for 6G communication networks: State-of-the-art and vision for the future. *IEEE Access*, 7, 46317–46350.

Shah, A. S. (2022, January). A survey from 1G to 5G including the advent of 6G: Architectures, multiple access techniques, and emerging technologies. In *2022 IEEE 12th Annual Computing and Communication Workshop and Conference (CCWC) Las Vegas, NV, USA* (pp. 1117–1123). IEEE.

Shahjalal, M., Kim, W., Khalid, W., Moon, S., Khan, M., Liu, S.,... & Jang, Y. M. (2022). *Enabling Technologies for AI Empowered 6G Massive Radio Access Networks*. ICT Express.

Shehzad, M. K., Rose, L., Butt, M. M., Kovács, I. Z., Assaad, M., & Guizani, M. (2022). Artificial intelligence for 6G networks: Technology advancement and standardization. *IEEE Vehicular Technology Magazine*, 17 (3), 16–25.

Shen, X. S., Gao, J., Li, M., Zhou, C., Hu, S., He, M., & Zhuang, W. (2023). Toward immersive communications in 6G. *Frontiers in Computer Science*, 4, 1068478.

Wang, C., & Rahman, A. (2022). Quantum-enabled 6g wireless networks: Opportunities and challenges. *IEEE Wireless Communications*, 29 (1), 58–69.

Zhao, Y., Zhao, J., Zhai, W., Sun, S., Niyato, D., & Lam, K. Y. (2021). A survey of 6G wireless communications: Emerging technologies. In *Advances in Information and Communication: Proceedings of the 2021 Future of Information and Communication Conference (FICC)* Vancouver, BC, Canada, Volume 1 (pp. 150–170). Springer International Publishing.

Chapter 4

Advancing the frontiers of connectivity

A comprehensive overview of 6G communication networks

*Rachit Jain, Vandana Vikas Thakare,
Pramod Kumar Singhal, Ramya Radhakrishnan,
and Shilpi Gupta*

4.1 INTRODUCTION

In the relentless march of technological progress, the realm of wireless communication stands as a testament to humanity's insatiable appetite for connectivity. As we navigate the intricate evolution from 1G to 5G, the horizon of communication technology stretches ever further, beckoning towards the next frontier – 6G communication networks. This multidimensional journey takes us through the architectural foundations of 6G, its diverse and transformative applications, the paramount significance of security, and the symbiotic synergy between antennas and machine learning [1–4].

The architectural underpinnings of 6G communication networks serve as the cornerstone upon which the future of wireless communication is erected. In a world where terahertz frequencies become the canvas for data transmission, the architecture of 6G ushers in an era of unprecedented possibilities. It embraces massive MIMO technology, satellite integration, and edge computing, forging a network infrastructure that transcends the boundaries of imagination. Our journey begins here, at the very heart of 6G, as we unravel the intricacies of its architectural innovations, infrastructural components, and the collaborative global efforts that make it possible. Yet, the significance of 6G extends far beyond its architectural marvels. It is a technology poised to revolutionize industries, societies, and the human experience itself. As we venture deeper into this exploration, we encounter a vast landscape of applications that promise to redefine how we live, work, and connect. From immersive extended reality experiences that blur the lines between the physical and digital realms to the precision of holographic communication, 6G applications span a spectrum as diverse as human imagination itself. Our journey continues, traversing the myriad domains where 6G's transformative potential is poised to leave an indelible mark [5,6].

However, amidst this era of boundless connectivity, the specter of security looms large. In an environment where data is the currency of

DOI: 10.1201/9781003522003-5

communication, safeguarding its integrity, privacy, and reliability is paramount. As we delve into the security imperatives of 6G, we confront the challenges and innovations that must converge to ensure the trustworthiness of these networks. Quantum-resistant encryption, AI-driven threat detection, and the ethical dimensions of security become guiding stars in this chapter of our exploration. Antennas, often unsung heroes in the narrative of wireless communication, undergo a metamorphosis in the age of 6G. They are not merely conduits of signals, but intelligent entities that adapt, optimize, and evolve. The integration of antennas with machine learning algorithms represents a pinnacle of this transformation. Together, they navigate complex design spaces, mitigate interference, and offer predictive maintenance, ushering in an era where antennas become dynamic architects of connectivity. Our journey nears its conclusion in the realm where technology and intelligence converge, unlocking the true potential of 6G communication networks [1–6].

As we embark on this journey through the architecture, applications, security, antennas, and machine learning of 6G communication networks, we peer into the future of connectivity. It is a future where the boundaries of what is possible in communication technology are continually pushed, where the fusion of human ingenuity and technological prowess charts a course toward uncharted horizons. It is a future that promises a world of connectivity and possibilities beyond our current comprehension, and it is a future that we are poised to unravel in the pages that follow.

4.2 RELATED WORK

This chapter offers a comprehensive overview of prior research on 6G antenna design, covering aspects such as antenna types, frequency selection, material usage, fabrication processes, enhancement techniques, and demonstrated performance. It explores the shift of antennas from the microwave region to the THz frequency band for broader coverage and incorporates new materials and metamaterials to enhance performance. The discussion also delves into the implementation of three-dimensional (3D) multiple input multiple output (MIMO) for evolving antenna propagation channels. Emphasized is the necessity for gain improvements, additional antenna gain, or antenna arrays to enhance positioning accuracy. The proposed antenna design in this chapter utilizes graphene as a radiator and a flexible polyimide film as a substrate, with improved gain achieved through additional substrate and an air gap [7].

This chapter introduces an innovative systematic design of a high-aperture-efficiency and 2-D beam-scanning liquid-crystal embedded reflectarray antenna tailored for 6G FR3 and radar applications. The proposed LC-based reflectarray unit cell (LC-RUC) exhibits significantly low reflection loss compared to other LC-RUCs. The antenna achieves a beam scanning range of ±50 degrees on the *xoz* plane and 0° to −65° in the yoz plane at 9.55 GHz, with a lowest side lobe level (SLL) of –15.5 dB. Experimental measurements conducted in an anechoic chamber using a VNA demonstrate a maximum gain of 14.7–17.1 dBi and an SLL of –8.2 to –15.5 dB at 9.55 GHz [8].

The paper proposes a cost-effective PCB-manufactured antenna for 5G/6G AiP applications, incorporating a radiating choke to suppress common-mode currents and contribute to the main radiation. Derived from a wideband, high-efficiency electromagnetic structure (WHEMS), the antenna is processed on low-cost, two-layer PCBs, achieving a wide gain bandwidth of 60–75 GHz and a relatively flat gain of 8–10 dBi within the band. The radiating choke, designed with wide spacing between enclosing walls (EW) and the backing cavity, serves both as a radiating choke with common-mode current suppression and radiating functions [9].

The paper introduces a low-profile high-gain circularly polarized (CP) Fabry–Perot cavity (FPC) antenna designed for sub-terahertz (THz) frequencies, specifically at 300 GHz. Comprising seven metallic layers, including a ground layer, the antenna achieves a measured peak right-hand circularly polarized (RHCP) gain of 16.5 dBic at 292 GHz and an axial ratio (AR) bandwidth of 5.12 GHz. Fabricated using a low-cost laser cutting technique, the antenna design incorporates a fully metallic frequency-selective surfaceInterface (FSS) layer with periodic hexagonal-shaped apertures, demonstrating good agreement between simulation and experimental results [10].

The paper presents a wideband aperture-coupled stacked patch (ACSP) antenna on flexible material for wireless applications in the WR-08 band. Unlike conventional designs, this novel approach achieves a large bandwidth through broadband impedance matching and the suppression of surface waves induced in the WR-08 band. The fabricated antenna achieves a measured peak gain of 7.95 dBi and S11 ≤ –10 dB across a wide frequency range from 90 to 128.5 GHz [11].

The paper proposes a compact dual-polarized antenna for sub-6G 5G customer premise equipment (CPE), based on a double-loop-dipole structure and a co-planar-slot feeding concept. The antenna achieves a wide bandwidth of 86% from 2.2 to 5.5 GHz with a VSWR of less than 1.8 and high polarization isolation of 25 dB. The design is cost-effective, low-profile, and suitable for mass production, making it a strong candidate for 5G CPE and other applications with limited production costs [12].

The paper introduces a wideband differentially fed dual-polarized magnetoelectric (ME) dipole for millimeter-wave (mm-Wave) applications. Characteristic mode analysis is employed to investigate electric and magnetic characteristic modes, resulting in a stable broadside radiation pattern covering 5G frequency bands. The lifted ground (LGND) concept achieves a 57.1% impedance bandwidth, and differential feeding allows for more than 36 dB of port-to-port isolation across the entire operating band. The proposed ME dipole demonstrates promising attributes for mm-Wave Antenna-in-Package (AiP) applications, including favorable electrical performance, compact size, simple structure, and low-cost fabrication [13].

4.3 THE ARCHITECTURE OF A 6G COMMUNICATION

The architecture of a 6G communication network represents a visionary blueprint for the future of wireless communication. While 6G is still in the conceptual and early research stages as of my last knowledge update in September 2021, several key architectural features and goals have been proposed based on emerging technology trends and anticipated requirements. Here's an outline of the potential architecture for a 6G communication network [14] as shown in Figure 4.1:

Terahertz Spectrum and Beyond: 6G is expected to operate in the terahertz (THz) frequency spectrum and beyond, which opens up a vast amount of available spectrum. This high-frequency spectrum will enable unprecedented data rates and extremely low latency communication.

4.3.1 Cellular infrastructure evolution

Cell Types: 6G networks may incorporate a variety of cell types, including traditional macro cells, small cells, and possibly even femtocells, depending on the specific use case.

Massive MIMO: Massive Multiple-Input Multiple-Output (MIMO) technology will become more prevalent, with an increasing number

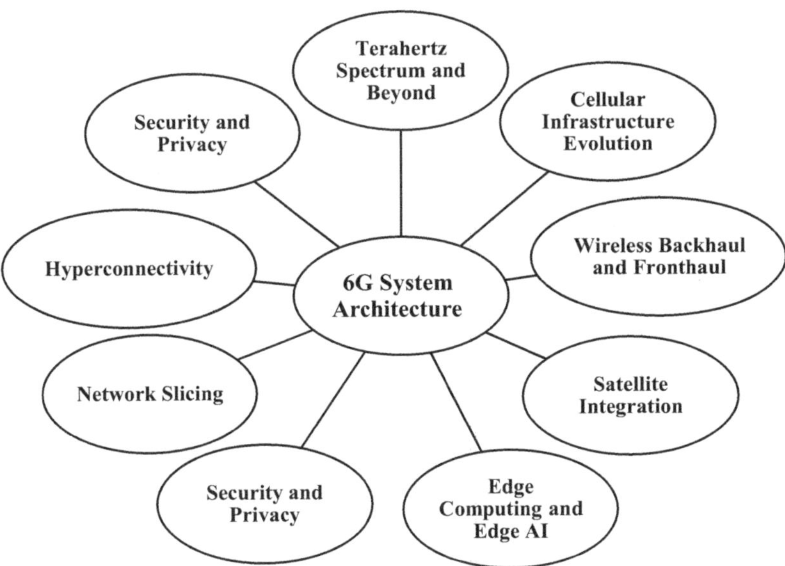

Figure 4.1 6G system architecture.

of antennas at base stations and user devices to improve spectral efficiency and coverage.

Wireless Backhaul and Fronthaul: High-capacity wireless backhaul and fronthaul links, potentially using optical wireless communication, will be essential to support the high data rates and low latency requirements of 6G.

Satellite Integration: 6G networks may incorporate low Earth orbit (LEO) satellites to extend coverage to remote and underserved areas as well as to provide a resilient and ubiquitous network.

Edge Computing and Edge AI: Edge computing and artificial intelligence (AI) will play a central role in 6G architecture, enabling real-time processing of data at the network edge. This will support applications like augmented reality (AR), virtual reality (VR), autonomous vehicles, and IoT with stringent latency requirements.

Network Slicing: To cater to diverse use cases, 6G networks are expected to implement network slicing. This technology allows the creation of customized, isolated network segments optimized for specific applications, such as industrial automation, healthcare, or entertainment.

Hyperconnectivity: Beyond connecting people and devices, 6G aims to connect everything, including machines, sensors, and objects, leading to a hyperconnected ecosystem often referred to as the "Internet of Everything" (IoE).

Security and Privacy: Enhanced security mechanisms, including quantum-resistant encryption and robust authentication, will be integrated into the architecture to protect sensitive data and ensure the privacy of users.

Energy Efficiency: Given the expected proliferation of devices and the increased energy demands, 6G networks will prioritize energy efficiency through advanced power management techniques and renewable energy sources.

Regulatory Considerations: 6G will require regulatory frameworks to allocate and manage the new spectrum bands, ensure compliance with international standards, and address potential interference issues.

Global Cooperation: Developing 6G technology will likely require international collaboration among governments, regulatory bodies, research institutions, and the private sector to establish common standards and ensure interoperability.

User-Centric Design: User experience will be a primary focus, with networks designed to deliver seamless connectivity and immersive experiences, such as holographic communication, tactile internet, and advanced telepresence.

It's important to note that the development of 6G is ongoing, and its architectural details are subject to change as research progresses. The transition from 5G to 6G will represent a significant leap in wireless communication capabilities, and the architecture will need to adapt to the evolving needs of society and technology [14,15].

4.4 SECURITY IN 6G COMMUNICATION

Security will be of utmost importance in the evolution and implementation of 6G communication networks. While 6G strives to offer unparalleled connectivity, extremely high data rates, and ultra-low latency, it concurrently introduces new security challenges. The following are key facets of security within 6G networks [16]:

Quantum-Safe Encryption: A critical security challenge involves ensuring that encryption methods are resilient against quantum threats. Traditional encryption algorithms are vulnerable to quantum computing advancements, necessitating the adoption of quantum-safe encryption techniques in 6G networks to safeguard transmitted data.

AI-Powered Threat Detection: With heavy reliance on AI and machine learning, 6G networks can harness AI for security purposes. AI algorithms can continuously monitor network traffic, detect anomalies, identify potential threats, and respond in real time to mitigate security risks.

Authentication and Access Control: Robust authentication and access control mechanisms are imperative in 6G networks. Multi-factor authentication, biometrics, and secure identity management systems will play a pivotal role in ensuring that only authorized devices and users can access the network.

Secure Device Identity: Ensuring the identity and integrity of devices connecting to 6G networks is vital. Secure device onboarding, certificate management, and hardware-based security solutions, such as hardware roots of trust, will be employed to prevent unauthorized access.

Privacy-Preserving Technologies: The proliferation of connected devices and massive data collection necessitates the implementation of privacy-preserving technologies in 6G networks. Techniques like federated learning and homomorphic encryption enable data analysis without compromising user privacy.

Network Slicing Security: As 6G networks are anticipated to support network slicing for various use cases, each network slice must have its security policies and isolation mechanisms. This ensures the prevention of cross-slice attacks and maintains the integrity and confidentiality of data within each slice.

Resilience against Physical Attacks: Robust physical security measures are required to protect 6G networks against physical attacks, including tampering with infrastructure or devices. Physical security audits, tamper-evident hardware, and secure installation practices are essential components.

Secure Edge Computing: While edge computing is a key component of 6G networks, it introduces new security concerns. Protection against unauthorized access to edge nodes and servers, as well as securing data processing at the edge during transmission, is imperative.

Regulatory Compliance: Mandatory adherence to international and regional regulations regarding data privacy and security is crucial for 6G network operators. Compliance with data protection laws and standards is necessary to avoid legal and financial repercussions.

Collaboration and Information Sharing: Given the global nature of 6G networks, collaboration between governments, regulatory bodies, network operators, and technology providers is essential. This collaborative effort facilitates the sharing of threat intelligence and coordination of security initiatives.

Security by Design: Security should be an integral part of the design and development of 6G networks from the outset, following the principles of security by design. This proactive approach helps identify and address security vulnerabilities early in the development lifecycle.

Ethical Considerations: The deployment of 6G networks will raise ethical questions related to surveillance, data ownership, and user consent. Establishing ethical frameworks is necessary to ensure that the deployment and use of 6G technology align with societal values and norms [16,17].

4.5 6G COMMUNICATION NETWORK APPLICATIONS

6G communication networks are expected to usher in a new era of connectivity, pushing the boundaries of what's possible in wireless communication. These networks will not only provide faster and more reliable connections but will also enable a wide range of transformative applications across various industries. Here are some of the potential applications of 6G communication networks [1,18], as shown in Figure 4.2:

4.5.1 Immersive extended reality (XR)

Augmented Reality and Virtual Reality: 6G will enable highly immersive and interactive AR and VR experiences with minimal latency,

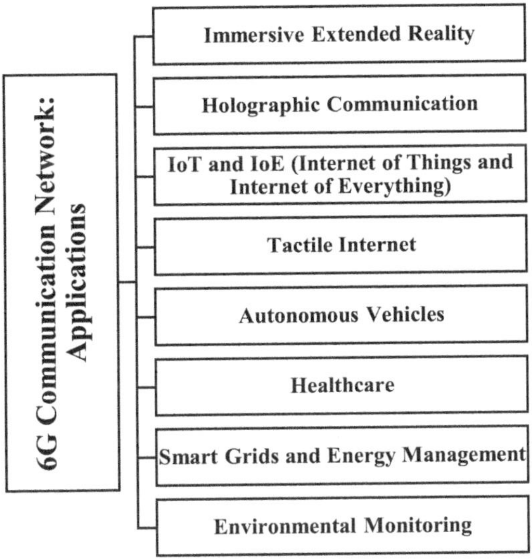

Figure 4.2 6G applications.

making it possible to seamlessly blend digital and physical worlds for applications like gaming, education, and remote collaboration.

4.5.2 Holographic communication

3D Telepresence: 6G networks will support real-time holographic communication, allowing people to interact as if they were physically present in the same location, which has applications in telemedicine, business meetings, and education.

4.5.3 IoT and IoE (Internet of Things and Internet of Everything)

Massive IoT Connectivity: 6G will accommodate a massive number of IoT devices, facilitating applications in smart cities, smart homes, agriculture, healthcare, and logistics.

IoT with AI: The integration of AI and 6G will enable autonomous decision-making by IoT devices, creating smart, adaptive ecosystems.

4.5.4 Tactile Internet

Real-time Remote Control: 6G will support the Tactile Internet, allowing for remote control of machinery and robots with an imperceptible delay. This is crucial for applications in industries like manufacturing, healthcare, and emergency response.

4.5.5 Ubiquitous and high-definition media

8K and 16K Streaming: 6G will enable seamless streaming of high-resolution, high-frame-rate content for entertainment, education, and telemedicine.

4.5.6 Autonomous vehicles

Vehicular Communication: 6G networks will provide the ultra-low latency and high reliability required for vehicle-to-vehicle (V2V) and vehicle-to-infrastructure (V2I) communication in autonomous driving, improving safety and traffic management.

4.5.7 Healthcare

Remote Surgery: Surgeons will be able to perform complex surgeries remotely with real-time feedback and precision, making healthcare more accessible globally.

IoT Health Monitoring: Continuous monitoring of patient's health through wearable IoT devices will become more accurate and proactive.

4.5.8 Smart grids and energy management

Efficient Energy Distribution: 6G can support the real-time monitoring and control of energy grids, optimizing energy distribution and consumption in smart cities.

4.5.9 Environmental monitoring

Precision Environmental Data: 6G networks will enable high-resolution environmental monitoring for climate research, disaster prediction, and wildlife conservation.

4.5.10 Space exploration and satellite communication

Interplanetary Communication: 6G could support communication between Earth and space probes, enhancing our understanding of the cosmos.

Global Satellite Internet: LEO (Low Earth Orbit) satellites in 6G networks may provide high-speed internet access to remote areas worldwide.

4.5.11 Industrial automation and robotics

Factory Automation: 6G will enable seamless communication between robots and machines, improving industrial efficiency and flexibility.

4.5.12 Smart agriculture

Precision Farming: IoT sensors and AI-powered analytics in 6G networks will enable precision agriculture practices for higher crop yields and reduced resource usage.

4.5.13 Emergency response and public safety

Disaster Management: 6G networks will enhance communication during natural disasters and emergencies, enabling rapid response and coordination.

4.5.14 Blockchain and secure transactions

Decentralized Finance (DeFi): 6G networks can support secure, real-time blockchain transactions, expanding the use of decentralized finance applications.

4.5.15 Education

Immersive Learning: 6G will revolutionize remote and online education with immersive, interactive, and personalized learning experiences.

These are just a glimpse of the countless possibilities that 6G communication networks can unlock. The seamless integration of ultra-fast, low-latency, and reliable connectivity into various industries and daily life is expected to drive innovation, improve efficiency, and create entirely new opportunities across the global economy [1,5,18].

4.6 ANTENNA FOR 6G COMMUNICATION NETWORKS

Antennas are a crucial component of 6G communication networks, as they enable the transmission and reception of signals over the wireless spectrum. In 6G networks, antennas will play an even more pivotal role due to higher frequencies, massive MIMO technology, and advanced beamforming techniques. Here are some key aspects of antennas in 6G communication networks [19–21]:

Higher Frequencies: 6G networks will operate in significantly higher frequency bands, including terahertz (THz) frequencies. Antennas for 6G must be designed to operate efficiently at these extremely high frequencies, where traditional antennas may not perform optimally.

Massive MIMO: Massive MIMO technology, which involves using a large number of antennas at both base stations and user devices, will be a core feature of 6G networks. These antennas will enable spatial multiplexing, improved coverage, and increased capacity.

Antenna Array Design: Antenna arrays with a multitude of elements will be used to form highly directional beams. These arrays will require advanced signal processing techniques to control beamforming and steering, ensuring efficient use of spectrum resources.

Beamforming and Beam Steering: Beamforming will be a fundamental technique in 6G networks to focus signal energy in specific directions. Dynamic beam steering will allow antennas to adapt to user locations and device mobility, optimizing signal strength and reducing interference.

MIMO and Diversity Techniques: MIMO techniques will be enhanced in 6G to improve channel capacity and reliability. Spatial diversity and precoding will be used to mitigate fading and multipath effects.

Metamaterial and Smart Antennas: Antennas made from metamaterials, which have unique electromagnetic properties not found in nature, may find applications in 6G. Smart antennas with reconfigurable elements will adapt to changing network conditions.

Miniaturization: Antennas for 6G will need to be compact and low-profile to accommodate the proliferation of IoT devices, wearables, and embedded sensors.

Energy Efficiency: Given the energy constraints of many IoT and sensor devices, 6G antennas will be designed with energy efficiency in mind. Techniques like energy harvesting and low-power electronics will be integrated into antenna systems.

Interference Mitigation: With the dense deployment of devices and antennas in 6G networks, interference management will be critical. Advanced interference cancellation and coordination algorithms will be employed.

Security Considerations: Antennas may also play a role in ensuring the security and privacy of communications in 6G networks. Directional transmissions can reduce the risk of eavesdropping.

Integration with Satellite Communication: In 6G, low Earth orbit (LEO) satellites may play a significant role in extending coverage. Antennas on the ground and in space will need to work together seamlessly.

Regulatory Compliance: Compliance with international regulations and spectrum allocations will be essential for 6G antennas, especially as they operate in new frequency bands.

Testing and Validation: Rigorous testing and validation processes will be necessary to ensure that 6G antennas meet performance, efficiency, and safety standards.

Antenna technology in 6G networks will be at the forefront of enabling the high data rates, low latency, and connectivity demanded by the next generation of wireless communication. The evolution of antennas in 6G will require interdisciplinary collaboration among researchers, engineers, and industry stakeholders to push the boundaries of what is possible in wireless communication.

4.7 MACHINE LEARNING IN 6G COMMUNICATION NETWORKS

In 6G communication networks, antennas and machine learning will work in tandem to address the unique challenges and opportunities presented by this next-generation wireless technology. Here's how antennas and machine learning are expected to converge in 6G networks [21,22]:

4.7.1 Antenna optimization with machine learning

Smart Beamforming: Machine learning algorithms will optimize beamforming techniques in real time. By analyzing data on user locations, movement patterns, and network conditions, ML can dynamically adjust beamforming to enhance signal strength and reduce interference.

Spectrum Management: Machine learning will play a critical role in managing the allocation of spectrum resources. It can adaptively allocate frequency bands to antennas based on usage patterns, optimizing spectrum utilization.

Energy Efficiency: ML algorithms will optimize the energy consumption of antennas, particularly important for IoT devices. By intelligently managing transmit power and antenna configurations, energy-efficient communication can be achieved.

4.7.2 Antenna health monitoring

Predictive Maintenance: Machine learning models can monitor the health of antennas by analyzing data from sensors embedded in the antennas themselves. Predictive maintenance algorithms can anticipate when an antenna is likely to fail or require maintenance, reducing downtime.

Anomaly Detection: ML algorithms can continuously analyze antenna performance data to detect anomalies or signs of interference. When anomalies are detected, automated responses can be triggered to address the issue promptly.

4.7.3 Adaptive antenna arrays

Dynamic Reconfiguration: Machine learning can optimize the configuration of adaptive antenna arrays to adapt to changing network conditions. ML models can analyze real-time data to determine the most effective antenna array configuration for a given situation.

Interference Mitigation: ML algorithms can identify sources of interference and dynamically adjust antenna array parameters to mitigate interference, ensuring reliable and high-quality communication.

4.7.4 Massive MIMO and precoding

Channel Estimation: Machine learning can improve channel estimation accuracy, which is crucial for massive MIMO systems. ML models can learn to predict channel conditions and optimize precoding matrices for efficient data transmission.

Spatial Processing: ML can assist in spatial processing tasks, such as spatial filtering, null steering, and interference cancelation, to enhance signal quality and reduce interference.

4.7.5 Security enhancement

Intrusion Detection: Machine learning can be used to detect unauthorized access attempts or physical tampering with antennas. ML models can analyze security logs and sensor data to identify security threats.

Secure Beamforming: ML-driven security mechanisms can ensure that beamforming is secure and resistant to eavesdropping or jamming attempts.

4.7.6 Antenna array calibration

Auto-Calibration: Machine learning can automate the calibration of antenna arrays, reducing the need for manual calibration and ensuring optimal performance.

4.7.7 Resource management

Network Slicing: Machine learning can assist in the allocation of network slices, ensuring that each slice's antenna resources are efficiently managed based on the specific requirements of different services and applications.

4.7.8 Regulatory compliance

Spectrum Regulations: Machine learning can help monitor and ensure compliance with spectrum regulations, automatically adjusting antenna parameters to stay within legal limits.

The integration of machine learning with antennas in 6G networks will empower these networks to adapt, self-optimize, and provide unprecedented levels of performance, reliability, and security. As antennas become more intelligent and dynamic, they will contribute significantly to the realization of the ambitious goals set for 6G communication networks.

4.8 CHALLENGES AND LIMITATIONS

6G communication networks hold immense promise, aiming to provide unprecedented levels of connectivity and performance. However, like any emerging technology, they also face a range of challenges and limitations. Here are some of the key challenges and limitations of 6G communication networks [23–25], as shown in Figure 4.3:

1. *Terahertz Frequency Challenges*:
   Propagation Loss: Terahertz frequencies used in 6G networks have significantly higher propagation losses compared to lower frequencies. Signals at these frequencies are easily absorbed by atmospheric gases and attenuated by obstacles, limiting coverage range.

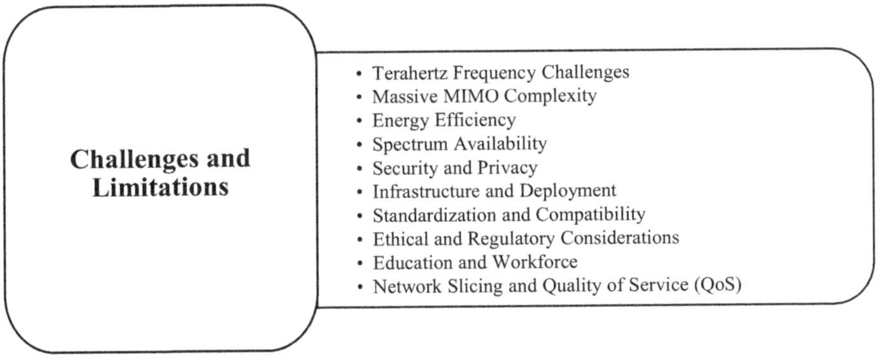

Figure 4.3 6G challenges and limitations.

Signal Blockage: Terahertz signals are sensitive to blockage by physical objects, including walls and even raindrops. This can result in signal interruptions, especially in urban environments.

2. *Massive MIMO Complexity*:

Signal Processing Overhead: Implementing massive MIMO technology, which involves a large number of antennas, introduces substantial signal processing complexities. Efficiently processing and coordinating signals from numerous antennas demand significant computational resources.

3. *Energy Efficiency*:

High Energy Consumption: Achieving high data rates and low latency in 6G networks may require more energy-intensive technologies. Balancing the need for performance with energy efficiency is a complex challenge, especially for battery-powered devices and IoT sensors.

4. *Spectrum Availability*:

Spectrum Crowding: The terahertz frequency bands proposed for 6G are limited in availability and highly sought after. Spectrum crowding can lead to interference and regulatory challenges.

5. *Security and Privacy*:

Quantum Threats: While 6G aims to incorporate quantum-resistant encryption, the advent of quantum computing may pose a threat to existing security measures over time.

Privacy Concerns: With the proliferation of connected devices and the collection of massive amounts of data, ensuring user privacy and data protection will be challenging. Ethical and legal frameworks will need to evolve accordingly.

6. *Infrastructure and Deployment*:

Cost and Infrastructure Deployment: Building the required infrastructure for 6G, including the deployment of massive MIMO antennas and terahertz-capable base stations, will be costly and time-consuming.

Backhaul and Fronthaul Networks: Ensuring high-capacity, low-latency backhaul and fronthaul networks to support 6G is a significant challenge, especially in remote or underserved areas.

7. *Standardization and Compatibility*:

Global Standardization: Developing global standards for 6G is a complex and time-consuming process that involves coordination among various stakeholders. Interoperability and compatibility with existing networks will also be essential.

8. *Ethical and Regulatory Considerations*:

Ethical Challenges: The deployment of 6G networks, particularly in sensitive applications like healthcare and surveillance, raises ethical questions related to surveillance, data ownership, and consent.

Regulatory Compliance: Ensuring compliance with evolving regulatory frameworks, including spectrum allocation and environmental impact assessments, will be essential.

9. *Network Slicing and Quality of Service (QoS)*:

Effective Network Slicing: Implementing network slicing to cater to diverse use cases with varying QoS requirements is technically challenging. Ensuring seamless transitions between slices and efficient resource allocation is complex.

10. *Education and Workforce*:

Skilled Workforce: Developing and maintaining 6G networks require a highly skilled workforce capable of designing, deploying, and managing the advanced technologies involved. Bridging the skills gap is crucial.

6G communication networks hold immense potential, but they are not without their share of challenges and limitations. Overcoming these challenges will require collaborative efforts from researchers, engineers, policymakers, and industry stakeholders. Addressing these limitations is essential to realize the vision of 6G as a transformative force in the world of wireless communication [1,3,23–25].

4.9 CONCLUSION

In the relentless pursuit of connectivity, the advent of 6G communication networks promises a transformative leap in the realm of wireless technology. This comprehensive exploration of 6G has taken us on a journey through its architectural foundations, applications that will redefine industries, the imperative of robust security, and the symbiotic relationship between antennas and machine learning. As we draw our journey to a close, several key takeaways emerge. 6G communication networks represent an unprecedented evolution in connectivity, where technology transcends boundaries to redefine how we communicate, interact, and innovate. As we stand on the cusp of this transformative era, the lessons and insights from this exploration serve as a beacon, guiding us toward a future where the potential of wireless communication knows no bounds. It is a future where the fusion of technology and human ingenuity propels us toward new horizons, promising a world of possibilities we can only begin to imagine.

REFERENCES

[1] M. Z. Chowdhury, M. Shahjalal, S. Ahmed, and Y. M. Jang, "6G Wireless Communication Systems: Applications, Requirements, Technologies, Challenges, and Research Directions," *IEEE Open Journal of the Communications Society*, vol. 1, pp. 1–1, 2020, doi: 10.1109/ojcoms.2020. 3010270.

[2] H. Viswanathan and P. E. Mogensen, "Communications in the 6G Era," *IEEE Access*, vol. 8, pp. 57063–57074, 2020, doi: 10.1109/access.2020.2981745.

[3] S. Nayak and R. Patgiri, "6G Communication: Envisioning the Key Issues and Challenges," *EAI Endorsed Transactions on Internet of Things*, p. 166959, Nov. 2020, doi: 10.4108/eai.11-11-2020.166959.

[4] I. F. Akyildiz, A. Kak, and S. Nie, "6G and Beyond: The Future of Wireless Communications Systems," *IEEE Access*, vol. 8, pp. 133995–134030, 2020, doi: 10.1109/ACCESS.2020.3010896.

[5] M. W. Akhtar, S. A. Hassan, R. Ghaffar, H. Jung, S. Garg, and M. S. Hossain, "The Shift to 6G Communications: Vision and Requirements," *Human-centric Computing and Information Sciences*, vol. 10, no. 1, Dec. 2020, doi: 10.1186/s13673-020-00258-2.

[6] R. Jain, K. Aole, S. Mittal, and P. Ranjan, "An Analysis on Wireless Communication in 6G THz Network and Their Challenges," *Terahertz Devices, Circuits and Systems*, pp. 167–181, 2022, doi: 10.1007/978-981-19-4105-4_10.

[7] S. N. Hafizah Sa'don *et al.*, "The Review and Analysis of Antenna for Sixth Generation (6G) Applications," *IEEE Xplore*, Dec. 01, 2020. https://ieeexplore.ieee.org/abstract/document/9344731/ (accessed Aug. 29, 2023).

[8] H. Kim, J. Kim, and J. Oh, "Communication A Novel Systematic Design of High-Aperture-Efficiency 2D Beam-Scanning Liquid-Crystal Embedded Reflectarray Antenna for 6G FR3 and Radar Applications," *IEEE Transactions on Antennas and Propagation*, vol. 70, no. 11, pp. 11194–11198, Nov. 2022, doi: 10.1109/TAP.2022.3209178.

[9] L. Chi, Z. Weng, Y. Qi, and J. L. Drewniak, "A 60 GHz PCB Wideband Antenna-in-Package for 5G/6G Applications," *IEEE Antennas and Wireless Propagation Letters*, vol. 19, no. 11, pp. 1968–1972, Nov. 2020, doi: 10.1109/LAWP.2020.3006873.

[10] B. Aqlan, M. Himdi, H. Vettikalladi, and L. Le-Coq, "A Circularly Polarized Sub-Terahertz Antenna with Low-Profile and High-Gain for 6G Wireless Communication Systems," *IEEE Access*, vol. 9, pp. 122607–122617, 2021, doi: 10.1109/access.2021.3109161.

[11] M. H. Maktoomi, Z. Wang, H. Wang, S. Saadat, P. Heydari, and H. Aghasi, "A Sub-Terahertz Wideband Stacked-Patch Antenna on a Flexible Printed Circuit for 6G Applications," *IEEE Transactions on Antennas and Propagation*, p. 1, 2022, doi: 10.1109/tap.2022.3185497.

[12] X. Tang, H. Chen, B. Yu, W. Che, and Q. Xue, "Bandwidth Enhancement of a Compact Dual-Polarized Antenna for Sub-6G 5G CPE," *IEEE Antennas and Wireless Propagation Letters*, vol. 21, no. 10, pp. 2015–2019, Oct. 2022, doi: 10.1109/LAWP.2022.3188751.

[13] J. Chen, M. Berg, K. Rasilainen, Z. Siddiqui, M. E. Leinonen, and A. Parssinen, "Broadband Cross-Slotted Patch Antenna for 5G Millimeter-Wave Applications Based on Characteristic Mode Analysis," *IEEE Transactions on Antennas and Propagation*, vol. 70, no. 12, pp. 11277–11292, Dec. 2022, doi: 10.1109/tap.2022.3209217.

[14] Z. Zhang et al., "6G Wireless Networks: Vision, Requirements, Architecture, and Key Technologies," *IEEE Vehicular Technology Magazine*, 2019, doi: 10.1109/MVT.2019.2921208.

[15] L. U. Khan, I. Yaqoob, M. Imran, Z. Han, and C. S. Hong, "6G Wireless Systems: A Vision, Architectural Elements, and Future Directions," *IEEE Access*, vol. 8, pp. 147029–147044, 2020, doi: 10.1109/access.2020.3015289.

[16] Y. Siriwardhana, P. Porambage, M. Liyanage, and M. Ylianttila, "AI and 6G Security: Opportunities and Challenges," *2021 Joint European Conference on Networks and Communications & 6G Summit (EuCNC/6G Summit)*, Jun. 2021, doi: 10.1109/eucnc/6gsummit51104.2021.9482503.

[17] S. Nayak and R. Patgiri, "6G Communication: A Vision on the Potential Applications," *Lecture Notes in Electrical Engineering*, pp. 203–218, 2022, doi: 10.1007/978-981-19-0019-8_16.

[18] Z. R. M. Hajiyat, A. Ismail, A. Sali, and M. N. Hamidon, "Antenna in 6G Wireless Communication System: Specifications, Challenges, and Research Directions," *Optik*, vol. 231, p. 166415, Apr. 2021, doi: 10.1016/j.ijleo.2021.166415.

[19] R. Jain, P. Singhal, and V. V. Thakare, "An Investigation on Unique Graphene-Based THz Antenna," *Advanced Structured Materials*, pp. 163–180, Jan. 2023, doi: 10.1007/978-3-031-28942-2_8.

[20] Y. J. Guo and R. W. Ziolkowski, *Advanced Antenna Array Engineering for 6G and beyond Wireless Communications*. John Wiley & Sons, 2021. [Online]. Available: https://books.google.com/books?hl=en&lr=&id=EKdJ EAAAQBAJ&oi=fnd&pg=PR9&dq=6g+communication+antenna&ots=58 2tF9p-YE&sig=zow32DBxSowCYD3vXGRrYOa8WmQ (accessed Aug. 29, 2023).

[21] A.Siddiqui, F. Qamar, A. Kazmi, R. Hassan, A. Arfeen, and Q. N. Nguyen, "A Study on Multi-Antenna and Pertinent Technologies with AI/ML Approaches for B5G/6G Networks," *Electronics*, vol. 12, no. 1, pp. 189–189, Dec. 2022, doi: 10.3390/electronics12010189.

[22] R. Jain, P. Ranjan, P. K. Singhal, and V. V. Thakare, "Estimation of S11 Values of Patch Antenna Using Various Machine Learning Models," *IEEE Xplore*, Dec. 01, 2022. https://ieeexplore.ieee.org/abstract/document/ 10119256/?casa_token=YlXcaTNrL2gAAAAA:JkgM9Za47dr14ma9kN9A esIBMzeMYMXMqGm28_5AiK07xyI6B9b5LMuO92lGzCTVvfJTISQNt 0AQ (accessed May 30, 2023).

[23] M. A. S. Sejan, M. H. Rahman, B.-S. Shin, J.-H. Oh, Y.-H. You, and H.-K. Song, "Machine Learning for Intelligent-Reflecting-Surface-Based Wireless Communication towards 6G: A Review," *Sensors*, vol. 22, no. 14, p. 5405, Jul. 2022, doi: 10.3390/s22145405.

[24] C. D. Alwis *et al.*, "Survey on 6G Frontiers: Trends, Applications, Requirements, Technologies and Future Research," *IEEE Open Journal of the Communications Society*, vol. 2, pp. 836–886, 2021, doi: 10.1109/ ojcoms.2021.3071496.

[25] H. Tataria, M. Shafi, A. F. Molisch, M. Dohler, H. Sjoland, and F. Tufvesson, "6G Wireless Systems: Vision, Requirements, Challenges, Insights, and Opportunities," *Proceedings of the IEEE*, vol. 109, no. 7, pp. 1166–1199, Jul. 2021, doi: 10.1109/jproc.2021.3061701.

Chapter 5

Digital twin architectures for 6G communication

Network and their future directions

Neelima K., C. Kavya, and Digvijay Pandey

5.1 INTRODUCTION: BACKGROUND AND DRIVING FORCES

The design and operation of 6G network requires real-time interaction and synchronization between physical systems and their virtual image that initiates the research towards Digital Twin Technology [1]. The Digital Twin offers flexibility and adaptability from elementary components to complex systems that enhance the efficiency of the design and the operational procedures in a deterministic fashion. It can provide the virtual image of a physical entity, validate a policy or assess the behaviour of an entity or a system in real-time environment. In 6G communication, it assesses the adequacy and the efficiency of quality-of-service policies as well as the design and the operation of innovative services. Maintenance costs and security risks raised by physical systems can also be better mastered but are subject to standardization activities that pave way to unprecedented challenges related to efficiency, accuracy, fault tolerance, and security. This chapter details the basic requirements of Digital Twin [1] focussed for 6G communication such as decoupling, scalable intelligent analytics, blockchain-based data management, scalability and reliability, latency, agility, interoperability, security, etc. Then the Digital Twin architectures are discussed for basic and improved architectures as projected in the literature with the comparisons between them for various aspects. This focuses on how the data is collected at what frequency by which mechanism, how to store, retrieve, and manage it. Further the details of the models used for digital twin along with interface types and their special features as suited for 6G Communications are discussed. The operational steps are investigated with a step-by-step procedure. Finally, this chapter concludes with the future directions with a focus on isolation between twins-based services for Internet of Everything, mobility management for Edge-based Twins, Digital Twin Forensics [1], Integrated Digital Twins, etc.

DOI: 10.1201/9781003522003-6

5.2 DIGITAL TWIN

A digital twin [1] is a virtual representation of elements and dynamics of a physical system. It uses machine learning, data analytics, and multiphysics simulation to model the dynamics of a given system.

The digital twins are categorized into

i. Monitoring digital twin which monitors the status of a physical system.
ii. Simulation digital twin predicts future states by using machine learning-based simulation tools.
iii. Operational digital twin performs system design, analysis and enables operators to interact with a cyber-physical system.

A reference 6G model is pictured in Figure 5.1. It consists of three layers: physical network layer, twin layer, and application layer. The 6G physical network layer refers to real-world 6G network that consists of physical network elements and their operating environment like a radio cell, a radio access network, a transport network, and a core network that exchange data and control messages with the 6G twin layer.

The 6G twin layer is the core of a 6G DTN which has three domains, namely data domain, model domain, and management domain. Data domain aims at data repository subsystem collection, storage, service, and management of data. Model domain represents a service-mapping subsystem comprising models to represent the real-world objects in the 6G

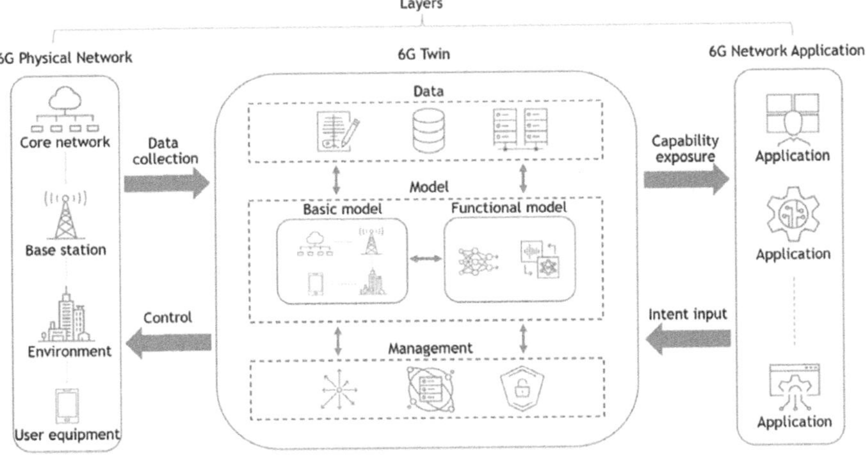

Figure 5.1 A reference 6G digital twin network architecture [1].

physical network based on the collected data. The model domain includes types such as basic and functional models. Basic models represent network elements and topology to describe the target physical network. Functional models refer to analytical models for extracting insights from the DTN like network planning, traffic analysis, fault detection, network emulation, and prediction. Management domain is responsible for the management function of the 6G twin layer used for model creation, configuration, update, monitoring, etc., and security management i.e., authentication, authorization, encryption, and integrity protection.

The 6G network application layer includes various network applications that exploit the capabilities exposed by the 6G twin layer that can also send control messages to the 6G physical network layer, i.e., network operations, administration, and maintenance (OAM), network optimization, and network visualization.

5.3 DTN USE CASES

For 6G networks and network operations, DTNs can support many use cases including network simulation and planning, network operation and management, data generation by simulation, AI training and interference, what-if analysis, etc.

5.3.1 Network simulation and planning

DTNs improve network simulation and planning process using large-scale, physically accurate digital replicas of city blocks that allow network deployment teams to test and optimize installation and placement of base stations and their configurations first in the virtual world to reduce resources and budget in the physical world.

5.3.2 Network operation and management

The softwarization of 6G changes networks through software in software-defined front haul network. To maintain system resilience and robustness, the DTN is used to create a virtual drive test prior to applying the configuration changes in the real network as per customer requirements.

5.3.3 Data generation by simulation

In AI-native 6G networks, real-world data access is very essential that reflects a diverse set of network operating conditions such as operating environment, weather, and network operational parameters. For optimization and network control, synthetic data is created in a cyber sibling of the physical world for statistics to train AI/ML models.

5.3.4 AI training and inference

AI/ML in image and video processing uses models that are trained offline with datasets such as ImageNet, Cityscapes, etc. Inference engines generate a priori model parameters e.g., deep reinforcement learning (DRL) based closed-loop control. DTN uses hardware accelerators with reprogramming capability to train and inference pipelines.

5.3.5 What-if-analysis

The dynamic nature of 6G networks requires network operators to modify key network parameters and architecture in the time scale of minutes and hours, rather than months or years. DTN is a virtual sandbox that identifies network misconfigurations, link bottlenecks, and security issues to enable prediction of future performance.

5.4 DTN REQUIREMENTS

The various requirements of 6G DTNs include reliability & latency, scalability, agility, generalizability, security, and interpretability, etc.

5.4.1 Reliability & latency

Trustworthiness with rendering high degree of stability for a DTN requires robustness in its operational infrastructure, including proactive handling of latency critical adverse situations involving human error, equipment failure or malicious attacks with real-time response, capability of ensuring robust data collection, storage, modeling, and information exchange between the DTN and other network entities, high degree of availability, and capabilities for disaster recovery with backup provision, and ability to restore critical historical states/data points.

5.4.2 Scalability

DTN can be scaled to vary over a wide range based on the dimension and complexity of its physical counterpart.

5.4.3 Agility

The DTN serves the various network applications on-demand with flexibility in its functionalities for cross-domain interaction, information exchange, and service cooperation between multiple DT entities.

5.4.4 Generalizability

DTNs require a certain degree of compatibility for data collection, storage, modelling, and interfaces need to be generalizable to support multi-vendor, multi-standard interoperability and ensure backward compatibility between updated and older versions.

5.4.5 Security

A robust DTN requires adequate protection from potential attacks on data, models, interactive interfaces, and overall network infrastructure to ensure integrity.

5.4.6 Interpretability

DTNs with easy-to-use asset management services aid user-friendly management of the overall lifecycle of the twin entities like network equipment, links, traffic flows, etc.

5.5 DIGITAL TWIN KEY DESIGN REQUIREMENTS

The key design requirements for DTs for wireless 6G systems are decoupling, scalable intelligent analytics, and blockchain-based data management with scalable and reliable architecture and algorithms as shown in Figures 5.2 and 5.3. The digital twin in 6G will virtually represent the

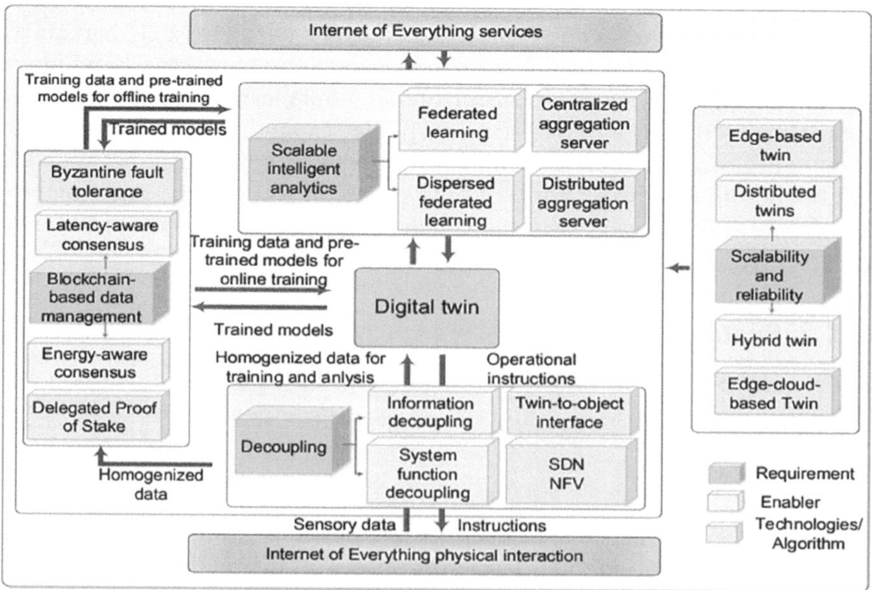

Figure 5.2 Digital twin enabled 6G key design requirements.

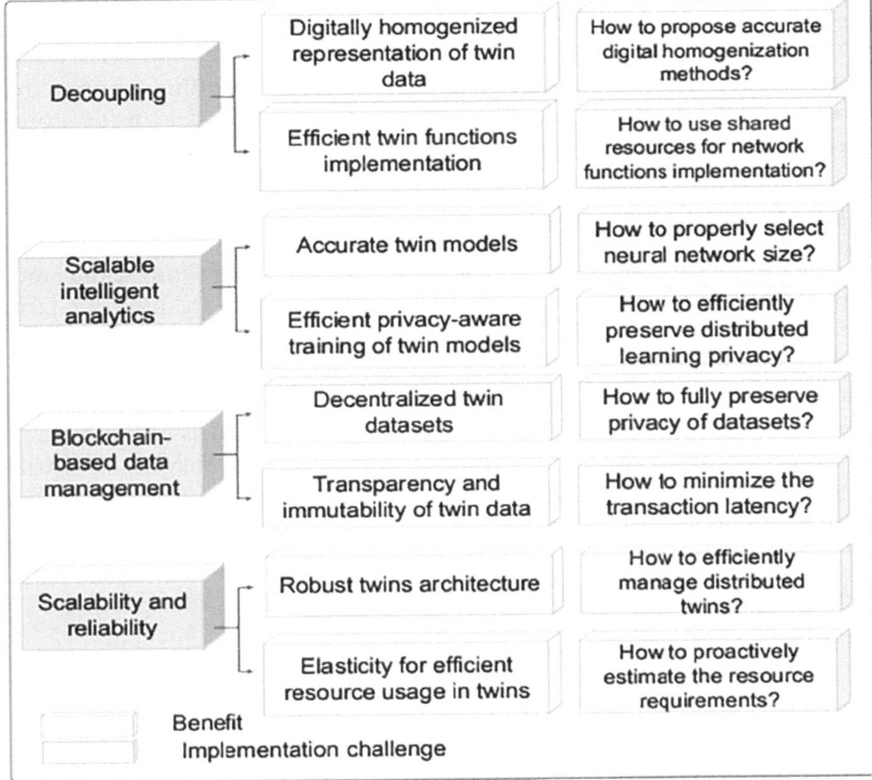

Figure 5.3 Benefits of digital twins-based architecture.

physical wireless system like intelligent reflecting surfaces with backhaul links, a typical 6G application physical system like autonomous driving car, Industry 4.0 plant, and ehealthcare system, and specific module such as edge caching module, computational offloading module at the edge.

5.5.1 Decoupling

Decoupling represents the transformation of a physical system into a digital twin [1, 2]. It can be of two types i.e., information decoupling and system functions decoupling. Information decoupling allows the transformation of physical system information or the system with dynamically changing states into a homogenized digital representation that will offer generality and ease the implementation of digital twins. A 6G physical interaction space consists of base stations (BSs), intelligent reflecting surfaces, smart devices/sensors, and edge/cloud servers whose information includes industrial process control sensory data and holographic images, haptics sensors, 6G spectrum usage data, edge servers resource management data, etc.

The system functions like mobility management, resource allocation, edge caching, etc. are to be taken care of from hardware to software. To enable digital twin [1, 2] SDN separates control and data planes and NFV offers system cost-efficient implementation. The DT performs offline or proactive analytics like data analytics and pre-training of twin models, using stored data on a blockchain and real-time control by network slicing.

5.5.2 Scalable intelligent analytics

To sustain heterogeneous system requirements, network structures, and hardware architectures, effective machine learning schemes are used for large datasets. DTs deal with a large size complex machine learning model and the high computing power which degrades performance in highly dynamic scenarios like mobility management, resource allocation, and edge caching. To overcome these, distributed deep learning-based twin models are used which are diverse in space and training time [13–15]. With the increase of distributed machines, the model computing time decreases and communication time increases. To remain scalable, Federated learning with sparsification is used.

5.5.3 Blockchain-based data management

To manage decentralized datasets in a transparent and immutable manner, DTs use blockchain as it can store unaltered data without collusion of the network majority and retroactive change of all subsequent blocks with seamless data transfer without loss in data integrity. To run a blockchain consensus algorithm, edge servers are used as miners. A few challenges include scalability, high latency, high-energy consumption, and privacy concerns.

5.5.4 Scalability and reliability

To implement massive ultra-reliable low latency communication (mURLLC) services, distributed twin architecture is used that reduces latency. Still distributed DTs suffer from higher management complexity. To balance computational power and latency, hybrid centralized and distributed DTs are used [10–12].

5.6 ARCHITECTURE OF DT ENABLED 6G

The DT-based 6G architecture components and their deployment possibilities are discussed in this section.

5.6.1 Twin objects

Single or multiple Twin objects perform optimization, training of a machine learning model, and control to enable 6G service which is created and terminated dynamically by using transient-based virtual machines (TVM). TVM-based twin objects offer the proactive customization of resources enabled by proactive intelligent analytics based on ML schemes and post-use cleanup [16–19].

5.6.2 Twin object deployment trends

Twin objects can be deployed based on architectures like edge-based digital twin, cloud-based digital twin, and edge cloud-based collaborative digital twin [1, 2], as shown in Figure 5.4. Edge-based twin objects like massive URLLC are more suitable for 6G applications with strict latency constraints, cloud-based twin objects are delay tolerant and are used in high computational power applications. The edge cloud-based twins use both edge and cloud resources to offer a tradeoff between latency and computational power. Various interfaces such as twins-to-things, twin-to-twin, and twin-to-service interfaces must be proposed for seamless, isolation-based, complex interaction in a scalable and reliable manner. The twin-to-object interfaces allow efficient decoupling of the IoE devices from the twin layer. A twin-to-twin interface will be used for communication among various twins at different levels (i.e., cloud and edge levels).

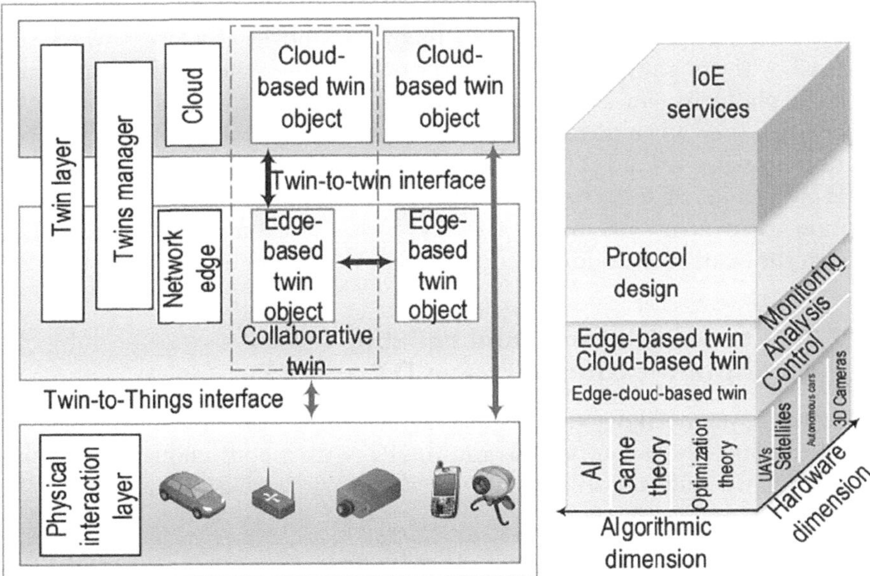

Figure 5.4 Twin objects deployment trends.

5.6.3 Digital twin operational steps

The twin operation happens in two phases, i.e., training and operation. In training phase, distributed machine learning is used, where the local learning models are sent to the twin layer for aggregation at the blockchain miner. The local learning models are updated in IoE devices after the global models are computed [8, 9]. The learning can happen either synchronously or asynchronously. In an asynchronous fashion, a device will send its local learning model only when getting a connection to miners, whereas the devices, must send their local learning models within a predefined time to the miner for global aggregation in case of synchronous fashion. In response to a service request from BS, first authentication takes place, after validation, the request is translated using semantic reasoning techniques, into a form understandable by the twin objects. The twin objects at the BS based on TVM are instantiated to associate with blockchain miners to enable the trustworthy sharing of data.

5.7 6G DIGITAL TWIN FUTURE DIRECTIONS

The 6G DTs have some research directions in setting up and transforming 6G networks, both as an enabler and a use-case. Some of them are detailed here.

5.7.1 DT ownership issues

DT ownership [7] arises technical, financial, and legal aspects which are caused by the potential difference in the ownership of the physical entity and the DT platform. For example, a fitness tracker device can be owned by an individual even when the generated data is owned by and stored on the application provider's cloud. The General Data Protection Rules (GDPR) protect this interest in IoT scenario. The owner of an appliance can opt for the viable option to own his/ her home gateway with storage and security capacity which they can rent cloud/fog/edge services to install and maintain the DT.

5.7.2 Ultra-low-latency and reliable communication between DT and PT

A critical system like a remote-surgery system needs a seamless real-time data exchange between the DT and the PT with a continuous and reliable exchange with ultra-low-latency.

5.7.3 Federated DT in the cloud/edge

Constrained resources such as power, storage, high-speed memory require significant resource management to sustain a technology like the DT, which includes communication, data analysis, and AI-based computation. The distribution or even replication of DTs may involve performance bottlenecks. Multiple copies of DTs can be distributed if the failures in the servers or network links might hamper the seamless connectivity between a PT and its DT [3, 4]. Then federated DT can be used to exchange data and/or train AI models to establish automated and intelligent operations for synchronized and collaborative AI algorithms over the nodes of the network.

5.7.4 DT of an entire network

The DT technology uses Software Defined Network (SDN) and Network Function Virtualisation (NFV) aided with AI for automated and autonomous telecommunication networks. The physical infrastructure like transceivers, antennas, optical fibers, filters, etc can be implemented as cloud native software. The several parameters can be managed, upgraded, and troubleshot using its DT [5, 6]. The research issues like network monitoring and troubleshooting using AI-based analytics and ownership issues using smart contracts hosted in a Blockchain.

5.7.5 Experimental investigation of DTs

The complete LTE network can be developed using commercially available software components such as AmarisoftTM LTE 100 eNodeB, UE from software radio systems (srsUETM) and a generic RF front end. This network on-off control can be done by using a python and Linux-based code. A DT has to collect data from the video for analysis in real time and plot various performance curves with an integrated Graphical User Interface (GUI) to visualize the operations of all components.

5.7.6 Isolation between twins-based services

For various twin-based IoE applications, efficient network resources utilization is necessary to improve performance by explicitly allocating resources (i.e., computation and communication resources). New optimization schemes that satisfy the isolation requirements can be developed for twin objects to efficiently use shared resources among many twin-based services.

5.7.7 Mobility management for edge-based twins

Interrupted service due to limited access point/ BS coverage with the twin can be overcome by using a backhaul link but still suffers from interrupted service and higher latency. Hence migrating services or new effective prediction schemes can be developed.

5.7.8 Digital twin forensics

A digital-twin-enabled 6G system players like end-devices, TVM-based twin-objects, communication interfaces, etc. are vulnerable to various security threats like evidence identification, evidence acquisition and preservation, and evidence presentation. To counteract them, effective forensic techniques can be investigated. The challenges faced would be attacks [20].

5.8 SUMMARY OF DT BASED 6G ARCHITECTURE NETWORKS

This chapter details digital twin [2] as a key enabler for 6G architecture networks and their future research. Using distributed deployment, the edge-based digital twins ensure scalability and reliability. The unique capabilities of DTNs make them attractive and powerful technology for the design, analysis, diagnosis, simulation, and control of 6G wireless networks. To improve reliability and speed, DT is integrated with AI and can also facilitate 6G network design, deployment, and operation which can achieve high network resilience. Further, the demand for DT ranging from aerospace to Industry 4.0 and healthcare is a major driver towards the development of 6G. The stringent performance requirements with higher security pave way to future innovative tools and platforms.

REFERENCES

[1] Y. Tao, J. Wu, X. Lin and W. Yang, "DRL-Driven Digital Twin Function Virtualization for Adaptive Service Response in 6G Networks," in *IEEE Networking Letters*, vol. 5, no. 2, pp. 125–129, Jun. 2023, doi: 10.1109/LNET.2023.3269766.

[2] Q. Guo, F. Tang and N. Kato, "Federated Reinforcement Learning-Based Resource Allocation for D2D-Aided Digital Twin Edge Networks in 6G Industrial IoT," in *IEEE Transactions on Industrial Informatics*, vol. 19, no. 5, pp. 7228–7236, May 2023, doi: 10.1109/TII.2022.3227655.

[3] P. Yang, J. Hou, L. Yu, W. Chen and Y. Wu, "Edge-Coordinated Energy-Efficient Video Analytics for Digital Twin in 6G," in *China Communications*, vol. 20, no. 2, pp. 14–25, Feb. 2023, doi: 10.23919/JCC.2023.02.002.

[4] N. P. Kuruvatti, M. A. Habibi, S. Partani, B. Han, A. Fellan and H. D. Schotten, "Empowering 6G Communication Systems with Digital Twin Technology: A Comprehensive Survey," in *IEEE Access*, vol. 10, pp. 112158–112186, 2022, doi: 10.1109/ACCESS.2022.3215493.

[5] S. Mihai et al., "Digital Twins: A Survey on Enabling Technologies, Challenges, Trends and Future Prospects," in *IEEE Communications Surveys & Tutorials*, vol. 24, no. 4, pp. 2255–2291, Fourthquarter 2022, doi: 10.1109/COMST.2022.3208773.

[6] Y. Dai and Y. Zhang, "Adaptive Digital Twin for Vehicular Edge Computing and Networks," in *Journal of Communications and Information Networks*, vol. 7, no. 1, pp. 48–59, Mar. 2022, doi: 10.23919/JCIN.2022.9745481.

[7] R. Zhao, K. Zhang and Y. Zhang, "Energy-Efficient Edge Association in Digital Twin Empowered 6G Networks," *2022 IEEE 22nd International Conference on Communication Technology (ICCT)*, Nanjing, China, 2022, pp. 869–874, doi: 10.1109/ICCT56141.2022.10073211.

[8] L. U. Khan, W. Saad, D. Niyato, Z. Han and C. S. Hong, "Digital-Twin-Enabled 6G: Vision, Architectural Trends, and Future Directions," in *IEEE Communications Magazine*, vol. 60, no. 1, pp. 74–80, Jan. 2022, doi: 10.1109/MCOM.001.21143.

[9] M. O. Ozdogan, L. Carkacioglu and B. Canberk, "Digital Twin Driven Blockchain Based Reliable and Efficient 6G Edge Network," *2022 18th International Conference on Distributed Computing in Sensor Systems (DCOSS)*, Marina del Rey, Los Angeles, CA, USA, 2022, pp. 342–348, doi: 10.1109/DCOSS54816.2022.00062.

[10] F. Tang, X. Chen, T. K. Rodrigues, M. Zhao and N. Kato, "Survey on Digital Twin Edge Networks (DITEN) Toward 6G," in *IEEE Open Journal of the Communications Society*, vol. 3, pp. 1360–1381, 2022, doi: 10.1109/ OJCOMS.2022.3197811.

[11] T. Q. Duong, D. Van Huynh, Y. Li, E. Garcia-Palacios and K. Sun, "Digital Twin-Enabled 6G Aerial Edge Computing with Ultra-Reliable and Low-Latency Communications: (Invited Paper)," *2022 1st International Conference on 6G Networking (6GNet)*, Paris, France, 2022, pp. 1–5, doi: 10.1109/6GNet54646.2022.9830363.

[12] Y. Lu, S. Maharjan and Y. Zhang, "Adaptive Edge Association for Wireless Digital Twin Networks in 6G," in *IEEE Internet of Things Journal*, vol. 8, no. 22, pp. 16219–16230, 15 Nov., 2021, doi: 10.1109/JIOT.2021.3098508.

[13] Y. Lu, X. Huang, K. Zhang, S. Maharjan and Y. Zhang, "Low-Latency Federated Learning and Blockchain for Edge Association in Digital Twin Empowered 6G Networks," in *IEEE Transactions on Industrial Informatics*, vol. 17, no. 7, pp. 5098–5107, Jul. 2021, doi: 10.1109/TII.2020.3017668.

[14] H. Ahmadi, A. Nag, Z. Khar, K. Sayrafian and S. Rahardja, "Networked Twins and Twins of Networks: An Overview on the Relationship Between Digital Twins and 6G," in *IEEE Communications Standards Magazine*, vol. 5, no. 4, pp. 154–160, Dec. 2021, doi: 10.1109/MCOMSTD.0001.2000041.

[15] J. Deng et al., "A Digital Twin Approach for Self-Optimization of Mobile Networks," *2021 IEEE Wireless Communications and Networking Conference Workshops (WCNCW)*, Nanjing, China, 2021, pp. 1–6, doi: 10.1109/WCNCW49093.2021.9420037.

[16] Y. Wu, K. Zhang and Y. Zhang, "Digital Twin Networks: A Survey," in *IEEE Internet of Things Journal*, vol. 8, no. 18, pp. 13789–13804, 15 Sep. 15, 2021, doi: 10.1109/JIOT.2021.3079510.

[17] M. Pengnoo, M. T. Barros, L. Wuttisittikulkij, B. Butler, A. Davy and S. Balasubramaniam, "Digital Twin for Metasurface Reflector Management in 6G Terahertz Communications," in *IEEE Access*, vol. 8, pp. 114580–114596, 2020, doi: 10.1109/ACCESS.2020.3003734.

[18] W. Sun, H. Zhang, R. Wang and Y. Zhang, "Reducing Offloading Latency for Digital Twin Edge Networks in 6G," in *IEEE Transactions on Vehicular Technology*, vol. 69, no. 10, pp. 12240–12251, Oct. 2020, doi: 10.1109/TVT.2020.3018817.

[19] G. Liu et al., "Vision, Requirements and Network Architecture of 6G Mobile Network beyond 2030," in *China Communications*, vol. 17, no. 9, pp. 92–104, Sep. 2020, doi: 10.23919/JCC.2020.09.008.

[20] H. Viswanathan and P. E. Mogensen, "Communications in the 6G Era," in *IEEE Access*, vol. 8, pp. 57063–57074, 2020, doi: 10.1109/ACCESS.2020.2981745.

Device to device communication for 6G communication

Neelash Patel, Hameed Khan, Kamal Kumar Kushwah, Monika Sahu, and Rekha Garg Solanki

6.1 INTRODUCTION

Device-to-device (D2D) communication in 6G is anticipated to be very important to improve connection and make new applications possible. It seeks to enable direct contact between gadgets without depending entirely on centralized infrastructure. Increased productivity, less latency, and support for several upcoming technologies may result. Since 6G is still in its early phases of development, further study will be needed to determine the precise specifications of the infrastructure and protocols for D2D communication [1]. Its main goal is to lessen dependency on centralized infrastructure by promoting direct connection between more effective devices with lower latency. As shown in Figure 6.1, 6G develops, new protocols and technologies will probably be introduced to serve a broader range of applications and improve overall connection.

Wireless communication technology has undergone development and regeneration about every ten years to satisfy the constantly growing needs of networks [2]. Up to now, 5G has been standardized in 2018 and will begin to be commercially deployed by the end of 2019. But it's anticipated

Figure 6.1 Introducing 6G technology.

DOI: 10.1201/9781003522003-7

that 5G won't be able to meet the demands of new Internet of Things (IoT) applications like mixed reality (MR), augmented reality (AR), and virtual reality (VR), which call for the convergence of processing, sensing, control, and communication capabilities. Furthermore, existing 5G network development cannot effectively provide new services like photorealistic communications, high-precision manufacturing, ecologically conscious and smart surroundings, or better energy efficiency. A unique network design for the next sixth-generation (6G) mobile network is needed to meet the above demanding objectives. Although it may still be too early to define 6G precisely, most experts agree it will be a tremendously dense, highly dynamic, ultra-heterogeneous, and inherently intelligent network that links everything. In more specific terms, 6G is anticipated to have end-to-end latency of less than 1 milliseconds (ms) and handle data rates that are 100–1,000 times faster than 5G. Current research trends indicate that the 6G structure of networks has the following features: more frequency bands, ultra-dense differing green energy, security and privacy, artificial intelligence (AI) driven space-air-terrestrial-sea integrated (SATSI), and higher density than current 5G architecture. Trillion-level devices in the earthly, aerial, space, and oceanic domains will be connected flexibly and effectively thanks to the SATSI network architecture, enabling ubiquitous connectivity. To support network systems with extensive data and low latency, 6G will use a much broader spectrum, such as visible light communications (VLC) and sub-terahertz and terahertz (THz) bands. The significant route loss will significantly reduce data transfer over distance in higher frequency bands. Thus, the 6G network will become very dense. Moreover, the 6G network will need ultra-massive, fast speeds, and low-latency D2D communication to overcome the limited range. Thus, the super-dense inconsistent network is considered to be one of the critical components of 6G [3]. D2D communication and ultra-dense networks have been advantageous for 5G; however, further network densification in 6G will encounter numerous significant challenges and obstacles, such as extreme interference, exceedingly complex resource management, a substantial amount of signalling, high-cost and energy consumption, etc. [4]. Researchers have discovered that network managers may use AI techniques to provide adaptive automated network administration, operation, and maintenance to achieve this. Artificial intelligence is predicted to be the most creative approach for 6G networks. Complex 6G networks will become more intelligent using AI methods like machine learning (ML) and network considerable data training through sensing. This will enable high-level network management and optimization, including proactive configuring, dynamic enhancing, self-healing, network environment sensing, and status prediction. Additionally, AI will progressively move from the cloud to the network edge, where multi-level AI will be used. It has been noted that the combination of high-end CPUs, large amounts of storage, long-lasting power sources, and many sensors is making prevalent UEs like smart phones more powerful. They may be considered personal portable workstations that allow for future intelligent networking using light-level on-device AI computing. Numerous studies have shown that on-device AI can effectively and optimally alter network settings by dynamically sensing the local environment to 2 Full-Frequency Radio Access THz Wave IR UV 300 MHz 30 GHz 300 GHz 3 THz 300 THz, Intra-Plane Two-Hop D2D, Intra-Ship D2D, NFV, SDN, IR, AI, Internet Edge-Core Cloud [5]. AI and 6G connectivity have the potential to transform contemporary living and alter our perception of technology completely. As a result, 6G communication technology will be very successful in the 2030–2040 market. Like many others, the commercial dawn of this idea will take decades to arrive. 6G communication technology will be wholly integrated with artificial intelligence. AI will be a crucial element of communication infrastructure in the 6G communication network. Furthermore, it is anticipated to facilitate both Augmented Reality (AR) and Extended Reality (XR). The network system may be advanced to a new level using these technologies alone. Everything will, in essence, be built using cloud-based architecture, such as Edge Computing. The consumer does not need to set up servers, software, or hardware; they only need a fast internet connection. Cloud technology will gain every network link and flexibility, creating a data-driven, real-time ecosystem [6]. Because of its vast capacity and low latency, 6G can run far larger network systems with minimal latency and power consumption.

The telecoms sector is changing dramatically on a worldwide scale. It gains additional strength with the aid of a decentralized network structure. Large-scale machine-type cross-connection, ultra-reliable, low-latency communications, and improved mobile broadband are the goals of the dynamic network system. Several modern pieces of hardware can also handle a significant degree of data integration. Cellular wireless networks include cutting-edge techniques, including virtualization, heterogeneous networks, and software-defined networking to maximize connection [7]. Network management is driven by AI to handle the quick monitoring system. Implementing several cloud services, virtual networks, and other independent components may simplify specific jobs. However, it leads to increased fragmentation over time and increases the requirement for tool control. We can create and modify that advanced toolbox with the help of 6G. Edge AI would handle the complex integration of the management system hierarchically.

6.1.1 Data availability in D2D

In D2D communication, "data availability" refers to the accessibility and preparedness of data for transfer between devices. It entails ensuring that the data can be effectively transferred between devices in a network and is regularly available. Reliability of the network, appropriate protocols, and device responsiveness to data transfers affect data availability. D2D communication may occur in several ways throughout the design and deployment phases. The following are some possible contexts in which conflicts of interest might arise:

- *Standardization Bodies*: Conflicts may emerge if organizations engaged in creating D2D communication standards have conflicting interests in promoting their proprietary technology or influencing standards to favour their goods.
- *Patents and intellectual property*: Conflicts may arise when businesses or individuals own patents about D2D technology. A disagreement over intellectual property rights may cause tensions between interested parties.
- *Service Providers*: Companies offering D2D communication services may experience issues with their buses.

6.1.2 Acknowledgements in terms of D2D

The D2D communication initiative may acknowledge the following:

- *Development Team*: Credit those who worked on the D2D communication system's conception, execution, and testing.
- *Collaborating Organizations*: Give credit to any outside groups or partners that contributed to the project in any way—through contributions, financing, or teamwork.
- *Technical Support*: Express gratitude to the people or organizations who offered technical advice, assistance, or knowledge while creating the D2D communication system.
- *Testing and Validation*: If appropriate, thank those who helped ensure the D2D system's functioning and dependability via testing and validation procedures.
- *Mentors and Advisers*: Show appreciation to mentors or advisers who helped with the project by offering advice and insightful commentary.
- *Funding Sources*: If the project was financially supported, credit the grant agencies or funding sources that enabled the D2D communication effort.

6.1.3 Domains for D2D communication and possible optimization factors

D2D communication includes several domains with possible parameters for optimization; Figure 6.2 shows some kinds of potential optimizing parameters such as:

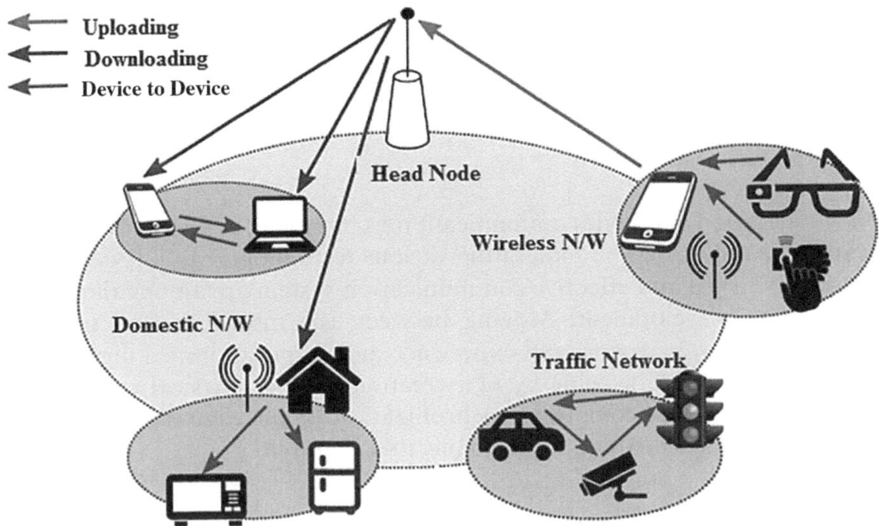

Figure 6.2 D2D communication domains and potential optimization parameters.

- *Spectrum Allocation*: Effective distribution of frequency bands to reduce interference. This is part of radio resource management.
- *Handover Optimization in Mobility Management*: Reducing interference while moving devices across cells.
    Location-Based Services: Sending customized messages depending on the user's location.
- Latency optimization, or Quality of Service (quality of service), minimizes communication latency for real-time applications.
    Reliability: Using error control techniques to guarantee dependable data flow.
- *Energy Efficiency*: Power Consumption: Reducing energy use to extend the battery life of devices [8].
- *Privacy and Security*: Ensure devices communicate securely.
- *Network Architecture and Topology Control*: Maximizing device configuration for effective communication.
- *Harmonization Time Alignment*: Guaranteeing that gadgets are in sync to promote unified correspondence. Harmonizing frequency features to improve interference control is known as frequency synchronization.
- *Interference Management*: Minimizing interference between adjacent devices that operate near one another. It may prevent congestion by responding to changes in the available spectrum using dynamic spectrum access. Predictive analysis by understanding and forecasting user behavior for proactive improvement is known as "user behavior modeling."

- *Traffic Offloading*: Redirecting traffic from portions of the network that are too crowded to enhance efficiency.
- *Burden Balancing*: Equitable distribution of the communication burden across the available devices. Improving D2D communication systems' overall efficiency and effectiveness is facilitated by optimizing these characteristics.

New generation of mobile communication technologies as seen in Figure 6.3, seamless connection and integrating various technologies and devices for a more networked and effective communication system are among the objectives of 6G development. Moving between the interconnected tangible realm of senses, behavior, and experience and its programmed digital representation is made feasible by 6G, creating a cyber-physical continuum. The network offers complete synchronization of the digital and physical worlds, intelligence, and infinite connection (Figure 6.4).

6.1.4 Technology is changing society

With the help of 6G, movement within a cyber-physical continuum is made possible, bridging the gap between the digitally programmable depiction of the natural world and its intricate web of perceptions, actions, and experiences. The network provides unlimited connectivity, intelligence, and perfect synchronization of the physical and digital worlds. Digital

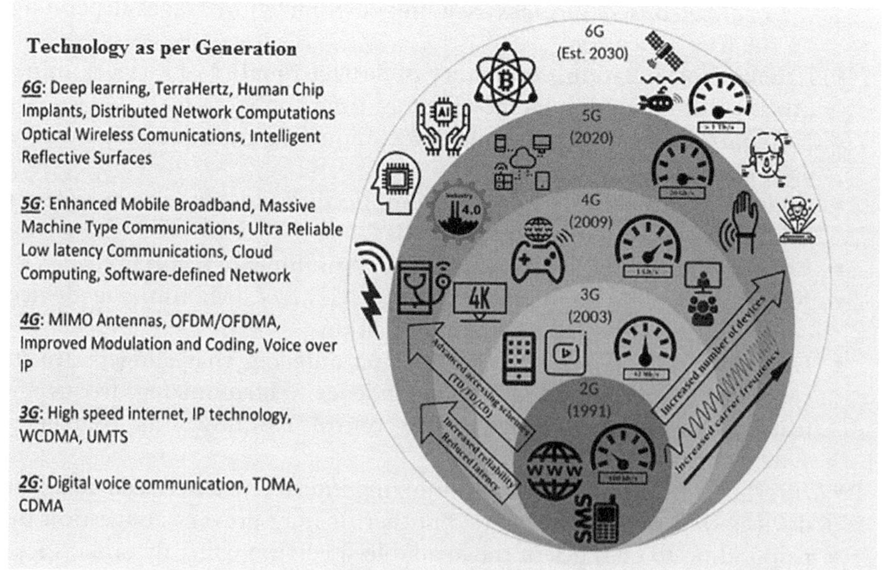

Figure 6.3 Generation of communication technologies.

Figure 6.4 Connecting everything is a goal of developing 6G.

representation is updated in real-time by a vast array of sensors implanted in the actual environment that sends data. Intelligent agents in the digital realm provide orders to actuators in the physical world. This opens up new possibilities for event analysis and tracing, real-time observation, action, modelling, forecasting, and programming. The cyber-physical continuum offers a tight connection to reality, where digital items are projected onto tangible objects encoded digitally, enabling them to cohabit as merged reality and enrich the actual world smoothly. This contrasts the VR and AR worlds, where avatars interact.

Future networks will need to have additional capabilities when new application areas arise. Addressing the apps now in use and closing any remaining digital gaps is equally crucial. A digitalized and programmable world will be able to provide interactive, accurate, real-time, four-dimensional maps of whole cities that many people may see and alter at once, together with intelligent devices, to allow for meticulous activity planning. Large-scale steerable systems, such as waste management, water, heating management, or public transportation, may receive orders from these cyber-physical service platforms, resulting in improved control, resilience, and resource efficiency. The emergence of precision healthcare, facilitated by small-scale nodes monitoring physiological processes and apparatuses providing drugs and aid, will be underpinned by an online digital representation constantly being examined. People's lives are highly reliant on technology, which highlights the need for reliability in terms of accessibility, security, and data privacy. It also calls for new devices that use efficient and distributed processing and administration, which can be safely installed almost anywhere, don't need maintenance, and communicate securely across local body networks [9].

An automated society would profit from AI support to raise people's standard of living and make their lives easier. Collaborative AI partners, for example, might help in both sectors and our homes, do various difficult jobs, including physical labor, more securely and effectively, operate autonomously, and adjust to human action. These high-trust cyber-physical systems are indispensable when it comes to low-latency communication, accurate location and sensing, high reliability and resilience, AI trust and integration, and seamless human-machine interaction. Intelligent identification and preference management will help individuals personally by facilitating interactions with the linked world and adjusting it to suit their preferences. Building a sustainable world requires massive social initiatives, with networks supporting global digital inclusion. A wide range of topics are covered, such as broad access to digital personal healthcare, high-end services offered to educational and medical institutions worldwide, connectivity for global sensors monitoring the welfare of forests and oceans, resource-efficient associated agriculture, and intelligent automation services provided everywhere on Earth. Autonomous supply chains may accelerate a fully circular resource economy via the worldwide, end-to-end monitoring of items' life cycles. Digital asset monitoring helps automate recycling and save waste. This requires highly available and secure network infrastructure, integrated autonomous devices and sensors, and worldwide coverage with exceptional energy, material, and cost efficiency [10].

Through immersive communication, engagement will be possible regardless of distance, thanks to the telepresence experience. For extended reality (XR) technological advances with human-grade sensory input, high data rates and capacity, minimal latency end-to-end with edge cloud computing, and spatial mapping from precise position and sensing are required. One example is the extensive usage of mixed reality on public transit. Passengers may complete virtual errands, seek XR guidance, and overlay games from all around the globe with this unique virtual experience. Furthermore, when sensations and holograms can be transmitted between the real and digital worlds, communication will become closer to a unified reality. In addition to adding entirely new communication modes with strict control over access and identities, personal immersive devices with precise body interaction will enable access to experiences and actions that are far away, ensuring an immersive perception and supporting people's communication needs even better—the significance of which has become particularly obvious during the COVID-19 pandemic.

6.1.5 6G: The network platform of the future

Growing aspirations provide a clear objective for the scientific community and industry. Through ubiquitous intelligent communication, 6G is expected to foster an effective, sociable, and sustainable society. In comparison to today's networks, the capabilities of future wireless access networks must be expanded and improved in several ways to support an extensive array of novel and developing services. This encompasses new and traditional capabilities, some of which may be more quantitative. Classic capabilities include possible data rates, latency, and system capacity. Future wireless networks should be able to support services that are not yet envisioned and support use cases that are now expected. Future networks should provide reduced latency and excellent attainable data rates in all critical contexts, starting with traditional capabilities. This includes the potential to provide several hundred gigabits per second and, in some situations, end-to-end latency of less than 1 ms. The ability to provide a high-speed connection with predictable low latency and low jitter rate is just as significant, if not more so [11].

Future wireless access networks need to handle traffic demand that is increasing exponentially in an affordable manner. One aspect is the improved spectral efficiency of essential radio access technologies; another is the inherent ability to access more spectrums. However, making it possible for very dense networks to be deployed cost-effectively is even more crucial [12]. The digital gap for rural places must be closed by advancing wireless communication toward complete worldwide coverage while enabling a much larger number of gadgets that will be integrated into society. Ensuring that the total cost is at a level that is sustainable for service providers as well as customers is a crucial part of this. High network energy performance was a critical need for creating 5G, which will only become more vital for following wireless access technologies [13]. The anticipated sharp rise in traffic must differ from an equivalent surge in energy use. Increased energy use should be further from an increase in traffic. Additionally, when there is no congestion within a node, the energy consumption must be almost negligible.

Resilience and security skills are vital as wireless networks become increasingly essential parts of society. The network must have strong defences against intentional harmful assaults and continue operating even if a portion of its structure is rendered inoperable by natural catastrophes or social breakdowns. Networks should be able to use emerging technologies for private computing, increase service availability, and provide better security credentials and protocols with end-to-end assurance in terms of trustworthiness [14]. These networks will require data and compute acceleration services that can be delivered through the web with performance guarantees, infrastructure that allows distributed applications and network functions to be developed and deployed quickly, and dependable compute and AI integration.

Figure 6.5 demonstrates the phases of 6G system. Finally, networks need their surroundings' exact locations and comprehensive sensing capabilities to support society's complete digitization and automation. Based on an

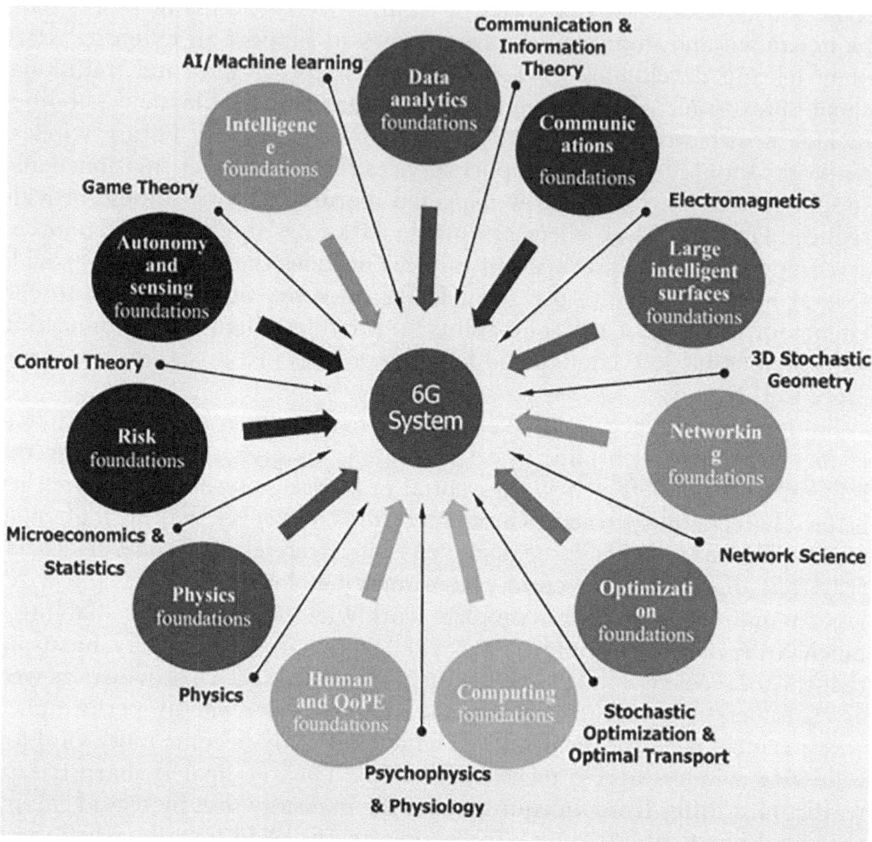

Figure 6.5 Phases of 6G system.

understanding of the effects of the environment on radio wave propagation, sensing is a new kind of capability. For instance, rainfall impacts microwave links—information necessary for weather forecasting. Additionally, radio signals may be actively delivered for sensing, enabling the communication network to perform a role similar to that of a radar. Reusing cellular networks for sensing may lead to a more affordable sensing system and a wider coverage area than what specialized sensing systems can provide. For example, sensing may detect road traffic, model the surroundings, or trigger alarms when someone enters a restricted area in a production hall. Future networks must make effective use of radio resources for sensing and communication. AI-based result interpretation, scalable methods for information distribution, and security measures to protect data privacy are also required. From Device to Device D2D Transmission in 5G and 6G Networks Concept Applications: Obstacles and Prospects. Introduced in 3rd Generation Partnership Project (3GPPs), D2D communication is a viable solution to handle growing traffic demand and reduce the burden on core networks. When nearby mobile users establish a communication group with a common interest in digital information, the D2D communication mode takes over [15]. Instead of downloading desired material via the base station from the Internet, a user may do it via peers in the same communication group. A persistent trend in the delivery and sharing of information is mobile communication. Traffic on mobile devices is increasing at an exponential rate. IP traffic is anticipated to reach 396 EB monthly by 2022 overall. Along with the development of mobile communication, several technologies have been invented to accommodate more significant data rates for a given spectrum [16]. The spectrum efficiency has significantly increased. However, we continue to look for innovative ways to achieve even greater spectrum efficiency in the face of rising traffic demand. Besides novel modulation techniques, emerging frameworks like Dense Small Cell Networks and D2D Communication provide encouraging answers to the problem. A setting that 3GPPs handle specifically for the D2D communication method [17]. A communication group is formed by users who are nearby. Members of communication groups often have a shared interest in certain digital information. If a group member already has a piece of material, other members may download it from them instead of using a base station to access the Internet. The D2D communication mode is intended to allow users to communicate directly, without the need for base stations, with adjacent UE. It suggests reduced base station loading, increased data rate, and improved spectrum efficiency.

6.1.6 Generations of advancements in mobile radio communication networks

This special issue is devoted to bringing together contributions from academic researchers, industry practitioners, and people working on this emerging, exciting research area in light of the recent massive number of activities surrounding future 6G communication systems. These individuals will share their innovative concepts and latest findings and identify and discuss possible application cases, open research issues, technical challenges, use cases, scenarios, and solution methods in this context [18]. Technology includes AI-driven networks, improved antenna technology, and terahertz frequencies. Advances in data speed, bandwidth utilization, latency, and the launch of new services and applications are commonplace throughout the generational shift. Every generation improves on the one before, correcting flaws and bringing new technology to satisfy consumers' increasing needs and new use cases. Research into the possible characteristics and technologies of generations beyond 6G is already in progress as part of the continuous process of developing mobile communication networks [19].

6.1.7 Distribution of spectrum

Two main categories exist for D2D communication in terms of spectrum consumption. Both In-band and Out-band exist.

6.1.8 D2D communication in-band

Here, the spectrum granted to the cellular operator is shared by D2D and cellular communication [20]. The licensed spectrum may be left undivided or split into non-overlapping segments for D2D and cellular transmission. Although overlay schemes are more straightforward to execute, underlay schemes encourage opportunistic behavior, increasing operator profit and facilitating more effective spectrum use.

6.1.9 D2D connection via out-band

In this case, D2D communication takes advantage of unlicensed spectrum in areas where cellular connectivity is impossible, such as the unrestricted 2.4 GHz ISM band or the 38 GHz millimeter Wave band. While other electrical devices (like Bluetooth and WiFi) running in this band still cause interference, it helps to eliminate interference between D2D and cellular users. When employing licensed spectrum, operators can regulate interference, but this is not possible with outbound schemes. There are two more categories for outbound technology: autonomous and controlled. In the former scenario, the cellular network manages the radio interface for D2D communication as shown in Figure 6.6 [21]. In contrast, the cellular network regulates cellular communication in the later system, leaving users in charge of D2D communication.

6.1.10 Networks with single hop and multi hops

A D2D connection often establishes a single-hop communication between a transmitter and its intended recipient UE. A multi-hop network of D2D connections is another option, similar to a mobile ad hoc network (MANET). The intermediate UEs in a multi-hop D2D network serve as relays between two UEs or between a BS and a UE. A cooperative cluster of UEs is one possible variation on the first scenario, where the BS sends a data item to the cluster head, who then group distributes it to other UEs in the ensemble. While 3GPP supports D2D-based vehicular communication, it also facilitates UE-to-network relay.

6.1.11 Synchronization

UEs in a typical cellular network need periodic broadcasts from the BS to synchronize time and frequency. As long as the devices in D2D

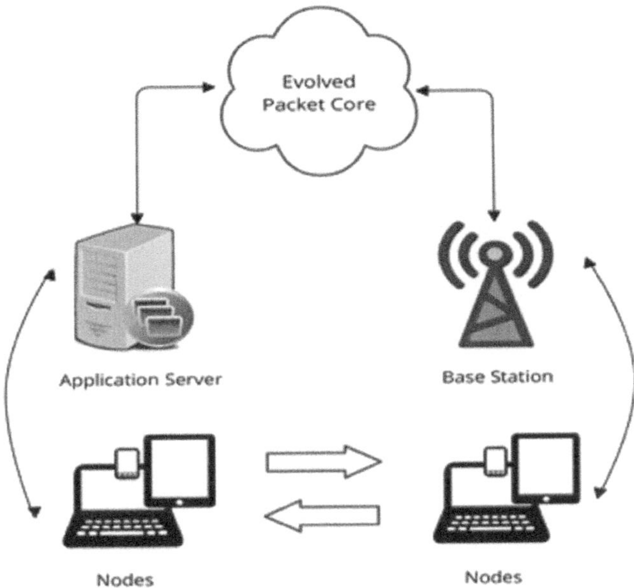

Figure 6.6 D2D communication model simplified using 3GPP& ProSe architecture.

communication are part of the same BS, they may also synchronize with the same broadcasts. In the following situations, things get more complicated: UEs may be divided into three categories: (1) all UEs are outside network coverage; (2) some UEs are within network coverage while others are outside; or (3) UEs belong to separate BSs that might not be synchronized themselves. For D2D communication, UE synchronization is advantageous because it enables a UE to find and communicate with its peers at the optimal time and frequency, resulting in a more energy-efficient touch [22]. Remember that local synchronization between nearby devices suffices for D2D communication, and global synchronization across all UEs in a network may not be necessary. While one can adopt the synchronization protocols recommended for MANETs and wireless sensor networks, D2D communication generally requires more precise synchronization and permits more complex algorithms in UEs, unlike the case in resource-constrained sensors. Researchers have developed various physical and MAC layer synchronization strategies for D2D communication.

6.1.12 Peer discovery

Given the demand for D2D networks, an effective way to find peers should exist. This implies that a UE should have little power consumption and be able to find other neighboring UEs quickly. Restricted and open peer discovery strategies are the two categories from which users may choose. In the first scenario, end users cannot find gadgets without consent. In the second scenario, devices are always discoverable near other users. From a network viewpoint, BS may exert loose or strict control over peer discovery. Peer discovery in a multi-cell network is a complex task since eliciting cooperation from nearby BSs is extremely hard. As a likely solution, incentive-based methods might be researched.

6.1.13 Selection of mode

A pair of self-discovery UEs may be good candidates for a D2D conversation. However, if the direct path is noisier, for example, cellular communication could work better. The process of selecting the appropriate communication mode between two UEs, such as cellular or D2D, to meet specific performance goals like high spectral efficacy, low latency, or low transmit power is known as mode selection. You may choose either the network or the UEs mode. Several goals and constraints may be included to define the mode selection issue, and each UE can have a decision variable associated with it that represents the chosen mode. A straightforward plan may be for the selected mode's channel gain to be greater than the other viable modes. More complex goals might also be used, such as weighted-sum-rate maximization and optimum spectrum reuse. Limitations may include maximum transmission power, minimal quality of service at the receiver, etc. Thus, power control and mode selection are often combined. The analysis might be performed using statistical system information, which would result in judgments that are optimum in the long run, or immediate system information, which would be challenging to get in reality.

6.1.14 Allocation of resources

In a cellular network, allocating radio resources—such as subcarriers—is essential to creating and maintaining direct links between D2D pairs, primarily in in-band mode. A simple yet general resource allocation framework is proposed for in-band multi-cell architecture: in overlay, the uplink range is divided into two symmetric portions with fractions assigned to cellular and D2D communication; in overlay, the spectrum is divided into bands that D2D UEs can access independently and randomly; etc. Assuming that the UEs are randomly distributed in accordance with a random global Poisson point process, and that the modes are selected depending on a UE's proximity to its intended recipient, the ideal values of and are computed. Subsequently, some performance goals are refined, including a combined assessment of the rates reached by both cellular and prospective D2D users.

By changing the optimization objectives and introducing various limits, several resource allocation strategies may be developed.

6.1.15 Management of interference

Due to the way they share frequencies, cellular and D2D connections may interfere with one another in in-band communication. D2D connections experience interference from other devices running in the same band and from one another during outbound communication. Lower transmission power levels from UEs may minimize interference, but this may impact the receiver's quality of service. Therefore, resource management that considers interference is a complex optimization issue. It is sometimes formulated as a transmit power minimization issue with a minimal quality of service constraint or as a weighted-sum-rate maximization problem with maximum transmit power and minimum quality of service requirements. One of the objectives of next-generation wireless networks is energy-efficient operation, which is facilitated by power management in addition to influence reduction. Transmissions should be carefully scheduled to reduce interference. Appropriate coding and modulation techniques and hybrid automated repeat requests strengthen the transmitted signal's resistance to noise. The optimization of mode selection, resource allocation, and interference reduction is often done in tandem. Numerous centralized, distributed, and hybrid methods have been developed for these three issues

6.1.16 D2D with mobility

While most D2D research has focused on static consumers, cellular networks mainly serve mobility users. More research is necessary to understand how performance gains rise in dynamic scenarios and what disturbance management and handover processes are required when UEs move within and across cells. D2D multi-hop communication is not without difficulties.

6.1.17 Costing

This constitutes one of the most pressing issues for mobile network operators. It might be difficult to regulate the direct link between users' devices and chargers. For example, operators may use UEs as switches for other users, paying the relay UEs with cash. Moreover, operators may charge for services like security provided during D2D connection. In the framework of D2D communication, many economic models include the following: D2D UEs in a cluster may purchase or sell data items; cellular users can sell their bandwidth to D2D UEs; and D2D-capable UE pairs may auction their cell phones to other queuing cellular users, and potentially opt to participate in D2D communication.

6.1.18 Safety

D2D communication offers better anonymity and data privacy than a traditional cellular connection since data is not kept centrally. D2D connections, however, may be rendered inoperable by several frequent assaults, including virus attacks, man-in-the-middle attacks, surveillance, denial of service attacks, and node impersonation. Additionally, users want to safeguard their privacy by, for example, limiting who may access their sensitive personal information. Implementing the same absence of a central authority complicates the implementation of security and privacy safeguards. The authors use three dimensions to represent threats:

1. The attacker's identity (internal vs external)
2. Their activity level (active vs passive)
3. The extent of the assault (local vs network-wide)

6.1.19 Applications and situations for D2D-6G environments

D2D connectivity is anticipated to be essential in allowing multifarious scenarios and applications in a 6G communication environment. Several situations and uses for D2D communication in a 6G environment are envisaged, even if 6G technologies are still in the conceptual and early development stages:

Massive Internet of Things (IoT): D2D communication can enable many IoT devices by allowing effective direct connections between devices.

Ultra-Reliable Low Latency Communication (URLLC): For crucial applications like industrial automation, healthcare monitoring, and driverless cars, D2D communication can provide ultra-low latency and excellent dependability [23].

Enhanced Mobile Broadband (eMBB): By facilitating direct connection between adjacent devices and lessening the strain on cellular networks, D2D communication may improve data rates and capacity for mobile broadband.

Smart Cities: Applications such as intelligent transportation systems, environmental monitoring, and smart grids may be made possible via D2D communication.

Immersive Virtual Reality (VR) and Augmented Reality (AR): D2D communication may provide smooth communication and cooperation between AR and VR devices, improving the user experience.

Edge Computing: By combining D2D communication with edge computing, latency may be decreased for various applications by supporting distributed processing and storage.

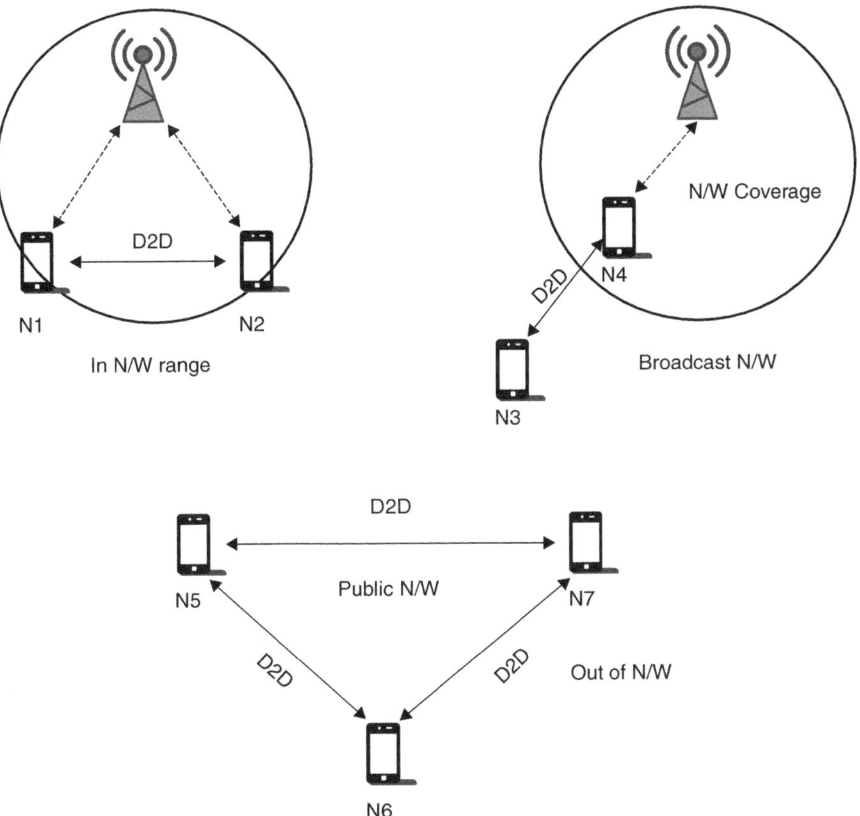

Figure 6.7 Network ranges.

Collaborative Edge Learning: Devices may exchange data and increase their collective intelligence using D2D connectivity, which enables collaborative machine learning and AI processing at the edge.

Energy-Efficient Communication: By allowing devices to speak with each other directly and minimizing the need for continuous contact with central network components as shown in Figure 6.7, D2D communication may help reduce energy consumption.

6.2 CONCLUSION

We intend to have D2D communications in the framework of future 6G. Our focus is on implementing intelligent D2D for future 6G communication systems, including D2D-enabled intelligent network splitting, D2D-enhanced edge computing for mobile devices, NOMA, and D2D-based intellectual networking, based on the trend of future 6G designs and the development of upcoming terminal user equipment. Our discussion needs to act as a blueprint for upcoming 6G D2D advancements.

REFERENCES

1. Akhtar, M.W., Hassan, S.A., Ghaffar, R. et al.; The shift to 6G communications: Vision and requirements. *Human-centric Computing and Information Sciences* 2020, 10, 53. https://doi.org/10.1186/s13673-020-00258-2.
2. Asghar, M.Z., Memon, S.A., Hämäläinen, J.; Evolution of wireless communication to 6G: Potential applications and research directions. *Sustainability* 2022, 14(10), 6356. https://doi.org/10.3390/su14106356.
3. Zhang, H., Song, L., Zhang, Y.J.; Load balancing for 5G ultra-dense networks using device-to-device communications. *IEEE Trans Wireless Communication* 2018, 17(6), 4039–4050. https://doi.org/10.1109/TWC.2018.2819648.
4. Zhang, S., Liu, J., Guo, H. et al.; Envisioning device-to-device communications in 6G. *IEEE Network* 2019, 34, 86–91.
5. Qureshi, S., Hassan, S.A., Jayakody, D.N.K.; Divide-and-allocate: An uplink successive bandwidth division NOMA system. *Transactions on Emerging Telecommunications Technologies* 2018, 29, e3216. https://doi.org/10.1002/ett.3216.
6. Bhat, J.R., Alqahtani, S.A.; 6G ecosystem: Current status and future perspective. *IEEE Access* 2021, 9, 43134–43167. https://doi.org/10.1109/ACCESS.2021.3054833.
7. Gozalvez, J., Lucas-Estañ, M.C., Sanchez-Soriano, J.; Joint radio resource management for heterogeneous wireless systems. *Wireless Network* 2012, 18, 443–455. https://doi.org/10.1007/s11276-011-0410-3.
8. Mishra, S.; Artificial intelligence assisted enhanced energy efficient model for device-to-device communication in 5G networks. *Human-Centric Intelligent Systems* 2023. https://doi.org/10.1007/s44230-023-00040-4.
9. Jayakumar, S., Nandakumar, S.; Reinforcement learning based distributed resource allocation technique in device-to-device (D2D) communication. *Wireless Network* 2023, 29, 1–16. https://doi.org/10.1007/s11276-023-03230-x.
10. Kuthadi, V.M., Selvaraj, R., Baskar, S. et al.; Optimized energy management model on data distributing framework of wireless sensor network in IoT system. *Wireless Personal Communications* 2022, 127(2), 1377–1403. https://doi.org/10.1007/s11277-021-08583-0.
11. Es-Saqy, A., Abata, M., Fattah, M. et al.; Terahertz VCO design for high-speed wireless communication systems. *Terahertz Wireless Communication Components and System Technologies* 2022, 1–16. https://doi.org/10.1007/978-981-16-9182-9_1.
12. Shah, S.W.H., Rahman, M.M.U., Mian, A.N., et al.; On the impact of mode selection on effective capacity of device-to-device communication. *IEEE Wireless Communications Letters* 2019, 8(3), 945–948. https://doi.org/10.1109/LWC.2019.2901460.
13. Chowdhury, M.Z., Shahjalal, M., Hasan, M. et al.; The role of optical wireless communication technologies in 5G/6G and IoT solutions: Prospects, directions, and challenges. *Applied Sciences* 2019, 9(20), 4367.

14. Puspitasari, A.A., An, T.T., Alsharif, M.H. et al.; Emerging technologies for 6G communication networks: Machine learning approaches. *Sensors* **2023**, 23(18), 7709. https://doi.org/10.3390/s23187709.

15. Ali, S., Sohail, M., Shah, S.B.H. et al.; New trends and advancement in next generation mobile wireless communication (6G): A survey. *Wireless Communications and Mobile Computing* **2021**, Article ID 9614520, 14 pages. https://doi.org/10.1155/2021/9614520.

16. Hakeem, S.A., Hussein, H.H., Kim, H.; Vision and research directions of 6G technologies and applications. *Journal of King Saud University Computer and Information Sciences* **2022**, 34(6), 2419–2442. https://doi.org/10.1016/j.jksuci.2022.03.019.

17. Kar, U.N., Sanyal, D.K.; A critical review of 3GPP standardization of device-to-device communication in cellular networks. *SN Computer Science* **2020**, 1, 37. https://doi.org/10.1007/s42979-019-0045-5.

18. Jameel, F., Hamid, Z., Jabeen, F. et al.; A survey of device-to-device communications: Research issues and challenges. *IEEE Communications Surveys & Tutorials* **2018**, 20(3), 2133–2168. https://doi.org/10.1109/COMST.2018.2828120.

19. You, X., Wang, C.X., Huang, J. et al.; Towards 6G wireless communication networks: Vision, enabling technologies, and new paradigm shifts. *Science China Information Sciences* **2021**, 64, 110301. https://doi.org/10.1007/s11432-020-2955-6.

20. Feng, D., Lu, L., Yuan-Wu, Y. et al.; Device-to-device communications in cellular networks. *IEEE Communications Magazine* **2014**, 52(4), 49–55. https://doi.org/10.1109/MCOM.2014.6807946.

21. Zong, B., Fan, C., Wang, X. et al.; 6G technologies: Key drivers, core requirements, system architectures, and enabling technologies. *IEEE Vehicular Technology Magazine* **2019**, 14(3), 18–27. https://doi.org/10.1109/MVT.2019.2921398.

22. Sultana, A., Woungang, I., Anpalagan, A. et al.; Efficient resource allocation in SCMA-enabled device-to-device communication for 5G networks. *IEEE Transactions on Vehicular Technology* **2020**, 69(5), 5343–5354. https://doi.org/10.1109/TVT.2020.2983569.

23. Du, J., Jiang, C., Wang, J. et al.; Machine learning for 6G wireless networks: Carrying forward enhanced bandwidth, massive access, and ultrareliable/low-latency service. *IEEE Vehicular Technology Magazine* **2020**, 15(4), 122–134. https://doi.org/10.1109/MVT.2020.3019650.

Chapter 7

Intelligent reflecting surface assisted wireless system design and analysis

Sanjeev Sharma and Dharmendra Dixit

7.1 INTRODUCTION

The intelligent reflecting surface (IRS)-assisted system design is considered for next-generation wireless systems to enhance their energy and spectral efficiencies. The IRS is also known as reconfigurable intelligent surface (RIS) or smart reflecting surface (SRS) in the literature [1–3]. The deployment of IRS in communication networks opens up new possibilities for advanced wireless services, including 5G and beyond. IRS technology represents a paradigm shift in wireless communication system since it changes the signal propagation environment with transmitter (Tx) and receiver (Rx) optimization. IRS consists of a large number of passive reflecting elements, such as meta-atoms or meta-surfaces, that can be electronically controlled to manipulate the propagation of electromagnetic waves. By adjusting the phase and amplitude of the reflected signals, IRS can enable beamforming, signal focusing, interference suppression, and other advanced signal processing techniques [3].

The concept of IRS leverages the principles of wave interference and beamforming to enhance wireless communication performance. Unlike traditional relay or amplification techniques, the IRS does not require its own power source or complex signal processing capabilities. Instead, it utilizes the existing signals and intelligently reflects them to achieve desired outcomes. By tuning the phases of the impinged signals, such that the constructive addition of reflected signals, it enhances the signal strength at the desired user [4,5]. IRS is also used to improve coverage, increased data rates, and enhanced security of a wireless system. It can be deployed in various environments, including indoor, outdoor, and urban areas, to overcome signal blockages, reduce interference, and extend coverage range [6,7].

The IRS can adapt its reflection coefficients and beamforming weights in real time based on the changing wireless channel conditions. This allows for dynamic channel reconfiguration to optimize the signal quality, capacity, and coverage area. Further, by actively manipulating the reflected signals, the IRS can cancel or mitigate interference coming from other sources at the desired user. For example, in Figure 7.1, IRS can be used to create virtual

DOI: 10.1201/9781003522003-8

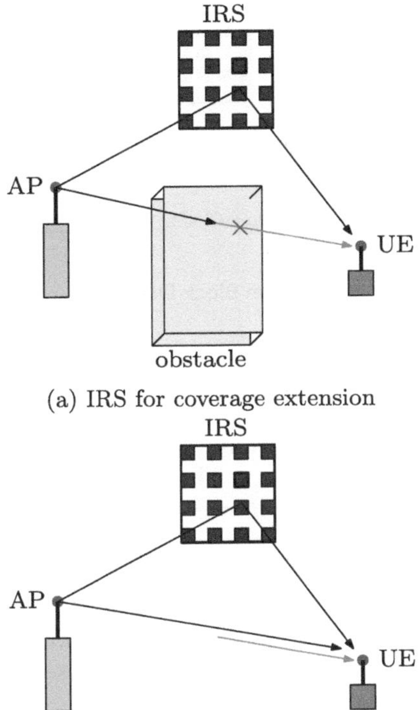

(a) IRS for coverage extension

(b) IRS for improving channel rank

Figure 7.1 Functions of IRS for wireless channel reconfiguration.

line-of-sight (LOS) channel between access point (AP) and user equipment (UE) when direct LOS path is blocked due to an obstacle. Additionally, the IRS can enhance signal paths in a specific direction to optimize the channel rank condition. It achieves this by improving the distribution of channel statistics, such as converting Rayleigh or fast fading to Rician or slow fading, thereby ensuring ultra-high reliability. Furthermore, the IRS can effectively suppress or nullify co-channel or inter-cell interference, among other benefits.

7.2 IRS-ASSISTED WIRELESS SYSTEM

In this section, we describe the analog and discrete time representation of an IRS-assisted wireless system. In Figure 7.2, IRS-assisted system is shown with Tx and Rx. Received signal is coming from the direct (Tx-Rx) and cascaded (Tx-IRS-Rx) links, as shown in Figure 7.2.

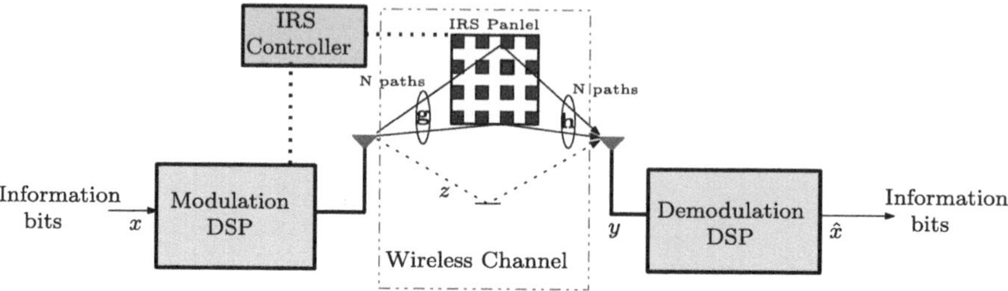

Figure 7.2 IRS-assisted Wireless System Block Diagram

7.2.1 IRS system modeling

To simplify the explanation, we consider a scenario where both the Tx and Rx have a single antenna. Communication system operates in a narrow band at a specific carrier frequency. The carrier frequency is represented by f_c and the system bandwidth by B in Hertz (Hz), with the constraint that B is less than or equal to f_c, i.e., $B \ll f_c$.

Let's consider the baseband transmit signal as $x(t)$, where $x(t)$ is a complex-valued signal. For simplicity, we initially focus on the propagation of the signal from the Tx to the Rx through a specific reflecting element of the IRS, represented by n (where n belongs to the set $\{1, ..., N\}$). We denote the complex channel coefficient from the Tx to the IRS element n as $h_n e^{j\theta_n}$, where h_n and θ_n represent the amplitude attenuation and phase shift of the frequency-flat channel in the narrow-band system, respectively. The passband signal impinging on the IRS element n can then be expressed as follows:

$$y_{n,\text{in}}(t) = \text{Re}\left[h_n e^{j\theta_n} x(t) e^{j2\pi f_c t} \right]. \tag{7.1}$$

Let's define the amplitude attenuation and time delay caused by IRS element n as a_n, which ranges from 0 to 1, and t_n, which ranges from 0 to $1/f_c$, respectively. The reflected signal by IRS element n is given as

$$y_{n,\text{out}}(t) = a_n y_{n,\text{in}}(t - t_n)$$

$$= \text{Re}\left[a_n h_n e^{j\theta_n} x(t - t_n) e^{j2\pi f_c(t - t_n)} \right] \tag{7.2}$$

$$\approx \text{Re}\left[a_n h_n e^{j\theta_n} e^{j\phi_n} x(t) e^{j2\pi f_c t} \right],$$

where we have considered that $x(t - t_n) \approx x(t)$ due to the fact that $t_n \leq 1/f_c \ll 1/B$, and $\phi_n = -j2\pi f_c t_n \in [0, 2\pi)$ is the phase shift induced by IRS element n.

Further, we consider $s_{n,\text{in}}(t) = h_n e^{j\theta_n} x(t)$ and $s_{n,\text{out}}(t) = a_n h_n e^{j\theta_n} e^{j\phi_n} x(t)$ which is the equivalent baseband signals corresponding $y_{n,\text{in}}(t)$ and $y_{n,\text{out}}(t)$, respectively. Thus, the IRS signal reflection model in the baseband is expressed as

$$s_{n,\text{out}}(t) = a_n e^{j\phi_n} s_{n,\text{in}}(t). \tag{7.3}$$

Here, ϕ_n within the range of $[0, 2\pi]$. For the sake of convenience, we consider the IRS phase shift within the interval $[0, 2\pi)$. According to Equation (7.3), we observe that in the baseband signal model, the output or reflected signal from IRS element n can be obtained by multiplying the corresponding input or impinging signal by a complex reflection coefficient $a_n e^{j\phi_n}$.

When the reflected signal from IRS element n reaches the receiver, it experiences a similar equivalent narrow-band frequency-flat channel, represented by $g_n e^{j\psi_n}$. Consequently, the passband signal received by the receiver through the reflection of IRS element n can be expressed as follows [8]

$$y_n(t) = \text{Re}\left[\left(g_n e^{j\psi_n} a_n e^{j\phi_n} h_n e^{j\theta_n} x(t)\right) e^{j2\pi f_c t}\right]. \tag{7.4}$$

Therefore, we have successfully modeled the cascaded channel from the transmitter to the receiver through IRS element n. We can denote the cascaded channel as h_n and g_n. The corresponding baseband signal model for Equation (7.4) can be expressed as follows

$$y_n(t) = a_n e^{j\phi_n} g_n h_n^* x(t). \tag{7.5}$$

By examining Equation (7.5), we can observe that the IRS reflected channel is obtained through the multiplication of three components. These components are the transmitter-to-element n channel (h_n), the IRS reflection coefficient $(a_n e^{j\phi_n})$, and the IRS element n-to-receiver channel (g_n).

The received signal, considering the contributions from all IRS elements, can be represented as a superposition of their respective reflected signals. Consequently, the baseband signal model, which takes into account all N IRS elements, can be expressed as follows

$$y(t) = \sum_{n=1}^{N} y_n(t) = \left(\sum_{n=1}^{N} a_n e^{j\phi_n} g_n h_n^*\right) x(t). \tag{7.6}$$

Here, h_n^* denotes the conjugate of the h_n. It is important to note that an IRS with N elements essentially functions as a linear mapping from the incident (input) signal to the reflected (output) signal.

7.2.2 Digital baseband system modeling

Here, the input and output relation of IRS-assisted system is formulated by considering discrete-time representation of the baseband channel as vectors. We consider a single-input single-output (SISO) based wireless system assisted by an IRS panel, as shown in Figure 7.2. $\mathbf{g} \in C^{N \times 1}$ and $\mathbf{h} \in C^{N \times 1}$ are the channel vectors between Tx and IRS, and between IRS and Rx, respectively. In addition, a direct path between Tx and Rx is denoted as $z \in C$. The received signal y is expressed as

$$y = \left(\mathbf{h}^T \mathbf{\Phi} \mathbf{g} + z \right) x,$$

where x is the transmitted symbol and $\mathbf{\Phi}$ is the IRS panel response. $\mathbf{\Phi}$ can be expressed as [9] $\mathbf{\Phi} = \mathrm{diag}[a_1 e^{j\phi_1}, a_2 e^{j\phi_2}, ..., a_N e^{j\phi_N}]$. Here, a_i and ϕ_i $(i = 1, 2, ..., N)$ are the amplitude and phase of the ith reflecting surface of the IRS, respectively. The term "$\mathrm{diag}[\cdot]$" denotes a diagonal matrix, where the elements of the vector within the brackets are placed along the diagonal of the matrix.

The channel vector $\mathbf{h}$ is expressed as $\mathbf{h} = \left[h_1 e^{j\theta_1} \cdots h_N e^{j\theta_N} \right]^T$, where h_i and θ_i $(i = 1, 2, ..., N)$ denote the amplitude and phase of ith component of the channel vector $\mathbf{h}$. Similarly, the channel vector $\mathbf{g}$ is expressed as $\mathbf{g} = \left[g_1 e^{j\psi_1} \cdots g_N e^{j\psi_N} \right]^T$, where g_i and ψ_i $(i = 1, 2, ..., N)$ denote the amplitude and phase of ith component of the channel vector $\mathbf{g}$. Therefore, received signal (without direct path) is expressed as [10]

$$y = \mathbf{h}^T \mathbf{\Phi} \mathbf{g} = \left[h_1 e^{j\theta_1} \cdots h_N e^{j\theta_N} \right] \begin{bmatrix} a_1 e^{j\phi_1} & \cdots & 0 \\ \vdots & \ddots & \vdots \\ 0 & \cdots & a_N e^{j\phi_N} \end{bmatrix} \begin{bmatrix} g_1 e^{j\psi_1} \\ \vdots \\ g_N e^{j\psi_N} \end{bmatrix}$$

$$= \left[a_1 g_1 h_1 e^{j(\phi_1 + \theta_1 + \psi_1)} + a_2 g_2 h_2 e^{j(\phi_2 + \theta_2 + \psi_2)} + \cdots + a_N g_N h_N e^{j(\phi_N + \theta_N + \psi_N)}. \right] \quad (7.7)$$

To maximum value of the received signal can be expressed as

$$y_{\max} = \max_{\{\phi_i \in (0, 2\pi]\}_{i=1}^N} \{y\}. \quad (7.8)$$

From the Equation (7.7) is clear that y is maximum when all the paths are added constructively at the receiver. Therefore,

$$\phi_1 = -\left(\theta_1 + \psi_1 \right), \phi_2 = -\left(\theta_2 + \psi_2 \right), ..., \phi_N = -\left(\theta_N + \psi_N \right) \quad (7.9)$$

and

$$y = a_1 g_1 h_1 + a_2 g_2 h_2 + \cdots + a_N g_N h_N = \sum_{i=1}^N a_i g_i h_i. \quad (7.10)$$

When the direct path is also present, each ϕ_i to optimize the received signal is expressed as $\phi_i = \angle z - (\theta_i + \psi_i)$, $i = 1, \ldots, N$, where $\angle z$ represents the phase of direct path z, and the maximized received signal is given as

$$y = \sum_{i=1}^{N} a_i g_i h_i e^{j\angle z} + z. \tag{7.11}$$

7.2.3 Impact of distance

In general, the channel coefficients in $\mathbf{h}$ and $\mathbf{g}$ generally depend on distance-related path loss, large-scale shadowing, and small-scale multipath fading. In particular, the path loss of the IRS-reflected channel plays a crucial role in analyzing the link budget and evaluating the performance of IRS-aided communications.

To illustrate this, let's consider an IRS element n, which is assumed to be located sufficiently far from both the Tx and Rx, with distances $d_{1,n}$ and $d_{2,n}$, respectively. Under the far-field propagation condition, we can assume that $d_1 = d_{1,n}$ and $d_2 = d_{2,n}$ $\forall n$.

Based on this assumption, it follows that $\mathbb{E}(|h_n|^2)$ is proportional to $c_1(d_1/d_0)^{-\alpha_1}$ and $\mathbb{E}(|g_n|^2)$ is proportional to $c_2(d_2/d_0)^{-\alpha_2}$, where c_1 and c_2 represent the corresponding path loss at the reference distance d_0, while α_1 and α_2 denote the path loss exponents with typical values ranging from 2 (in free-space propagation) to 6 [2]. Considering Equation (7.5), we can deduce that the average received signal power via the reflection by IRS element n, denoted as $P_{r,n}$, is inversely proportional to distance and is expressed as

$$P_{r,n} \propto \frac{1}{d_1^{\alpha_1} d_2^{\alpha_2}} \tag{7.12}$$

The reflected channel via IRS element n experiences double path loss, which is commonly known as the product-distance path loss model. Consequently, in practical scenarios, a significant number of IRS reflecting elements are required to counteract the substantial power loss caused by the double attenuation. This is achieved by jointly designing the reflection amplitudes and/or phases of the IRS elements to attain high passive beamforming gains.

In Figure 7.3(b), the IRS is substituted with an infinitely large perfect electric conductor (PEC), which can be represented as a metallic plate. Assuming free-space propagation and utilizing the image theory, it can be demonstrated that the received signal power at the receiver through the reflection from the PEC, denoted as P_r, is inversely proportional to the square of the sum of the distances of the two-hop links, as

$$P_r \propto \frac{1}{(d_1 + d_2)^2} \tag{7.13}$$

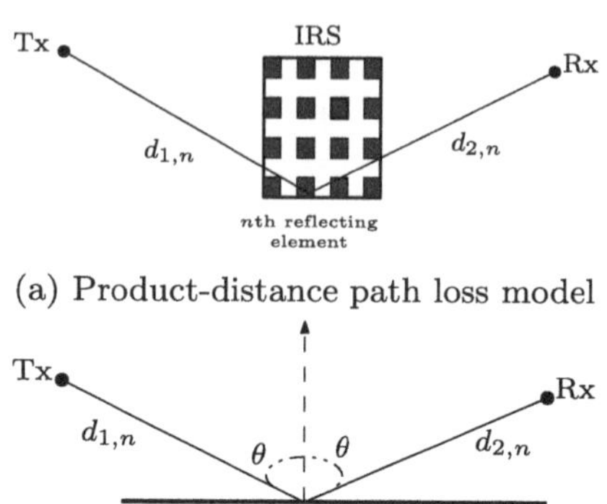

(a) Product-distance path loss model

(b) Sum-distance path loss model

Figure 7.3 Depiction for the path loss models of IRS-reflected channel.

This particular model is commonly known as the sum-distance path loss model. In this model, the received signal at the receiver is as if it originated from an equivalent transmitter located at the image point of the original transmitter, as shown in Figure 7.3(b). This effect is achieved due to the reflection from the infinitely large PEC, resulting in specular reflection.

However, it is essential to note that this model is specifically valid for free-space propagation with an infinitely large PEC and does not directly apply to the IRS-reflected channel modeled at the element level, as described in Equations (7.6)–(7.12). Therefore, it is inappropriate to utilize the sum-distance model in scenarios involving one or more finite-size tunable PECs and assume that the received signal power scales with the number of PECs by exploiting their multiplicative passive beamforming gains and benefiting from the more favorable sum-distance based path loss, even under free-space propagation.

7.3 OUTAGE PROBABILITY ANALYSIS OF IRS-ASSISTED SYSTEMS

In this section, outage probability (OP) of SISO based IRS-assisted system is derived over Rician fading channels. If we assume that the channel phases are accessible to every IRS panel controller and the optimal co-phasing matrix Φ is applied, the instantaneous optimal signal-to-noise ratio (SNR) γ of the received signal can be expressed as follows [11]:

$$\gamma = \gamma_0 \left(\sum_{i=1}^{N} |h_i| |g_i| \right)^2 . \tag{7.14}$$

In the above equation, the term $\gamma_0 = \dfrac{P}{N_0}$ represents the transmit SNR. The OP of a SISO-based communication system is given as

$$\begin{aligned} P_{\text{out}} &= \Pr\left(\gamma \le \gamma_{\text{th}}\right) \\ &= F_\gamma\left(\gamma_{\text{th}}\right), \end{aligned} \tag{7.15}$$

where $\Pr(\cdot)$ represents a probability operator, $F_\gamma(\cdot)$ denotes the cumulative distribution function (CDF) of the instantaneous SNR γ, and γ_{th} is a predefined threshold SNR. Therefore, it is necessary to derive an expression for $F_{\gamma_m}(\cdot)$ in order to obtain the expression for the OP.

The expression for the instantaneous SNR of the received signal can be reformulated as follows

$$\gamma = \mathcal{A}^2 \gamma_0. \tag{7.16}$$

Here, $\mathcal{A} = \displaystyle\sum_{i=1}^{N} z_i$ is defined as the sum of N terms, where each term $z_i = |h_i||g_i|$ represents the absolute value of the product of $|h_i|$ and $|g_i|$. It is assumed that the random variables (RVs) z_i are independent and identically distributed (i.i.d.) RVs. It is important to note that both $|h_i|$ and $|g_i|$ are independent and identically distributed (i.i.d.) Rician distributed RVs. Despite extensive research, it is not currently feasible to derive an exact, analytically tractable closed-form expression for the cumulative distribution function (CDF) $F_\gamma(\cdot)$. However, one can approximate $F_\gamma(\cdot)$ using the Central Limit Theorem (CLT)-based method.

For a large value of N, i.e., $N \gg 1$, with the aid of the CLT method, the RV $\mathcal{A}$ can be modeled as a normal distributed RV with the following mean μ and standard deviation σ [13]

$$\mu = \mathbf{E}\left[\mathcal{A}\right] = \frac{N\pi L_{0.5}\left(-K_{\text{h}}\right) L_{0.5}\left(-K_{\text{g}}\right)}{4\sqrt{d_{\text{h}}^{\alpha_{\text{h}}} d_{\text{g}}^{\alpha_{\text{g}}} \left(K_{\text{h}} + 1\right)\left(K_{\text{g}} + 1\right)}}, \tag{7.17}$$

$$\sigma^2 = \mathbf{Var}\left(\mathcal{A}\right) = \frac{N}{d_{\text{h}}^{\alpha_{\text{h}}} d_{\text{g}}^{\alpha_{\text{g}}}} \left[1 - \frac{\pi^2 L_{0.5}^2\left(-K_{\text{h}}\right) L_{0.5}^2\left(-K_{\text{g}}\right)}{16\left(K_{\text{h}} + 1\right)\left(K_{\text{g}} + 1\right)}\right], \tag{7.18}$$

where $\mathbf{E}[\cdot]$ is the statistical averaging operator, $\mathbf{Var}(\cdot)$ is the statistical variance operator, and $L_m(\cdot)$ denotes the Laguerre polynomial of degree m and K denotes the Rician factor. Hence, the CDF of $\mathcal{A}$ is expressed as [12]

$$F_{\mathcal{A}}(x) = 1 - \Theta Q\left(\frac{x - \mu}{\sigma}\right), \tag{7.19}$$

where $\Theta = \left(0.5 + 0.5\,\mathrm{erf}\left(\sqrt{\dfrac{\mu}{2\sigma^2}}\right)\right)^{-1}$, $\mathrm{erf}(\cdot)$ is the error function, and the Gaussian Q-function is denoted as $Q(\cdot)$ [12]. Now, one can get the CDF of γ as the following

$$F_\gamma(x) = \Pr(\gamma \le x) = \Pr\left(\mathcal{A}^2\gamma_0 \le x\right) = 1 - \Theta Q\left(\frac{\sqrt{\dfrac{x}{\gamma_0}} - \mu}{\sigma}\right) \tag{7.20}$$

Hence, the approximate OP of a SISO wireless system assisted by an IRS can be given as [13]

$$P_{\mathrm{out}} = \Pr(\gamma \le \gamma_{\mathrm{th}})$$

$$= F_\gamma(\gamma_{\mathrm{th}}) = 1 - \Theta Q\left(\frac{\sqrt{\dfrac{\gamma_{\mathrm{th}}}{\gamma_0}} - \mu}{\sigma}\right). \tag{7.21}$$

7.4 MIMO-BASED IRS SYSTEM

We consider a multiple-input multiple-output (MIMO) based wireless system assisted by an IRS panel, as shown in Figure 7.4. $\mathbf{G} \in \mathcal{C}^{N \times N_t}$ and $\mathbf{H} \in \mathcal{C}^{N \times N_r}$ are the channel matrices between Tx and IRS, and between IRS and Rx, respectively. In addition, a direct path between Tx and Rx is denoted as $\mathbf{Z} \in \mathcal{C}^{N_r \times N_t}$. The received signal $\mathbf{y}$ is expressed as

$$\mathbf{y} = \left(\mathbf{H}^T \boldsymbol{\Phi} \mathbf{G} + \mathbf{Z}\right)\mathbf{w}x,$$

where $\mathbf{w}$ is a precoding vector [2,14] and $\boldsymbol{\Phi}$ is the IRS panel response. In MIMO-based system, each IRS element has N_t incoming signal links and N_r outgoing signal link. Therefore, the optimal phase of IRS element can be set by considering all the incoming and outgoing channel links. In literature

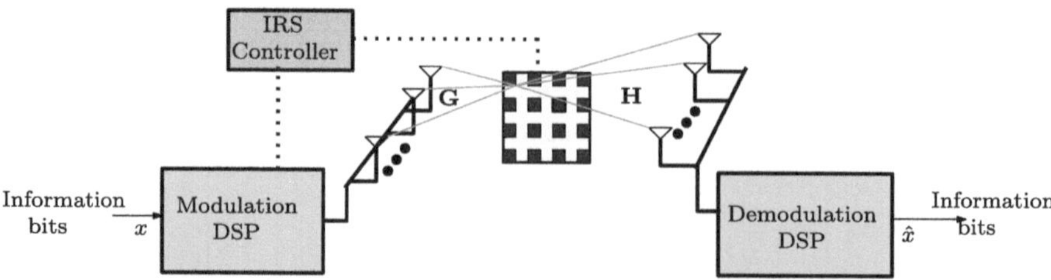

Figure 7.4 MIMO-based IRS-assisted Wireless System Block Diagram

[2,15], some low complexity algorithms are suggested to optimize the phase of IRS elements in MIMO-based system design. Further, joint beamforming (active beamforming at the BS and passive beamforming at the IRS panel) techniques are considered in MIMO system design.

7.5 IRS DESIGN AND DEVELOPMENT

The architecture of an IRS, also known as reconfigurable intelligent surface (RIS) or software-controlled metasurface, typically consists of a large number of passive reflecting elements [16]. These elements are designed to manipulate the electromagnetic waves by adjusting their reflection coefficients based on control signals.

In essence, a metasurface is a flat array consisting of carefully designed reflecting elements, also known as meta-atoms. These elements possess an electrical thickness typically smaller than the wavelength of the signal being considered. By manipulating various characteristics such as shape (e.g., square, triangle, or split-ring), size, orientation, and arrangement, it becomes possible to achieve the desired response from each element, such as controlling reflection amplitude or phase shift. However, in wireless communication, the channel tends to change over time due to the movement of transmitters, receivers, and surrounding objects. Therefore, it is necessary for an IRS to dynamically adjust its response in real time based on these channel variations. This requires manufacturing IRS elements with adjustable reflection coefficients and establishing a connection between the IRS and the wireless network. By learning and adapting to the external communication environment, the IRS can enable adaptive reflection, ensuring optimal performance [17].

Figure 7.5 depicts a typical architecture of an IRS, consisting of three layers. The outermost layer is made up of a multitude of tunable metallic patches printed on a dielectric substrate. These patches directly manipulate the incident signals. The intermediate layer employs a copper plate to minimize any leakage of signal energy during the reflection process. Following that is the innermost layer, which serves as a control circuit board. This layer is responsible for exciting the reflecting elements and dynamically adjusting their reflection amplitudes and/or phase shifts in real time.

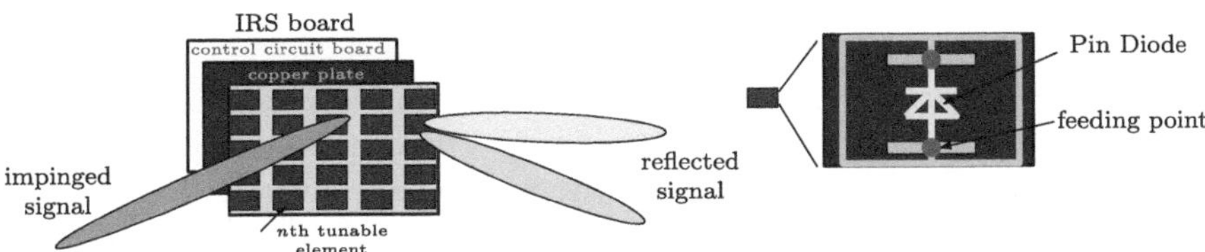

Figure 7.5 IRS architecture using tunable reflecting element based on PIN diode.

The reflection adaptation is initiated and regulated by a smart controller attached to each IRS. In practical applications, dedicated sensors can be deployed in the first layer, interlaced with the reflecting elements of the IRS. These sensors are used to sense the surrounding radio signals of interest, enhancing the IRS's ability to learn about the environment. The information obtained by the sensors facilitates the smart controller in designing the reflection coefficients.

Hardware implementation of IRS involves various components such as substrate, reflecting elements, sensors, phase shifters, and control circuitry. The substrate forms the base of the reflecting element surface. It provides structural support and can be made of dielectric or other suitable materials. Reflecting elements can be metallic patches, resonators, or other structures, depending on the desired functionality. Tunability can be achieved through various mechanisms, such as using varactors, phase shifters, or microelectromechanical systems technology. In some designs, an intermediate layer is incorporated between the substrate and the reflecting elements. This layer, often made of materials like copper, serves to minimize signal energy leakage during the reflection process and enhance overall performance.

The control circuit board is responsible for controlling and coordinating the operation of the reflecting elements. It provides the necessary signals and power to excite the elements and adjust their reflection properties dynamically. The control circuit board can include circuitry, microcontrollers, or programmable logic devices. In certain scenarios, dedicated sensors can be integrated into the reflecting element surface. These sensors, often placed within or alongside the reflecting elements, are used to gather information about the surrounding radio signals or environmental conditions. The sensor data can assist the smart controller in optimizing the reflection coefficients and adapting to the specific communication environment.

7.6 POSSIBLE CHALLENGES IN IRS-ASSISTED SYSTEM IMPLEMENTATION

We highlight some challenges in IRS-assisted system design. Some of the possible challenges are [5,15,17]:

Hardware Complexity: Designing and deploying IRS with a large number of reflecting elements can be technically challenging and costly. The complexity increases with the need for precise beamforming and control of each reflecting element.

Channel Estimation: Accurate channel estimation is crucial for optimizing the IRS performance. However, estimating the channel between the transmitter, IRS, and receiver accurately can be challenging due to multipath propagation, dynamic environment, and limited feedback resources.

Power Efficiency: Efficient power allocation and utilization are essential in IRS-assisted systems. Maximizing the signal power at the receiver while minimizing the power consumption at the IRS can be a challenging optimization problem.

Mobility Support: Handling mobility of users and devices in IRS-assisted systems poses challenges. The system needs to adapt to fast-changing channel conditions and ensure continuous connectivity and seamless handover without compromising performance.

Interference Management: Coexistence with neighboring systems and mitigating interference from other wireless devices operating in the same frequency bands can be a significant challenge in IRS-assisted systems. Ensuring interference control and coordination mechanisms are essential for maintaining system performance.

Deployment and Scalability: Deploying IRS in real-world environments, such as indoor or outdoor scenarios, may require careful planning, infrastructure modifications, and coordination with existing wireless networks. Additionally, ensuring scalability to accommodate a large number of users and multiple IRS deployments can be a challenge.

Complexity of System Optimization: Optimizing the overall system performance, including IRS phase configuration, beamforming, power allocation, and resource allocation, involves dealing with a highly complex optimization problem. Developing efficient algorithms and protocols for system optimization is a significant challenge.

7.7 RESULTS

We present BER, sum-rate, and OP results for an IRS-assisted wireless system. Simulation setup of IRS-assisted system is shown in Figure 7.6 by considering direct path is blocked, path only via IRS, and only the direct path exists between Tx and Rx, as mentioned in Figure 7.6. Average BER performance of an IRS-assisted system is shown in Figure 7.7 over Rayleigh fading channels by varying reflecting elements N. As we observed in Figure 7.7, BER improves as N increases. Further, in the presence of both links (direct and reflected via IRS), BER performance is better (around 2 dB at BER $= 10^{-3}$ for $N = 64$) as compared to only reflected link. Furthermore, BER improves using the IRS as shown in Figure 7.7.

Sum-rate performance of system is shown in Figure 7.8 by considering only the reflected link (direct link is blocked). As N increases, sum-rate improves due to higher receiver signal strength. Sum-rate R is calculated using the following expression

$$R = \log_2\left(1 + \mathcal{A}^2 \gamma_0\right) \text{ bps/Hz.} \tag{7.22}$$

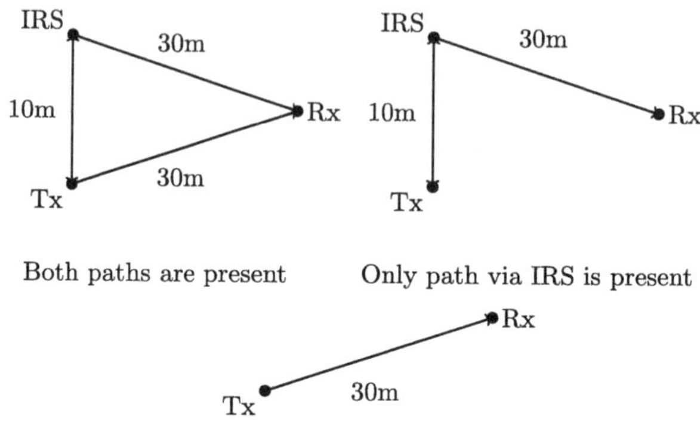

Both paths are present Only path via IRS is present

Only direct path is present

Figure 7.6 Simulation setup for BER plot.

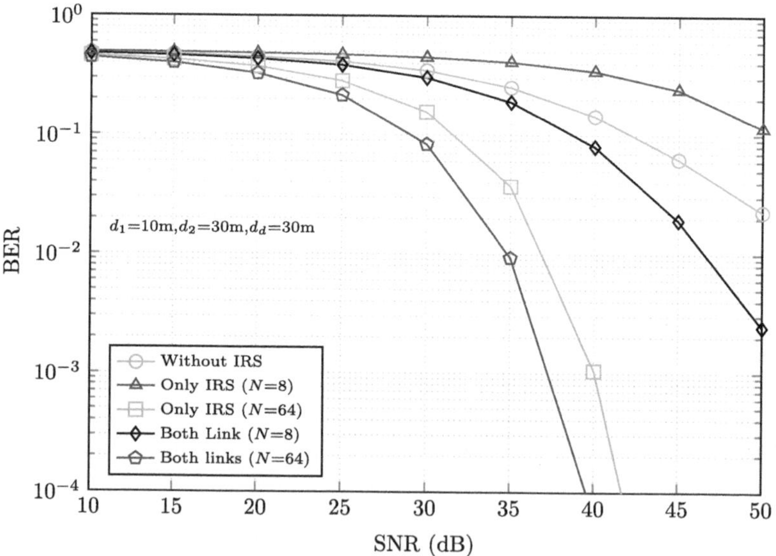

Figure 7.7 BER performance of IRS-assisted system by considering different links and reflecting elements at the IRS panel.

Further, we observe that IRS is more effective at low SNR as compared to high SNR region.

Next, we have shown the OP performance of IRS-assisted system over Rayleigh fading channels by varying N using the reflected link only in Figure 7.9. For OP, we consider a sub-6G scenario with the bandwidth of system is 180 kHz, the power spectral density of noise is -173 dBm/Hz, the distances (in meters) are taken to be $d_1 = 100$ m and $d_2 = 150$ m. Path-loss

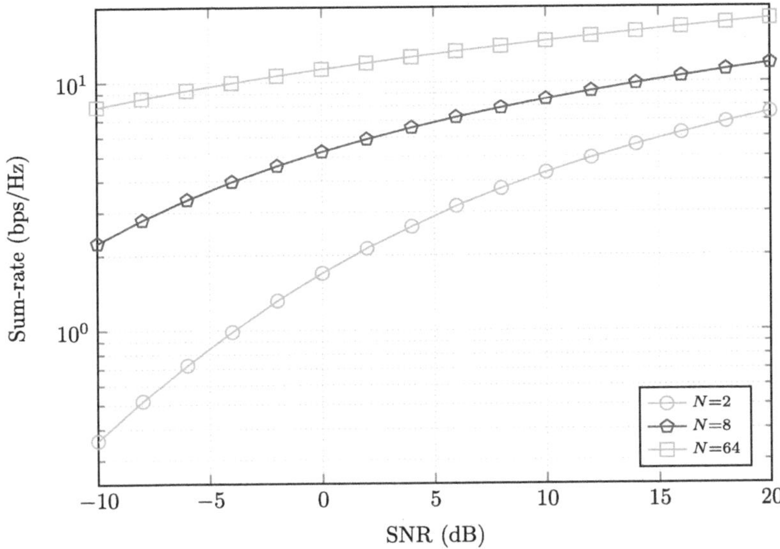

Figure 7.8 Sum-rate performance of IRS-assisted system by considering different reflecting elements at the IRS panel.

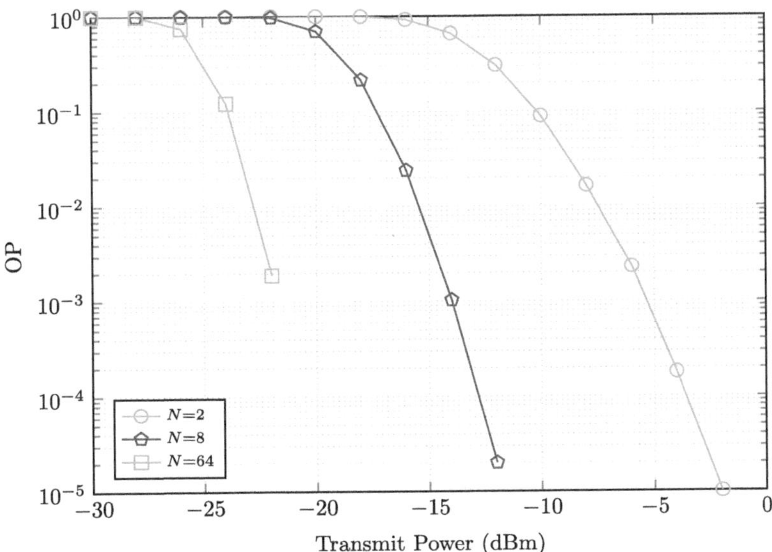

Figure 7.9 OP performance of IRS-assisted system by considering different reflecting elements at the IRS panel.

exponent is 2.7 and the reference path loss at the reference distance $d_0 = 1$ m is taken to be -30 dB. The threshold SNR is taken to be $\gamma_{\mathrm{th}} = 10$ dB. The OP performance is plotted against transmitted power P (dBm). As N increases, OP performance improves, as observed in Figure 7.9.

7.8 EXAMPLES

Example 7.1 For SISO IRS-assisted communication system channels h and g form BS to IRS, and form IRS to Rx for a SISO system are defined as

$$\mathbf{h} = [-0.9247 - 0.9545i \ -0.3066 + 2.1460i \ 0.2423$$

$$+0.5129i \ 2.5303 - 0.0446i \ 1.9583 + 0.5054i]$$

$$\mathbf{g} = [-0.1449 + 0.4748i \ -0.0878 - 0.8538i \ 1.0534$$

$$+ 0.5072i \ 0.9963 + 1.1528i \ 1.0021 + 0.3457i].$$

Find the optimal received signal amplitude.

Solution: The received signal is expressed as $y = \mathbf{h}\Phi\mathbf{g}^T$, where Φ is a diagonal matrix of size 5×5. Φ is expressed as

$$\Phi = \begin{bmatrix} a_1e^{j\phi_1} & 0 & 0 & 0 & 0 \\ 0 & a_2e^{j\phi_2} & 0 & 0 & 0 \\ 0 & 0 & a_3e^{j\phi_3} & 0 & 0 \\ 0 & 0 & 0 & a_4e^{j\phi_4} & 0 \\ 0 & 0 & 0 & 0 & a_5e^{j\phi_5} \end{bmatrix}.$$

Let consider $a_i = 1$, i from 1 to 5, i.e., reflection coefficients are one. Further, h and g are written as

$$\mathbf{h} = \begin{bmatrix} 1.3290e^{-2.3403j} & 2.1678e^{1.7127j} & 0.5673e^{1.1295j} \\ 2.5307e^{-0.0176j} & 2.0225e^{0.2526j} \end{bmatrix}$$

$$\mathbf{g} = \begin{bmatrix} 0.4964e^{1.8671j} & 0.8583e^{-1.6732j} & 1.1691e^{0.4487j} \\ 1.5237e^{0.8581j} & 1.0601e^{0.3322j} \end{bmatrix}.$$

The received signal can be expressed as

$$y = \begin{bmatrix} 1.3290e^{-2.3403j} & 2.1678e^{1.7127j} & 0.5673e^{1.1295j} \\ 2.5307e^{-0.0176j} & 2.0225e^{0.2526j} \end{bmatrix}$$

$$\times \begin{bmatrix} e^{j\phi_1} & 0 & 0 & 0 & 0 \\ 0 & e^{j\phi_2} & 0 & 0 & 0 \\ 0 & 0 & e^{j\phi_3} & 0 & 0 \\ 0 & 0 & 0 & e^{j\phi_4} & 0 \\ 0 & 0 & 0 & 0 & e^{j\phi_5} \end{bmatrix} \begin{bmatrix} 0.4964e^{1.8671j} \\ 0.8583e^{-1.6732j} \\ 1.1691e^{0.4487j} \\ 1.5237e^{0.8581j} \\ 1.0601e^{0.3322j} \end{bmatrix}$$

$$= 0.6598e^{j(\phi_1 - 0.4733)} + 1.8607e^{j(\phi_2 + 0.0395)} + 0.6632e^{j(\phi_3 + 1.5782)}$$

$$+ 3.8559e^{j(\phi_4 + 0.8404)} + 2.1439e^{j(\phi_5 + 0.5848)}$$

To maximize the y, phase values are set as $\phi_1 = 0.4733$, $\phi_2 = -0.0395$, $\phi_3 = -1.5782$, $\phi_4 = -0.8404$, and $\phi_5 = -0.5848$. Thus, y is written as

$$y = 0.6598 + 1.8607 + 0.6632 + 3.8559 + 2.1439 = 9.1835 \, \text{unit}.$$

The received signal is real due to phase cancellation using IRS panel, and results in enhancement of the received signal strength.

Example 7.2 For a MIMO based IRS-assisted system channels H and G form BS to IRS, and form IRS to Rx are defined as

$$\mathbf{H} = \begin{bmatrix} 0.7316 - 0.0723i & 0.2078 + 0.2212i & -0.8111 - 0.1166i \\ 0.5140 - 0.1707i & -0.5567 - 0.6116i & -0.7558 + 0.4439i \\ -0.2146 + 0.2257i & 0.6282 - 0.0212i & -0.5724 + 0.7731i \\ \\ -2.0819 + 0.7844i & -0.5338 - 0.8585i \\ 1.0171 - 0.6107i & 0.9689 - 0.7874i \\ 0.2299 + 0.0547i & -1.2102 - 0.0048i \end{bmatrix}$$

$$\mathbf{G} = \begin{bmatrix} 1.0837 + 0.0608i & 0.2626 - 0.5249i & 0.7901 + 1.6620i \\ -0.5442 - 1.0547i & -0.1595 - 0.7507i & -0.7701 - 0.4353i \\ \\ 0.0230 + 0.5290i & 0.7782 + 0.6283i \\ 0.3907 - 0.1361i & 1.0919 - 0.5408i \end{bmatrix}$$

Find the optimal received signal amplitude using maximum energy channel between BS-IRS and IRS-Rx: The channel is expressed as $\tilde{\mathbf{h}} = \mathbf{G}\boldsymbol{\Phi}\mathbf{H}^T$, where $\boldsymbol{\Phi}$ is a diagonal matrix of size 5×5. $\boldsymbol{\Phi}$ is expressed as

$$\boldsymbol{\Phi} = \begin{bmatrix} a_1 e^{j\phi_1} & 0 & 0 & 0 & 0 \\ 0 & a_2 e^{j\phi_2} & 0 & 0 & 0 \\ 0 & 0 & a_3 e^{j\phi_3} & 0 & 0 \\ 0 & 0 & 0 & a_4 e^{j\phi_4} & 0 \\ 0 & 0 & 0 & 0 & a_5 e^{j\phi_5} \end{bmatrix}.$$

Let consider $a_i = 1$, i from 1 to 5, i.e., reflection coefficients are one. Further, norm of each row of **H** as 2.6974, 2.1706, 1.7140. Therefore, we chooses first row of **H** to set the phase in maximum energy path based phase optimization. Similar, norm of each row of **G** as 2.4880, 2.1062. Thus we select first row of **G** to set the phase pf the IRS panel as

$$\mathbf{H}(1,:) = \begin{bmatrix} 0.7352 e^{-0.0985j} & 0.3035 e^{0.8167j} & 0.8194 e^{-2.9988j} \\ 2.2248 e^{2.7813j} & 1.0109 e^{-2.1271j} \end{bmatrix}$$

$$\mathbf{G}(1,:) = \begin{bmatrix} 1.0854 e^{0.0560j} & 0.5869 e^{-1.1069j} & 1.8403 e^{1.1270j} \\ 0.5295 e^{1.5273j} & 1.0002 e^{0.6792j} \end{bmatrix}.$$

The phases of IRS elements are set as $\phi_1 = 0.0985 - 0.0560 = 0.0425$, $\phi_2 = -0.8167 + 1.1069 = 0.2902$, $\phi_3 = 2.9988 - 1.1270 = 1.8718$, $\phi_4 = -2.7813 - 1.5273 = -4.3086$, and $\phi_5 = 2.1271 - 0.6792 = 1.4479$. Thus, Φ is expressed as

$$\Phi = \begin{bmatrix} e^{j\phi_1} & 0 & 0 & 0 & 0 \\ 0 & e^{j\phi_2} & 0 & 0 & 0 \\ 0 & 0 & e^{j\phi_3} & 0 & 0 \\ 0 & 0 & 0 & e^{j\phi_4} & 0 \\ 0 & 0 & 0 & 0 & e^{j\phi_5} \end{bmatrix}$$

$$= \begin{bmatrix} 2.7134 + 0.1153i & 0 & 0 \\ 0 & 2.5009 + 0.7359i & 0 \\ 0 & 0 & 0.4294 + 0.6069i \\ 0 & 0 & 0 \\ 0 & 0 & 0 \\ & & \\ 0 & 0 \\ 0 & 0 \\ 0 & 0 \\ 0.4092 + 0.5369i & 0 \\ 0 & 0.6180 + 0.9466i \end{bmatrix}$$

The $\tilde{\mathbf{h}}$ is written as

$$\tilde{\mathbf{h}} = \begin{bmatrix} 2.3444 - 2.0443i & 1.1827 + 2.3716i & 1.5023 - 2.8653i \\ -2.6572 - 2.0213i & 1.4391 + 1.9048i & -2.0759 - 0.7085i \end{bmatrix}$$

In MIOM-based system multiple ways to set the phase of the IRS panel. Therefore, phase can be optimized to get the best performance of a system.

7.9 CONCLUSION

A comprehensive design of an IRS-assisted wireless system is considered. Use cases of IRS are highlighted. Optimal phase shift design methods of the IRS are illustrated in the chapter. The IRS hardware design is also explained in detail. Further, an end-to-end IRS-assisted system design is mentioned with received and transmitted signals using mathematical models. BER and OP performance of the IRS-assisted system design are shown with some examples.

APPENDIX

MATLAB code of IRS-assisted system BER plot in Rayleigh fading:

```matlab
%% Sample Matlab code
clc;
clear all;
SNR_dB=-10:2:30;  % Range of SNR values in dB
SNR=10.^(SNR_dB/10);  % Convert SNR to linear scale
numBits=1e4;  % Number of bits to transmit
% Generate random bits
    bits=randi([0 1], numBits, 1);
    M=2;
    s1=qammod(bits,M); % M=2 BPSK, M=4->QPSK
    P=1; %unit
    s=sqrt(P)*s1;
    N=8; % number of reflecting surfaces
    nErr=zeros(1,length(SNR_dB));
for ii=1:length(SNR)
    n=sqrt(1/2)*[randn(numBits,1)+1j.*randn(numBits,1)];
        % white gaussian noise, 0dB variance
      h1=sqrt(1/2)*[randn(numBits,N)+1j.*randn(numBits,N)];
          % h=|h|exp(-angle(h)) Rayleigh channel Soure--> RIS
      g1=sqrt(1/2)*[randn(numBits,N)+1j.*randn(numBits,N)];
          % g=|g|exp(-angle(g)) Rayleigh channel RIS-->
          Destination
      %hd=sqrt(1/2)*[randn(numBits,1)+1j.*randn(numBits,1)];
          % h=|h|exp(-angle(h)) Rayleigh channel Soure--> RIS
    % Determination of Phases;
    q1=exp(-1j*(angle(h1)+angle(g1)));
    ghq1=sum((g1.*h1).*q1,2);% perfect phase cancellation
    %ghq1=sum((g1.*h1),2);% Random phase
    Es=mean((abs(s)).^2);
    sigma=sqrt(Es/(SNR(ii)));
    y=ghq1.*s++sigma.*n; % Channel and noise Noise addition
    yHat=y./(sqrt(P)*ghq1); % channel equalization
    xHat=qamdemod(yHat,M); % demodulation
    %%%%%
    nErr(ii)=sum(bits~=xHat); %bit errors calculation
end
simSer=nErr/numBits;% simulated BER
semilogy(SNR_dB,simSer ,'->k','LineWidth',2); hold on;
```

Here, we can observe $ghq1 = [7.1676 + 0.0000i 4.8691 - 0.0000i, ..., 6.0685 + 0.0000i]$ is pure real signal die to perfect phase cancellation.

MATLAB code of IRS-assisted system sumrate plot in Rayleigh fading:

```matlab
%% Sample Matlab code
clc;
clear all;
SNR_dB=-10:2:20;   % Range of SNR values in dB
SNR=10.^(SNR_dB/10);   % Convert SNR to linear scale
numBits=1e4;   % Number of bits to transmit
% Generate random bits
    bits=randi([0 1], numBits, 1);
    M=2;
    s1=qammod(bits,M); % M=2 BPSK, M=4->QPSK
    P=1; %unit
    s=sqrt(P)*s1;
    N=16; % number of reflecting surfaces
    Sumrate=zeros(length(SNR_dB), numBits);
for ii=1:length(SNR)
    n=sqrt(1/2)*[randn(numBits,1)+1j.*randn(numBits,1)];
        % white gaussian noise, 0dB variance
    h1=sqrt(1/2)*[randn(numBits,N)+1j.*randn(numBits,N)];
        % h=|h|exp(-angle(h)) Rayleigh channel Soure--> RIS
    g1=sqrt(1/2)*[randn(numBits,N)+1j.*randn(numBits,N)];
        % g=|g|exp(-angle(g)) Rayleigh channel RIS-->
        Destination
     %hd=sqrt(1/2)*[randn(numBits,1)+1j.*randn(numBits,1)];
          % h=|h|exp(-angle(h)) Rayleigh channel Soure--> RIS
     % Determination of Phases;
    q1=exp(-1j*(angle(h1)+angle(g1)));
    ghq1=sum((g1.*h1).*q1,2);% perfect phase cancellation
    %ghq1=sum((g1.*h1),2);% Random phase
    %sigma=sqrt(Es/(SNR(ii)));
    % Channel power gain
    channelGain=abs(ghq1).^2;
    Sumrate(ii,:)=Sumrate(ii,:)+ (log2(1+(P*SNR(ii).
        *channelGain)))';
end
sum_rate=sum(Sumrate,2)/numBits;% simulated BER
semilogy(SNR_dB,sum_rate ,'->k','LineWidth',2); hold on;
```

For example, $N = 8$ sumrate vector is $[2.2, 2.7, 3.3, 3.9, 4.5, 5.2, 5.8, 6.5, 7.1, 7.8, 8.5, ..., 9.1, 9.8, 10.4, 11.1, 11.8]$.

REFERENCES

1. E. Basar, M. Di Renzo, J. De Rosny, M. Debbah, M.-S. Alouini, and R. Zhang, "Wireless commun. through reconfigurable intelligent surfaces," *IEEE Access*, vol. 7, pp. 116753–116773, 2019.
2. Q. Wu, S. Zhang, B. Zheng, C. You, and R. Zhang, "Intelligent reflecting surface-aided wireless communications: A tutorial," *IEEE Transactions on Communications*, vol. 69, no. 5, pp. 3313–3351, 2021.
3. M. H. Kumar, S. Sharma, K. Deka, and V. Bhatia, "Intelligent reflecting surface assisted terahertz communications," in *2022 IEEE International Conference on Signal Processing and Communications (SPCOM)*. IEEE, 2022, pp. 1–5. Bangalore, India.
4. A. Thomas, K. Deka, S. Sharma, and N. Rajamohan, "IRS-assisted OTFS system: Design and analysis," *IEEE Transactions on Vehicular Technology*, vol. 72, no. 3, pp. 3345–3358, 2022.
5. Q. Wu and R. Zhang, "Towards smart and reconfigurable environment: Intelligent reflecting surface aided wireless network," *IEEE Communications Magazine*, vol. 58, no. 1, pp. 106–112, 2019.
6. M. Di Renzo, A. Zappone, M. Debbah, M.-S. Alouini, C. Yuen, J. De Rosny, and S. Tretyakov, "Smart radio environments empowered by reconfigurable intelligent surfaces: How it works, state of research, and the road ahead," *IEEE Journal on Selected Areas in Communications*, vol. 38, no. 11, pp. 2450–2525, 2020.
7. M. H. Kumar, S. Sharma, K. Deka, and M. Thottappan, "Reconfigurable intelligent surfaces assisted hybrid NOMA system," *IEEE Communications Letters*, vol. 27, no. 1, pp. 357–361, 2022.
8. R. K. Hindustani, S. Sharma, D. Dixit, and M. Sharma, "Intelligent reflecting surfaces assisted downlink UWB system," in *2023 IEEE Global Conference on Artificial Intelligence and Internet of Things (GCAIoT)*. IEEE, 2023, pp. 115–120. Dubai, United Arab Emirates.
9. S. Sharma, K. Deka, Y. Hong, and D. Dixit, "Intelligent reflecting surface-assisted uplink SCMA system," *IEEE Communications Letters*, vol. 25, no. 8, pp. 2728–2732, 2021.
10. M. H. Kumar, S. Sharma, K. Deka, and M. K. Sharma, "RIS-assisted user pairing NOMA system for THz communications," in *2023 National Conference on Communications (NCC)*. IEEE, 2023, pp. 1–6. Guwahati.
11. Qurrat-Ul-Ain Nadeem, Abla Kammoun, Anas Chaaban, Mérouane Debbah, and Mohamed-Slim Alouini. "Asymptotic max-min SINR analysis of reconfigurable intelligent surface assisted MISO systems," *IEEE Transactions on Wireless Communications*, vol. 19, no. 12, pp. 7748–7764, 2020.
12. J. G. Proakis and M. Salehi, *Digital Communications*, 4th ed. McGraw-Hill Companies, Inc., New York, 2001.
13. R. Hindustani, D. Dixit, S. Sharma, and V. Bhatia, "Outage probability of multiple-IRS-assisted SISO wireless communications over rician fading," *Physical Communication*, vol. 59, p. 102102, 2023.

14. A. Sirojuddin, D. D. Putra, and W.-J. Huang, "Low-complexity sum-capacity maximization for intelligent reflecting surface-aided MIMO systems," *IEEE Wireless Communications Letters*, vol. 11, no. 7, pp. 1354–1358, 2022.

15. W. Tang, J. Y. Dai, M. Z. Chen, K.-K. Wong, X. Li, X. Zhao, S. Jin, Q. Cheng, and T. J. Cui, "MIMO transmission through reconfigurable intelligent surface: System design, analysis, and implementation," *IEEE Journal on Selected Areas in Communications*, vol. 38, no. 11, pp. 2683–2699, 2020.

16. W. Tang, M. Z. Chen, J. Y. Dai, Y. Zeng, X. Zhao, S. Jin, Q. Cheng, and T. J. Cui, "Wireless communications with programmable metasurface: New paradigms, opportunities, and challenges on transceiver design," *IEEE Wireless Communications*, vol. 27, no. 2, pp. 180–187, 2020.

17. H. Zhang, B. Di, L. Song, and Z. Han, "Reconfigurable intelligent surfaces assisted communications with limited phase shifts: How many phase shifts are enough?" *IEEE Transactions on Vehicular Technology*, vol. 69, no. 4, pp. 4498–4502, 2020.

Chapter 8

Survey on IoT-enabled 6G wireless systems

*Sridhar Iyer, Rahul J. Pandya, Rakhee Kallimani,
Krishna Pai, Rajashri Khanai, Dattaprasad A. Torse,
and Swati Mavinkattimath*

8.1 INTRODUCTION

Over the past decade, technological advancements in Internet of Things (IoT) have revolutionized wireless systems. In specific, the technology of smart devices enabled by advanced computing and sensing has interconnected the Internet with many physical objects [1]. The enabling IoT technology results in continuous communication between many different devices without requiring human intervention simultaneously benefiting the society as the created system is intelligent and self-regulated. Indeed, the predictions indicate that IoT will result in significant development of the future wireless systems [2–4]. Further, Nano-Things' connection to the Internet is gaining immense research attention in view of creating advances in the existing ecosystems [5]. Overall, IoT is now viewed as a key technology capable to deliver quality services to customers demanding the latest lifestyles [6].

Within the IoT, the Internet is an enabler that supports the IoT networks and the applications, and in this regard, the first to the fifth generation (5G) of mobile technology has been proposed and commercially deployed. Specifically, the latest 5G wireless technology will be able to provision multiple advanced services to the advanced IoT systems for higher performance in regard to throughput(high), delay(low) and power efficiency [7,8]. Within the 5G networks, researchers have been investigating new methods of communicating the connecting devices in sophisticated applications viz., Wireless Brain-Computer Interfaces (WBCI), five sense communications, holographic communications, etc., that will result in truly immersive experience into an isolated environment. Simultaneously, advances within personal communications result in the evolution of smart verticals within 5G technology to higher level. Overall, in times of this migration, connection of physical and cyberspace of the communication systems, it is noted that IoT will play a key part in enabling the aforementioned trending applications.

However, with rapidly expanding IoT networks and the emergence of multiple smart devices, 5G technology will face the daunting challenge of supporting the next generation applications with increased technical

DOI: 10.1201/9781003522003-9   117 requirements such as completely dynamic and intelligent services, autonomous devices, unmanned aerial vehicles, etc. In specific, the next-generation IoT enabled applications will (i) requisite superior performances in regard to the data rate, latency, coverage, and localization, (ii) be more data- and computation-intensive, which will exceed the URLLC and mMTC range of the 5G technology [7], (iii) face difficulty in efficiently managing the massive IoT devices, and (iv) face massive generated data which will be accompanied with serious security and privacy issues [8]. Hence, with the rapid evolution of the IoTs, the 5G technology will meet the limitations gradually and will not be able to provide any assistance to the majority of the recent modern applications. Hence, researchers are motivated to enable massive IoTs by extending the capabilities of 5G towards the sixth generation (6G) networks.

In view of the above, there has already begun the conceptualization of 6G wireless network architecture [9,10] and promising progress has been made towards the concept of IoT with 6G for meeting the needs of 6G-enabled ubiquitous smart society. The major applications of 6G-IoT will include healthcare, vehicular, and autonomous driving, satellite and industrial IoT, and many more, requiring quality of service and enhanced experience of users from the systems that can be provisioned by the 6G technology owing to the superior features such as communications with ultra-low latency, throughput to be extremely high, customer services related to satellite-based, autonomous massive networks [11–13]. Further, the research community has already identified that the capacity levels required by the next-generation applications will be outstanding, in turn, increase applications and the organization of 6G-enabled IoT systems for sensing of data, connection of devices, wireless communication, and management of network. Hence, recently, after having identified the immense potential of the 6G-IoT, much effort has been conducted in research within this promising area [10,14–16]. During these efforts, it has been identified that the 6G technology, owing to the special features and enhanced capabilities, will be a crucial factor to support the future generation of IoT systems by providing complete spatial coverage with integration of all the functionalities ranging from sensing, computing, communication, to smart and self-organized autonomous control. Consequently, in comparison to 5G, 6G systems have been anticipated to contribute extensive range and improved expandability to promote connectivity of IoT devices and deliver quality service [17].

8.2 BREAKTHROUGH IOT TECHNOLOGIES

The key technology is to integrate variant devices with wireless communication systems, IoT targets at connecting various devices (things) to the Internet, thereby developing an environment connected; wherein the sensing of data, computations, and communications are carried out without any interventions of human automatically. In recent years, with the expanse of human activities towards extreme environments such as higher altitudes, outer space, under-sea/ocean, it is required to build a ubiquitous Internet of Everything (IoE) which is omniscient and omnipotent, and which can recognize the connection at any moment and in any place with differing needs. For this, a 4-tier network architecture enabled via edge computing is proposed for regulation by 6G [18–20]. From the users' context, edge computing will aid IoT devices in performing tasks directly thereby, demonstrating an outperformance in comparison to cloud computing. From the context of system, Machine Learning (ML) enabled edge intelligence, is ensured via computing at edge, which helps handle IoT-enabled systems through multiple intelligent methods.

6G will enhance multiple advanced technology(s) including ML and Blockchain. As an intelligence-enabling technology, ML is extensively employed in various forms of IoT-enabled applications. Recently, in wireless systems, ML is being adopted to deal the related challenges/issues and provide an organized way for future extensive IoT communications [21–23]. Regarding 6G system in IoT networking, ML algorithms will be extensively employed to solve various key issues [24]. Further, the current IoT network models are centralized, with cloud server for connection and IoT devices connected to an individual gateway for transferring the data. However, this will not comply with the needs of advanced IoT devices and hence, Blockchain technology is proposed to solve the many key issues existing in current IoT models [25,26]. The main advantage of Blockchain is the ability to dynamically manage the network via decentralization at much lower cost [27,28].

Also, along with intelligence in the 6G systems, the ML functions will be extendible to edge of network mainly due to the computational capabilities of the edge nodes [29]. The aforementioned is the novel paradigm known as edge intelligence [30,31], which uses the convergence provided by ML along with edge computation. Also, lately, Federated Learning (FL) has emerged as distributed collaborative ML method which is transforming the edge/fog intelligence architectures [32]. Another recent technology is Reconfigurable Intelligent Surfaces (RISs) technology in which there exist artificial planar structures [33]. Electronic circuits enable every element in a software-defined manner by reflecting the impinging electromagnetic wave [34]. RIS is a potential solution to support the design of wireless system owing to the configurability characteristic, which facilitates propagation of signal, modelling of channel, and acquisition, thereby ensuring the benefits of smart radio environments for the 6G-enabled applications. Further, the multiple diverse applications via 6G-IoT will require very large bandwidth and very low latency, for which Tera-Hertz (THz) communications have been envisioned as a driving technology [35]. A benefit of THz spectrum is that it can deal with issues related to scarcity of spectrum in wireless communications and can predominately improve the capacities of wireless system within the 6G-IoT. Owing to the aforementioned key benefits of THz communications, various studies related to IoT have been conducted in regard to the 6G technology [36,37].

Lastly, in line 6G IoE vision, the aim will be to provide complete connectivity for which unified communication platform will be required by 6G IoT. Further, in view of achieving a broad coverage and ubiquitous connectivity to support the various IoT applications, a de-cellular and 4-tier large-dimensional network will be necessary [10]. The 4-tiers will comprise of:

i. *Space Communication* will use various satellite types to provide wireless coverage for the terrestrial networks' areas. Further, 6G technology will introduce the concept of satellite-terrestrial communications [38].
ii. *Air Communication* will introduce the concept of base stations in the air with flying base stations constructed by UAVs and balloons to cover expanded coverage in remote areas [39].
iii. *Terrestrial Communication* will provision ground connectivity and coverage [40].
iv. *Underwater Communication* will provide connectivity services to underwater IoT devices [41].

8.3 TECHNOLOGICAL DRIVERS

The IoT-enabled 5G will transform and generate Industry 4.0 with many IoT-enabled applications emerging to create advances in technology. These advanced applications will require robust systems. In this section, we detail the major technological drivers applicable to 6G IoT.

1. *Holographic Communications*: Holographic communications manipulate motion 3D images and projects in real-time. It captures human/object images, and presents them in real-time, and transmits them with voices/sounds to the receiver [21]. A key requirement is that of high data rate for streaming-related media [10].
2. *Five-Sense Communications*: This will initiate novel remote human interactions [21], and will combine all data related to sense organs of human. The key will be to consider the engineering and perceptual needs during process of design [10].
3. *Wireless Brain-Computer Interfaces (WBCI)* will introduce interfaces to use human thoughts for interaction with machines and/or environments [42,43]. This technology will require very high data rates,

very low delays, and high reliability to guarantee required Quality of Service (QoS) and Quality of Experience (QoE).

4. *Smart Healthcare*: Smart healthcare systems will provide reliable remote monitoring system, diagnosis, guidance, and surgery [44]. 6G will facilitate this use case by provisioning high data rate, low delay, and ultra-reliability.

5. *Smart Education*: Smart education will use innovative techniques to allow learner view many structures and models in 3D, and aid instructor to deliver content remotely. The intelligent classes will allow for data collection via sensors to be sent to cloud or edge cloud for analysis.

6. *Industry Internet*: Here, many vertical industry types will be introduced which will automate process control and operation of systems/ devices. Remote maintenance and control will be possible resulting in low cost and enhanced safety [45].

7. *Self Driving*: This will involve the deployment of many high-definition cameras and high precision radar sensors for vehicles to perform all the driving tasks [46]. The key challenge will be that of strict safety demands [10].

8. *Smart City*: This will enable a city to operate in a smart and self-organized manner leading to enhanced QoS, and low public administrations' costs [47]. The key challenge will be that of connectivity and coverage [10].

9. *Intensive and Sensitive IoT*: This will result in systems that are highly intensive in regard to data and, very sensitive in view of to privacy/ security and associated delays [48]. This will however impose stringent delay requirements and rigorous data security/privacy protection [10].

10. *IoT Evolution*: This will mandate the implementation of 3C's viz., Communication, Caching, and Computing to converge to the 4CSL [49].

8.4 USE CASES

The technological drivers detailed in the previous section will aid in the realization of novel applications within the 6G-IoTs. In the present section, we review the various evolving use cases related to 6G-IoTs.

1. *Healthcare IoT (HIoT)*: The opportunity of Healthcare IoT (HIoT) is in possible impending usage of new sensor and actuator technologies. For example, the use of non-invasive technologies in sensor measurement can offer more accurate and faster sensing of various critical body parameters. The availability of such conveniently deployable technologies (e.g., non-invasive sensing and imaging) can radically raise their usage in clinical settings. The evolution in the design

of actuators can also advance the mechanization of routine medical tasks such as drug delivery. The diagnostic systems involved in diagnosing vital organ diseases such as heart and brain promise to create more accurate devices using advancements in actuator technologies. Further, automation of the healthcare system can promise substantial improvements in patient health monitoring and hospital management. HIoT promises advancement in the procedures followed for routine measurements of patients, drug consumption, and the customized drug based on individual patient's health needs. On the technology end, HIoT offers most healthcare institutions an opportunity to get connected to the Internet owing to increasing accessibility of high bandwidth connectivity, inexpensive cloud storage and computation, and large-scale data analytics. In this emerging scenario, HIoT technologies are important because of the personalization of clinical healthcare [50]. Further, IoT revolution will occur through the enabling technologies of 6G-IoT.

The 6G technology enabled with cloud computing offers healthcare services with cloud storage, computing and analysis of biomedical data. The biomedical data obtained through advanced non-invasive sensors are stored to the cloud by optimizing constraints in communication resources and bandwidth. This brings the end user devices closer to the data source using edge computing technology. 6G will depend on edge technology for provisioning high-speed Internet to smart biomedical devices. Thus, important healthcare data is collected, processed and analysed in real time at the edge technology's nodes which are located near the medical sensors. For example, in cardiology, edge node will receive patients' electrocardiogram (ECG) data which is transmitted from non-invasive biomedical sensors and determines whether patient is suffering from any cardiovascular disorder. The ECG signal data are recorded continuously and communicated to the edge nodes. These nodes process and analyse the biomedical data and communicate vital information to cloud for storage. 6G with cloud offers important advantages such as reliability, low latency, scalability, privacy, and adaptability. The authors in [51] focused on using 6G via URLLC and THz spectrum to transmit healthcare data which demands lowest delay, and to provide wearable device(s) supply in market. Further, in [52], the authors demonstrated the conduction of surgery via blockchain. Also, doctors can observe and handle surgical procedures via video streaming on a real-time basis with the aid of medical robotics and biomedical devices that are connected through the 6G networks [44]. In the context of intelligence, AI/ML techniques must be exploited in the healthcare domain to ensure data learning. As an example, the authors in [53] have used ML methods for analysing the patients' medical history. The relevant data is collected through the wearable sensors. The authors introduce an inter-disciplinary approach by connecting concepts radio resource optimization by developing 6G Heterogeneous Networks (HetNets). The authors have analysed stroke patients' medical data and recordings from non-invasive IoT medical sensors to predict the possibility of a future stroke [53]. In recent years, with COVID-19, major health concerns have been cited by various countries around the globe. In this regard, many aspects of computing at the edge (edge computing), and at cloud level (cloud computing) are implemented for combating COVID-19 [54]. Also, AI methods can be used for accurate and speedy disease diagnosis by processing the data using edge intelligence and computing techniques [55].

2. *Vehicular IoT (VIoT)*: With the advances in the 6G technology, intelligent transportation systems are being revolutionized. In [56], the authors have exploited mMTC in VIoT networks enabled by the 6G technology for enabling the vehicle-to-everything connectivity in view of transmitting shorter vehicular information payloads through many vehicles without any human interaction. The study clarifies that, in the future, within the 6G enable VIoTs, data rate prediction will be an open challenge for research owing to various complex interdependencies between different factors to manage the mobility, handle channels, and networking. Further, ML techniques provide efficient throughput prediction to mimick the attained network behaviours via 6G upon network training through load data which is obtained through the analytics of control channel [57]. For leveraging potential intelligence within VIoTs, edge computing with Artificial Intelligence (AI) can be implemented [58, 59]. Also, this can be ensured Deep Learning (DL) to obtain support for network management [60]. A similar focus on vehicular intelligence appears in [61]. The authors have adopted the DL techniques for autonomous data transmission scheduling via supervised, unsupervised, and reinforcement learning methods. The study also mentions that, in the future, cooperative DL approaches such as FL will be required to be developed since the 6G-enabled VIoT networks will be highly scalable and distributed. Also, at the vehicular data controller, intelligent software based on DL is deployed [62], thereby providing apt security solutions and protection in the 6G-VIoT networks. Further, regarding the Autonomous Vehicle (AV) applications, the 6G technology is envisioned to offer multiple opportunities for satisfying the required stringent service requirements to achieve reliability and high-speed communications [63]. For realizing AV in the 6G era, it will be mandatory to explore the performance vehicle-to-vehicle network communication since every AV in the interconnected system will be recognized as an individual entity having complete control access.

3. *Unmanned Aerial Vehicles (UAVs)*: Much research focus lately has been on the exploration of applications related to the 6G-enabled UAV networks. The study in [64] has considered a cell-free UAV network for 6G-enabled wide area IoT, focusing upon the UAV flight process-oriented optimization. The proposed method supports massive access for wide area IoT devices and is promising method in identifying the cell-free coverage patterns in the era of 6G era. The researchers in [65] have proposed IoT networks enabled by 6G with UAV-supported Clustered Non-Orthogonal Multiple Access (C-NOMA) method [66] to support wireless powered communications. The authors in [67] have characterized UAV to ground channel with arbitrary three-dimensional (3D) UAV trajectories for the UAV-based 6G networks, and the authors in [68] have studied a collaborative multi-UAV trajectory optimization and resource scheduling framework for a 6G IoT network wherein, multiple UAVs fly as the base stations for transferring the energy to multiple terrestrial IoT users. Further, techniques such as AI are used to contribute optimal solutions to UAV networks enabled by 6G via edge intelligence techniques [69], and FL methods have been implemented in [70] and contributed to privacy preserved intelligence for the UAV-based 6G networks. As a major challenge, in future UAVs within the 6G-IoT networks, regulations are required to contribute and the necessary guidelines/regulations for deploying the UAVs within the IoT systems must ensure safety and privacy [71].

4. *Satellite Internet of Things (SIoT)*: With the 6G technology, it will be mandatory to incorporate satellite communications with the existing deployed wireless networks in view of massive IoT coverage, which will result in the IoT [72]. With advanced satellite technologies and 6G, multiple satellites are deployable at multiple orbits above the earth, and in this regard, the low earth orbit systems will be most suitable to realize the global coverage and reuse frequency efficiently [73]. In [74], the authors have focused the research considering a low earth orbit satellite network able to reinforce UAV trajectories' navigation to collect IoT data. The study in [75] provides further insights into the SIoTs as the authors have built a comprehensive model and have highlighted the relevant technical discussions and challenges. Further, as demonstrated in [76], RIS technology and satellite communications can be integrated to improve power of signal transmission strength at the satellites. The study in [77] has considered an energy-aware massive random access scheme for satellite communications in the 6G-enabled global IoT. Further, spectrum sharing is another key issue to be addressed within satellite communications for 6G IoT networks [78].

5. *Industrial Internet of Things (IIoT)*: Lately, 6G within IIoT has been investigated in which ML is used [79]. Also, transfer learning is implemented to coordinate data distribution and transmission in

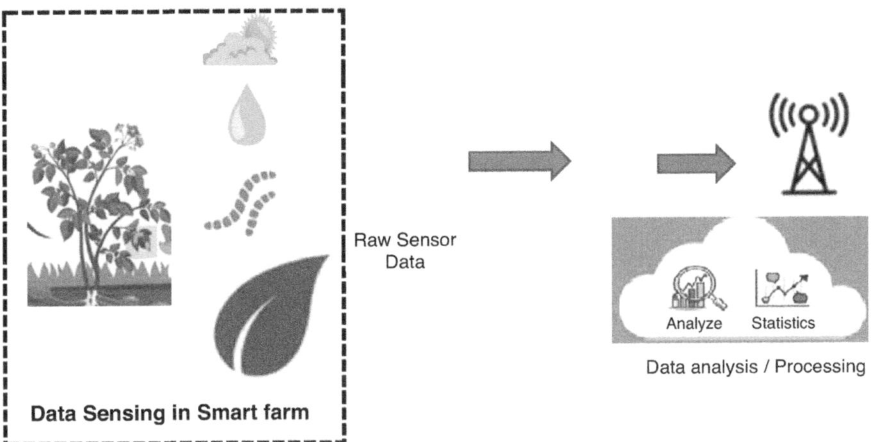

Figure 8.1 MEC and 5G for a smart farm.

blockchain-enabled 6G IIoT network [80]. ML is also used in [81] for tackling issues of detecting fault. AI techniques for intelligent agriculture is highlighted in [82], and security attacks and privacy issues to be addressed are mentioned in [83,84].

6. *Smart Agriculture*: Smart farming ensures that food supply and demand is met with safety and consistency. The productivity is improved via specific environmental control and management through IoT devices resulting in the use of rarer resources and fewer wastes. The agriculture sector comprises of use cases like sensing and monitoring and data analysis. The amalgamation of 5G/6G and edge computing in smart farming/agriculture is shown in Figure 8.1.

Finally, multiple use cases of edge considerations for vertical industries are summarized in Table 8.1.

8.5 OPEN PROBLEMS OF RESEARCH

In this section, we underline various research challenges and mention the possible perspectives in the domain of 6G-IoT.

1. In regard to ML for massive 6G-IoT, the existing research studies have only provided an introductory perspective, and much research considering the key issues still remains to be conducted [85]. These challenges include the issue of the incomplete dataset are various layers, security, and privacy, long time cost, computation, and storage. Further, existing centralized techniques will not be applicable to the large-scale IoT challenges of 6G, and hence, ML techniques such as

Table 8.1 Use-cases of edge considerations for vertical industries

Edge Considerations	Agriculture	ITS	Manufacturing	Energy	Video Analytics
Architecture	Multi server	Multi server	Single server	Singe/Multi server	Single/Multi server
Edge location	RAN	RSU or RAN	On-premise	On-premise	5G RAN
Storage capacity	Medium/High	Medium/High	Low	Low/Medium	Very High
Computing capacity	Medium	High	Low/Medium	Medium	Very High
Typical Latency Requirement	< 1 s	10 ms–1 s	< 100 ms	< 10 ms	< 1 s
Security Issues	DDoS and Data poisoning attacks	Data & ML model poisoning	Online adversarial attacks	Online adversarial and DDoS attacks	Data & ML model poisoning
Edge Intelligence	Anomalies detection, crop treatment recommendations	Object detection, tracking, traffic/vehicle prediction & control etc.	Anomaly detection, process automation & control etc.	Fault isolation, predictive maintenance etc.	Object detection, recommendation, summarization etc.
Real-world deployments	5G-ENSURE	5G-ENSURE	SONATA [4]	COGNET [5]	5G-ENSURE [3]

Distributed ML will be required, which needs much research attention. Lastly, considering the modifications in the ML techniques that will be required in the 6G-IoT, optimization of the new ML methods will be an open area for research.

2. Even though Blockchain technology is emerging and intended to play a crucial role in 6G-IoT, many challenges need timely solutions before the implementation of Blockchain for the massive IoTs [86]. The first issue includes the performance of Blockchain-enabled systems, which needs to guarantee simultaneous performance of both the overlaid and the underlaid layers within the IoT systems. A second issue is the blockchain technology requiring large amounts of computation, caching, and transmission to generate the block(s), verify, store the ledger, and consensus among the nodes. Specifically, the main advantage(s) of including Blockchain within the IoT requires investigation. The aforementioned also points towards the issue of allocating the resources between the Blockchain and the wireless system in view of ensuring the optimal utilization of the resources. Next, multiple security-related problems originate once large amounts of services are outsourced at the edge, which requires accurate investigation. Lastly, there are open problems for research within the domain of blockchain optimization. Specifically, an adaptive generic blockchain model for 6G-IoT needs to be developed to support multiple services. To meet heterogeneous demands of 6G-IoT, a novel consensus protocol must be designed for improving throughput, security, and privacy.

3. There are multiple issues regarding the security and privacy of 6G-IoTs, which require investigation and solutions [87]. This occurs mainly due to the integration of 6G within the IoT networks, which makes the 6G-IoT vulnerable to multiple threats that are associated with wireless interface attacks at computing units/servers due to unauthorized access of data, integrity threats within access network infrastructure, and Denial of Service (DoS) to shut down the software and data centres.

4. A major concern in the 6G-IoTs will be achieving high energy efficiency since next-generation applications such as autonomous driving, space, terrestrial communications, etc., will require efficient network operations with efficient energy resources. Hence, much research is necessary to explore the issues of energy efficiency within 6G-IoT networks [88].

5. Another challenge relates to constraints faced in choosing IoT devices due to extensive communications and computations in 6G-IoTs [89].

6. With the 6G technology integration within the IoT networks, there will be a transformation in the IoT markets, and the IoT ecosystems will be enabled via wireless systems. This will require the standardization of the stringent specifications through the collaboration of the various relevant stakeholders [90].

REFERENCES

[1] A. Al-Fuqaha, M. Guizani, M. Mohammadi, M. Aledhari, and M. Ayyash, "Internet of Things: A Survey on Enabling Technologies, Protocols, and Applications," *IEEE Communications Surveys & Tutorials*, vol. 17, no. 4, pp. 2347–2376, 2015.

[2] "Internet of Things 2016," 2016. [Online]. Available: https://www.cisco.com/c/dam/en/us/products/collateral/se/internetof-things/at-a-glance-c45-731471.pdf.

[3] "Number of Connected IoT Devices Will Surge to 125 Billion by 2030," 2021. [Online]. Available: https://news.ihsmarkit.com/prviewer/releaseonly/slug/number-connected-iot-devices-will-surge-125-billion-2030.

[4] "5G IoT Market by Connection, Radio Technology, Range, Vertical and Region - Global Forecast to 2025," 2019. [Online]. Available: https://www.globenewswire.com/fr/news-release/2019/04/19/1806975/0/en/Global-5G-IoT-Market-Forecast-to-2025-Market.html.

[5] A. O. Balghusoon and S. Mahfoudh, "Routing Protocols for Wireless Nanosensor Networks and Internet of Nano Things: A Comprehensive Survey," *IEEE Access*, vol. 8, pp. 200724–200748, 2020.

[6] K. Shafique, B. A. Khawaja, F. Sabir, S. Qazi, and M. Mustaqim, "Internet of Things (IoT) for Next-Generation Smart Systems: A Review of Current Challenges, Future Trends and Prospects for Emerging 5GIoT Scenarios," *IEEE Access*, vol. 8, pp. 23022–23040, 2020.

[7] L. Chettri and R. Bera, "A Comprehensive Survey on Internet of Things (IoT) toward 5G Wireless Systems," *IEEE Internet of Things Journal*, vol. 7, no. 1, pp. 16–32, Jan. 2020.

[8] A. Yadav and O. A. Dobre, "All Technologies Work Together for Good: A Glance at Future Mobile Networks," *IEEE Wireless Communications*, vol. 25, no. 4, pp. 10–16, Aug. 2018.

[9] W. Saad, M. Bennis, and M. Chen, "A Vision of 6G Wireless Systems: Applications, Trends, Technologies, and Open Research Problems," *IEEE Network*, vol. 34, no. 3, pp. 134–142, May 2020.

[10] Z. Zhang, Y. Xiao, Z. Ma, M. Xiao, Z. Ding, X. Lei, G. K. Karagiannidis, and P. Fan, "6G Wireless Networks: Vision, Requirements, Architecture, and Key Technologies," *IEEE Vehicular Technology Magazine*, vol. 14, no. 3, pp. 28–41, Sep. 2019.

[11] G. Gui, M. Liu, F. Tang, N. Kato, and F. Adachi, "6G: Opening New Horizons for Integration of Comfort, Security, and Intelligence," *IEEE Wireless Communications*, vol. 27, no. 5, pp. 126–132, Oct. 2020.

[12] L. Bariah, L. Mohjazi, S. Muhaidat, P. C. Sofotasios, G. K. Kurt, H. Yanikomeroglu, and O. A. Dobre, "A Prospective Look: Key Enabling Technologies, Applications and Open Research Topics in 6G Networks," *IEEE Access*, vol. 8, pp. 174792–174820, 2020.

[13] H. B. Eldeeb, S. Naser, L. Bariah, S. Muhaidat and M. Uysal, "Digital Twin-Assisted OWC: Towards Smart and Autonomous 6G Networks," *IEEE Network*, 2024. Early Access. doi: 10.1109/MNET.2024.3374370.

[14] "Nokia to Lead the EU's 6G Project Hexa-X," 2021. [Online]. Available: https://www.nokia.com/about-us/news/releases/2020/12/07/nokia-to-lead-the-eus-6g-project-hexa-x.

[15] Y. Yuan, Y. Zhao, B. Zong, and S. Parolari, "Potential Key Technologies for 6G Mobile Communications," *Science China Information Sciences*, vol. 63, no. 8, p. 183301, Aug. 2020.

[16] "South Korea to Launch 6G Pilot Project in 2026: Report," 2020. [Online]. Available: https://www.rcrwireless.com/20200810/asia-pacific/south-korea-launch-6g-pilot-project-2026-report.

[17] F. Guo, F. R. Yu, H. Zhang, X. Li, H. Ji, and V. C. M. Leung, "Enabling Massive IoT toward 6G: A Comprehensive Survey," *IEEE Internet of Things Journal*, Vol. 8, no. 15, pp. 11891–11915, 2021.

[18] Y. Xiao Hu, W. Cheng-Xiang, H. Jie, G. Xi Qi, Z. Zaichen, W. Michael, H. Yongming, Z. Chuan, J. Yan Xiang, J. Wang, Z. Min, S. Bin, W. Dong Ming, P. Zhi Wen, Z. Pengcheng, Y. Yang, L. Zoning, Z. Ping, T. Xiao Feng, L. Shao Qian, M. X. Chen Zhi, I. Chih-Lin, H. Shuang Feng, L. Ke, P. Chengkang, Z. Zhimin, H. Lajos, S. Xuemin, G. Y. Jay, D. Zhiguo, H. Harald, T. Wen, Z. Peiying, Y. Ganghua, W. Jun, L. E. G. N. Hien, H. Wei, W. Haiming, H. Debin, C. Jixin, C. Zhe, H. Zhang-Cheng, L. Geoffrey, T. Rahim, G. Yue, P. Vincent, F. Gerhard, and L. Ying-Chang, "Towards 6G Wireless Communication Networks: Vision, Enabling Technologies, and New Paradigm Shifts," *Science China Information Sciences*, vol. 64, no. 1, p. 110301, Jan. 2020. https://link.springer.com/article/10.1007/s11432-020-2955-6

[19] P. Porambage, J. Okwuibe, M. Liyanage, M. Ylianttila, and T. Taleb, "Survey on Multi-Access Edge Computing for Internet of Things Realization," *IEEE Communications Surveys and Tutorials*, vol. 20, no. 4, pp. 2961–2991, 2018.

[20] M. Ejaz, T. Kumar, M. Ylianttila, and E. Harjula, "Performance and Efficiency Optimization of Multi-Layer IoT Edge Architecture," in *Proceedings of the 2020 2nd 6G Wireless Summit (6G SUMMIT)*, Levi, 2020, pp. 1–5.

[21] E. Calvanese Strinati, S. Barbarossa, J. L. Gonzalez-Jimenez, D. Ktenas, N. Cassia, L. Maret, and C. Dehos, "6G: The Next Frontier: From Holographic Messaging to Artificial Intelligence Using Subterahertz and Visible Light Communication," *IEEE Vehicular Technology Magazine*, vol. 14, no. 3, pp. 42–50, Aug. 2019.

[22] M. Z. Chowdhury, M. Shahjalal, S. Ahmed, and Y. M. Jang, "6G Wireless Communication Systems: Applications, Requirements, Technologies, Challenges, and Research Directions," *CoRR*, Sep. 2019, arxiv: 1909.11315.

[23] Y. Zhao, G. Yu, and H. Xu, "6G Mobile Communication Network: Vision, Challenges and Key Technologies," *Scientia Sinica Informationis*, May 2019, arxiv: 1905.04983.

[24] S. K. Sharma and X. Wang, "Toward Massive Machine-Type Communications in Ultra-Dense Cellular IoT Networks: Current Issues and Machine Learning-Assisted Solutions," *IEEE Communications Surveys and Tutorials*, vol. 22, no. 1, pp. 426–471, 2020.

[25] F. R. Yu, J. Liu, Y. He, P. Si, and Y. Zhang, "Virtualization for Distributed Ledger Technology (vDLT)," *IEEE Access*, vol. 6, pp. 25019–25028, Apr. 2018.

[26] M. S. Ali, M. Vecchio, M. Pincheira, K. Dolci, F. Antonelli, and M. H. Rehmani, "Applications of Blockchains in the Internet of Things: A Comprehensive Survey," *IEEE Communications Surveys and Tutorials*, vol. 21, no. 2, p. 1, 2018.

[27] M. Wu, K. Wang, X. Cai, S. Guo, M. Guo, and C. Rong, "A Comprehensive Survey of Blockchain: From Theory to IoT Applications and Beyond," *IEEE Internet of Things Journal*, vol. 6, no. 5, pp. 8114–8154, 2019.

[28] J. Xie, H. Tang, T. Huang, F. R. Yu, R. Xie, J. Liu, and Y. Liu, "A Survey of Blockchain Technology Applied to Smart Cities: Research Issues and Challenges," *IEEE Communications Surveys and Tutorials*, vol. 21, no. 3, p. 1, 2019.

[29] S. Han, T. Xie, C.-L. I. L. Chai, Z. Liu, Y. Yuan, and C. Cui, "Artificial-Intelligence-Enabled Air Interface for 6G: Solutions, Challenges, and Standardization Impacts," *IEEE Communications Magazine*, vol. 58, no. 10, pp. 73–79, Oct. 2020.

[30] S. Deng, H. Zhao, W. Fang, J. Yin, S. Dustdar, and A. Y. Zomaya, "Edge Intelligence: The Confluence of Edge Computing and Artificial Intelligence," *IEEE Internet of Things Journal*, vol. 7, no. 8, pp. 7457–7469, Aug. 2020.

[31] Z. Zhou, X. Chen, E. Li, L. Zeng, K. Luo, and J. Zhang, "Edge Intelligence: Paving the Last Mile of Artificial Intelligence with Edge Computing," *Proceedings of the IEEE*, vol. 107, no. 8, pp. 1738–1762, Aug. 2019.

[32] D. C. Nguyen, M. Ding, P. N. Pathirana, A. Seneviratne, J. Li, and H. V. Poor, "Federated Learning for Internet of Things: A Comprehensive Survey," *IEEE Communications Surveys Tutorials*, vol. 23, no. 3, p. 1, 2021.

[33] M. Di Renzo, A. Zappone, M. Debbah, M.-S. Alouini, C. Yuen, J. de Rosny, and S. Tretyakov, "Smart Radio Environments Empowered by Reconfigurable Intelligent Surfaces: How It Works, State of Research, and The Road Ahead," *IEEE Journal on Selected Areas in Communications*, vol. 38, no. 11, pp. 2450–2525, Nov. 2020.

[34] C. Huang, S. Hu, G. C. Alexandropoulos, A. Zappone, C. Yuen, R. Zhang, M. D. Renzo, and M. Debbah, "Holographic MIMO Surfaces for 6G Wireless Networks: Opportunities, Challenges, and Trends," *IEEE Wireless Communications*, vol. 27, no. 5, pp. 118–125, Oct. 2020.

[35] C. Han, Y. Wu, Z. Chen, and X. Wang, "Terahertz Communications (TeraCom): Challenges and Impact on 6G Wireless Systems," Dec. 2019, arXiv: 1912.06040. https://doi.org/10.48550/arXiv.1912.06040

[36] C. She, C. Sun, Z. Gu, Y. Li, C. Yang, H. V. Poor, and B. Vucetic, "A Tutorial on Ultra-Reliable and Low-Latency Communications in 6G: Integrating Domain Knowledge into Deep Learning," Jan. 2021, arXiv: 2009.06010. https://doi.org/10.48550/arXiv.2009.06010

[37] S.-Y. Lien, S.-C. Hung, D.-J. Deng, and Y. J. Wang, "Efficient Ultra-Reliable and Low Latency Communications and Massive Machine-Type Communications in 5G New Radio," in *Proceedings of the 2017 IEEE Global Communications Conference*, Singapore, Dec. 2017, pp. 1–7.

[38] A. Saeed, O. Gurbuz, and M. A. Akkas, "Terahertz Communications at Various Atmospheric Altitudes," *Physical Communication*, vol. 41, p. 101113, Aug. 2020.

[39] Y. Huo, X. Dong, T. Lu, W. Xu, and M. Yuen, "Distributed and Multilayer UAV Networks for Next-Generation Wireless Communication and Power Transfer: A Feasibility Study," *IEEE Internet of Things Journal*, vol. 6, no. 4, pp. 7103–7115, Aug. 2019.

[40] S. Andreev, V. Petrov, M. Dohler, and H. Yanikomeroglu, "Future of Ultra-Dense Networks Beyond 5G: Harnessing Heterogeneous Moving Cells," *IEEE Communications Magazine*, vol. 57, no. 6, pp. 86–92, Jun. 2019.

[41] A. Celik, N. Saeed, B. Shihada, T. Y. Al-Naffouri, and M.-S. Alouini, "A Software-Defined Opto-Acoustic Network Architecture for Internet of Underwater Things," *IEEE Communications Magazine*, vol. 58, no. 4, pp. 88–94, Apr. 2020.

[42] J. R. Wolpaw, N. Birbaumer, D. J. McFarland, G. Pfurtscheller, and T. M. Vaughan, "Brain-Computer Interfaces for Communication and Control," *Clinical Neurophysiology*, vol. 113, no. 6, pp. 767–791, 2002.

[43] S. Guruprakash, R. Balaganesh, M. Divakar, K. Aravinth, and S. Kavitha, "Brain-Controlled Home Automation," *International Journal of Advanced Research in Biology Engineering Science and Technology (IJARBEST)*, vol. 2, no. 10, pp. 430–436, 2016.

[44] M. S. Kaiser, N. Zenia, F. Tabassum, S.A. Mamun, M. A. Rahman, Md. S. Islam, and M. Mahmud. "6G Access Network for Intelligent Internet of Healthcare Things: Opportunity, Challenges, and Research Directions," in M. S. Kaiser, A. Bandyopadhyay, M. Mahmud, and K. Ray (eds.) *Proceedings of International Conference on Trends in Computational and Cognitive Engineering. Advances in Intelligent Systems and Computing*, vol. 1309, Springer, Singapore, 2021. https://doi.org/10.1007/978-981-33-4673-4_25.

[45] J. Chen and J. Zhou, "Revisiting Industry 4.0 with a Case Study," in *2018 IEEE International Conference on Internet of Things (iThings) and IEEE Green Computing and Communications (GreenCom) and IEEE Cyber, Physical and Social Computing (CPSCom) and IEEE Smart Data (SmartData)*, Halifax, NS, Canada, 2018, pp. 1928–1932. https://doi.org/10.1109/Cybermatics_2018.2018.00319.

[46] S. Pendleton, H. Andersen, X. Du, X. Shen, M. Meghjani, Y. Eng, D. Rus, and M. Ang, "Perception, Planning, Control, and Coordination for Autonomous Vehicles," *Machines*, vol. 5, no. 1, p. 6, 2017.

[47] "6G and Smart Cities: Transformation of Communications, Services, Content, and Commerce 2025–2030," Report, Mind Commerce, 2021. https://www.asdreports.com/market-research-related-572020/g-smart-cities-transforma-tion-communications-services-content-commerce code: ASDR-572020.

[48] Y. Zhou, L. Liu, L. Wang, N. Huia, X. Cuia, J. Wua, Y. Peng, Y. Qi, and C. Xing, "Service-Aware 6G: An Intelligent and Open Network Based on the Convergence of Communication, Computing and Caching," *Digital Communications and Networks*, vol. 6, no. 3, pp. 253–260, Aug. 2020.

[49] M. Rasti, S. S. K. Taskou, H. Tabassum, and E. Hossain, "Evolution toward 6G Wireless Networks: A Resource Management Perspective," arxiv: 2108.06527, 2021. https://doi.org/10.48550/arXiv.2108.06527

[50] H. Habibzadeh, K. Dinesh, O. R. Shishvan, A. Boggio-Dandry, G. Sharma, and T. Toyota, "A Survey of Healthcare Internet of Things (IoT): A Clinical Perspective," *IEEE Internet of Things Journal*, vol. 7, no. 1, pp. 53–71, Jan. 2020.

[51] S. Nayak and R. Patgiri, "6G Communication Technology: A Vision on Intelligent Healthcare," in R. Patgiri, A. Biswas, and P. Roy (eds.) Health Informatics: A Computational Perspective in Healthcare. *Studies in Computational Intelligence*, vol. 932, pp. 1–18, Springer, Singapore, 2021. https://doi.org/10.1007/978-981-15-9735-0_1.

[52] R. Gupta, A. Shukla, and S. Tanwar, "BATS: A Blockchain and AI-Empowered Drone-Assisted Telesurgery System towards 6G," *IEEE Transactions on Network Science and Engineering*, vol. 8, no. 4, pp. 2958–2967, 2020.

[53] M. S. Hadi, A. Q. Lawley, T. E. H. El-Gorashi, and J. M. H. Elmirghani, "Patient-Centric HetNets Powered by Machine Learning and Big Data Analytics for 6G Networks," *IEEE Access*, vol. 8, pp. 85639–85655, 2020.

[54] Y. Siriwardhana, G. Gur, M. Ylianttila, and M. Liyanage, "The Role of 5G for Digital Healthcare against COVID-19 Pandemic: Opportunities and Challenges," *ICT Express*, vol. 7, no. 2, pp. 244–252, Nov. 2020.

[55] D. Nguyen, M. Ding, P. N. Pathirana, and A. Seneviratne, "Blockchain and AI-based Solutions to Combat Coronavirus (COVID-19)-like Epidemics: A Survey," *IEEE Access*, vol. 9, pp. 95730–95753, 2021.

[56] C. Kallas and J. Alonso-Zarate, "Massive Connectivity in 5G and Beyond: Technical Enablers for the Energy and Automotive Verticals," in *Proceedings of the 2020 2nd 6G Wireless Summit (6G SUMMIT)*, Levi, Finland, Mar. 2020, pp. 1–5.

[57] B. Sliwa, R. Falkenberg, and C. Wietfeld, "Towards Cooperative Data Rate Prediction for Future Mobile and Vehicular 6G Networks," in *Proceedings of the 2020 2nd 6G Wireless Summit (6G SUMMIT)*, Levi, Finland, Mar. 2020, pp. 1–5.

[58] W. Yuan, S. Li, L. Xiang, and D. W. K. Ng, "Distributed Estimation Framework for Beyond 5G Intelligent Vehicular Networks," *IEEE Open Journal of Vehicular Technology*, vol. 1, pp. 190–214, 2020.

[59] C. Li, W. Guo, S. C. Sun, S. Al-Rubaye, and A. Tsourdos, "Trustworthy Deep Learning in 6G-Enabled Mass Autonomy: From Concept to Quality-of-Trust Key Performance Indicators," *IEEE Vehicular Technology Magazine*, vol. 15, no. 4, pp. 112–121, Dec. 2020.

[60] D. C. Nguyen, P. N. Pathirana, M. Ding, and A. Seneviratne, "Privacy-Preserved Task Offloading in Mobile Blockchain with Deep Reinforcement Learning," *IEEE Transactions on Network and Service Management*, vol. 17, no. 4, pp. 2536–2549, Dec. 2020.

[61] B. Sliwa, R. Adam, and C. Wietfeld, "Client-Based Intelligence for Resource Efficient Vehicular Big Data Transfer in Future 6G Networks," *IEEE Transactions on Vehicular Technology*, vol. 70, no. 6, pp. 5332–5346, 2021.

[62] Z. Zhang, Y. Cao, Z. Cui, W. Zhang, and J. Chen, "A Many-Objective Optimization-Based Intelligent Intrusion Detection Algorithm for Enhancing Security of Vehicular Networks in 6G," *IEEE Transactions on Vehicular Technology*, vol. 70, no. 6, pp. 5234–5243, 2021.

[63] X. Chen, S. Leng, J. He, and L. Zhou, "Deep Learning-Based Intelligent Inter-Vehicle Distance Control for 6G-Enabled Cooperative Autonomous Driving," *IEEE Internet of Things Journal*, vol. 8, no. 20, pp. 15180–15190, 2020.

[64] C. Liu, W. Feng, Y. Chen, C.-X. Wang, and N. Ge, "Cell-Free Satellite-UAV Networks for 6G Wide-Area Internet of Things," *IEEE Journal on Selected Areas in Communications*, vol. 39, no. 4, pp. 1116–1131, 2020.

[65] Z. Na, Y. Liu, J. Shi, C. Liu, and Z. Gao, "UAV-supported Clustered NOMA for 6G-Enabled Internet of Things: Trajectory Planning and Resource Allocation," *IEEE Internet of Things Journal*, vol. 8, no. 20, pp. 15041–15048, 2020.

[66] S. M. R. Islam, N. Avazov, O. A. Dobre, and K.-s. Kwak, "Power-Domain Non-Orthogonal Multiple Access (NOMA) in 5G Systems: Potentials and Challenges," *IEEE Communications Surveys & Tutorials*, vol. 19, no. 2, pp. 721–742, 2017.

[67] H. Chang, C.-X. Wang, Y. Liu, J. Huang, J. Sun, W. Zhang, and X. Gao, "A Novel Non-Stationary 6G UAV-to-Ground Wireless Channel Model with 3D Arbitrary Trajectory Changes," *IEEE Internet of Things Journal*, vol. 8, no. 12, pp. 9865–9877, 2020.

[68] J. Wang, Z. Na, and X. Liu, "Collaborative Design of Multi-UAV Trajectory and Resource Scheduling for 6G-enabled Internet of Things," *IEEE Internet of Things Journal*, vol. 8, no. 20, pp. 15096–15106, 2020.

[69] C. Dong, Y. Shen, Y. Qu, Q. Wu, F. Wu, and G. Chen, "UAVs as a Service: Boosting Edge Intelligence for Air-Ground Integrated Networks," Mar. 2020, arXiv: 2003.10737. https://doi.org/10.48550/arXiv.2003.10737

[70] S. Tang, W. Zhou, L. Chen, L. Lai, J. Xia, and L. Fan, "Battery Constrained Federated Edge Learning in UAV-Enabled IoT for B5G/6G Networks," *Physical Communication*, vol. 47, p. 101381, 2021.

[71] Z. Ullah, F. Al-Tudjman, and L. Mostarda, "Cognition in UAV-Aided 5G and Beyond Communications: A Survey," *IEEE Transactions on Cognitive Communications and Networking*, vol. 6, no. 3, pp. 872–891, Sep. 2020.

[72] J. Chu, X. Chen, C. Zhong, and Z. Zhang, "Robust Design for NOMA Based Multibeam LEO Satellite Internet of Things," *IEEE Internet of Things Journal*, vol. 8, no. 3, pp. 1959–1970, Feb. 2021.

[73] C. Liu, W. Feng, X. Tao, and N. Ge, "MEC-Empowered Non-Terrestrial Network for 6G Wide-Area Time-Sensitive Internet of Things," Mar. 2021, arXiv: 2103.11907. https://doi.org/10.48550/arXiv.2103.11907

[74] Z. Jia, M. Sheng, J. Li, D. Niyato, and Z. Han, "LEO Satellite-Assisted UAV: Joint Trajectory and Data Collection for Internet of Remote Things in 6G Aerial Access Networks," *IEEE Internet of Things Journal*, vol. 8, no. 12, pp. 9814–9826, 2020.

[75] X. Fang, W. Feng, T. Wei, Y. Chen, N. Ge, and C.-X. Wang, "5G Embraces Satellites for 6G Ubiquitous IoT: Basic Models for Integrated Satellite Terrestrial Networks," *IEEE Internet of Things Journal*, vol. 8, no. 18, pp. 14399–14417, Nov. 2020.

[76] K. Tekbıyık, G. K. Kurt, and H. Yanikomeroglu, "Energy-Efficient RIS assisted Satellites for IoT Networks," *IEEE Internet of Things Journal*, vol. 9, no. 16, pp. 14891–14899, Jan. 2021.

[77] L. Zhen, A. K. Bashir, K. Yu, Y. D. Al-Otaibi, C. H. Foh, and P. Xiao, "Energy-Efficient Random Access for LEO Satellite-Assisted 6G Internet of Remote Things," *IEEE Internet of Things Journal*, vol. 8, no. 7, pp. 5114–5128, 2020.

[78] X. Liu, K.-Y. Lam, F. Li, J. Zhao, and L. Wang, "Spectrum Sharing for 6G Integrated Satellite-Terrestrial Communication Networks Based on NOMA and Cognitive Radio," *IEEE Network*, vol. 35, no. 4, pp. 28–34, 2021.

[79] A. Mukherjee, P. Goswami, M. A. Khan, L. Manman, L. Yang, and P. Pillai, "Energy-Efficient Resource Allocation strategy in Massive IoT for Industrial 6G Applications," *IEEE Internet of Things Journal*, vol. 8, no. 7, pp. 5194–5201, 2020.

[80] P. K. Deb, S. Misra, T. Sarkar, and A. Mukherjee, "Magnum: A Distributed Framework for Enabling Transfer Learning in B5G-Enabled Industrial-IoT," *IEEE Transactions on Industrial Informatics*, vol. 17, no. 10, pp. 7133–7140, 2020.

[81] A. Mohamed, H. Ruan, M. H. H. Abdelwahab, B. Dorneanu, P. Xiao, H. Arellano-Garcia, Y. Gao, and R. Tafazolli, "An Inter-Disciplinary Modelling Approach in Industrial 5G/6G and Machine Learning Era," in *2020 IEEE International Conference on Communications Workshops (ICC Workshops)*, Dublin, Ireland, Jun. 2020, pp. 1–6.

[82] D. C. Nguyen, P. N. Pathirana, M. Ding, and A. Seneviratne, "Integration of Blockchain and Cloud of Things: Architecture, Applications and Challenges," *IEEE Communications Surveys & Tutorials*, vol. 22, no. 4, pp. 2521–2549, 2020.

[83] H. Lin, S. Garg, J. Hu, G. Kaddoum, M. Peng, and M. S. Hossain, "A Blockchain-based Secure Data Aggregation Strategy Using 6G-enabled NIB for Industrial Applications," *IEEE Transactions on Industrial Informatics*, vol. 17, no. 10, pp. 7204–7212, 2020.

[84] C. Zhang, P. Patras, and H. Haddadi, "Deep Learning in Mobile and Wireless Networking: A Survey," *IEEE Communications Surveys and Tutorials*, vol. 21, no. 3, pp. 2224–2287, 2019.

[85] A. Jahid, M. H. Alsharif, and T. J. Hall, "The Convergence of Blockchain, IoT and 6G: Potential, Opportunities, Challenges and Research Roadmap," 2021, arxiv: abs/2109.03184. https://doi.org/10.48550/arXiv.2109.03184

[86] M. Wang, T. Zhua, T. Zhang, J. Zhang, S. Yu, and W. Zhou, "Security and Privacy in 6G Networks: New Areas and New Challenges," *Digital Communications and Networks*, vol. 6, no. 3, pp. 281–291, Aug. 2020.

[87] U. M. Malik, M. A. Javed, S. Zeadally, and S. ul Islam, "Energy-Efficient Fog Computing for 6G Enabled Massive IoT: Recent Trends and Future Opportunities," *IEEE Internet of Things Journal*, vol. 9, no. 16, pp. 14572–14594, 2021.

[88] Secure software updates for the Internet of Things "Constraints of Internet of Things devices". https://securingiot.projectsbyif.com/constraints-of-internet-of-things-devices/.

[89] A. Salim, H. Chaibi, A. Chehri, R. Saadane, and G. Jeon, "Toward 6G: Understanding Network Requirements and Key Performance Indicators," *Transactions on Emerging Telecommunications Technologies*, vol. 32, no. 3, p. e4201, Mar. 2021. https://doi.org/10.1002/ett.4201.

[90] J. Yeo, T. Kim, J. Oh, S. Park, Y. Kim, and J. Lee, "Advanced Data Transmission Framework for 5G Wireless Communications in the 3GPP New Radio Standard," *IEEE Communications Standards Magazine*, vol. 3, no. 3, pp. 38–43, Sep. 2019.

Part II

Security

6G communications – security issues and possible solutions

Neelima K., C. Kavya, and Digvijay Pandey

9.1 INTRODUCTION: BACKGROUND AND DRIVING FORCES

The 6G Communication network emerged with the developments in new generation Information and Communication Technologies such as Artificial Intelligence, Virtual Reality/Augmented Reality/Extended Reality, Internet of Things, and blockchain technology, etc. The development of 6G has a profound impact on the intelligence process of communication development that consists of intelligent connectivity, deep connectivity, holographic connectivity, and ubiquitous connectivity. Some of the security issues are portrayed in Figure 9.1. The security and privacy challenges may

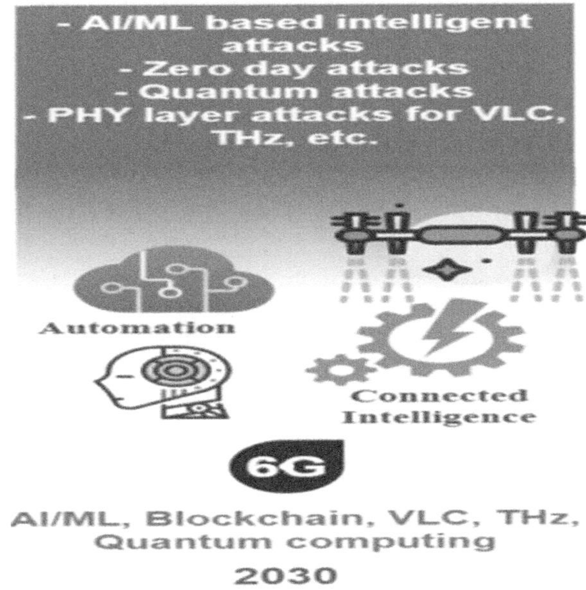

Figure 9.1 6G security issues. (Adapted from Porambage et al. 2021).

DOI: 10.1201/9781003522003-11

be encountered with the expecting 6G requirements, security key performance indicators (KPIs), novel network architecture, new applications, and enabling technologies. The challenges arise due to UAV-based mobility, Holographic Telepresence, Extended Reality, Connected Autonomous Vehicles, Smart Grid 2.0, Industry 5.0, Intelligent Healthcare, Digital Twin, etc. The security solutions in terms of distributed ledger technology (DLT), physical layer security, quantum communication, distributed AI/ML are provided. The Distributed Ledger Technology threats include Majority attack/51% attack, double spending attacks, Re-entrency Attack, Sybil attacks, Privacy Leakages, other attacks, etc. The possible solutions include proper access control and authentication mechanisms, selecting the proper blockchain/DLT type according to the 6G application and services, etc. The Physical Layer Security requires line of sight transmission which can be overcome by multipath transmission in Terahertz Technology, broadcast nature of visible light communications which can be overcome by enhancing secrecy performance by using multiple input multiple output technology, Quantum cloning attacks and quantum collision attacks in Quantum Computing which can be overcome by lattice-based, code-based, hash-based and multivariate-based cryptography, etc, poisoning attacks, evasion attacks, API – based attacks that can affect Distributed and Scalable AI/ML that can be overcome by adversarial training injects, defensive distillation, etc. Finally, the road map for materializing 6G security visions into a reality is provided [1–5].

9.2 6G SECURITY

6G Communication network is intended to facilitate a wide range of services to human kind which require stringent security measures. The key requirements are portrayed in Figure 9.2, which details the various sectors that need security at the respective levels. The details are summarized as per the applications

   a. UAV-based Mobility

   Unmanned aerial vehicles (UAVs) with AI-enabled service applications like medicine delivery, logistics, and military operations using 6G are prone to potential threats due to limited processing and power handling capabilities. To justify low latency requirements, lightweight security mechanisms are used. 6G supports path planning, route optimization, collision avoidance, etc. Specially, protected integrity of control data is a vital requirement for proper operation. The security issues that arise are physical attacks due to unmanned nature, organized attacks due to swarm of drones ranging from cyber-attacks to physical terrorist attacks.

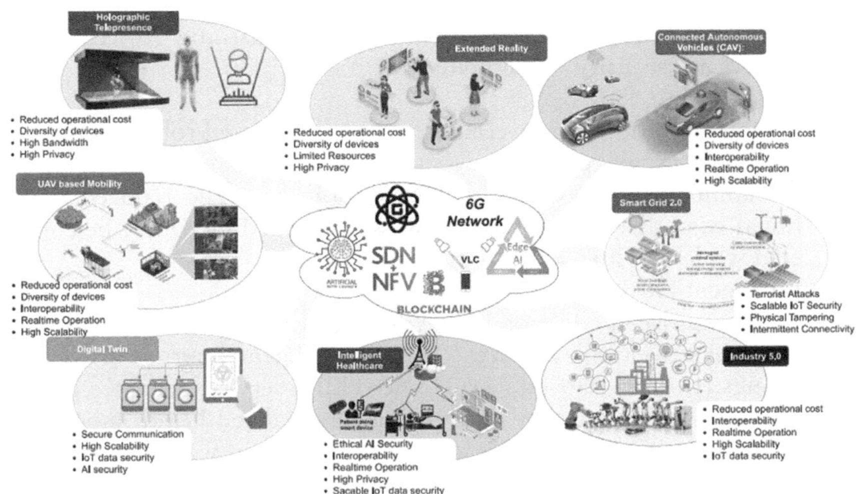

Figure 9.2 6G key security requirements. (Adapted from Porambage et al. 2021).

b. Holographic Telepresence

Holographic telepresence is used to project physical presence of distant people during 3D Video conferencing or news broadcasting in a realistic manner which requires very high bandwidth. The security requirement arises due to low operational cost of diverse devices interconnected for privacy if projected for a remote location due to lack of control over the prevailing environmental conditions.

c. Extended Reality

Extended reality (XR) combines Augmented Reality (AR), Virtual Reality (VR), Mixed Reality (MR), etc. used in applications like virtual tourism, online gaming, entertainment, online teaching, healthcare, robot control, etc. Personal data breach and fake or forged data are the potential threats in XR due to the usage of diverse devices with low overhead and high scalability requirements [6].

d. Connected Autonomous Vehicles (CAV)

The new era anticipates commercially viable autonomous vehicles for public transport that are reliable and safe. The security issues appear at vehicle level, CAV supply chain and data collecting stages. The V2X communications, hijacking of sensors, stealing physical controls can happen for vehicle-level attacks which require integrated auto-stop options with consistent monitoring. Privacy issues do arise for owners and passengers [7].

e. Smart Grid 2.0

Smart grid 2.0 features automated meter data analysis, intelligent dynamic pricing, intelligent line loss analysis, distribution grid management automation, reliable electric power delivery with self-healing capabilities, etc. Physical attacks, software-related threats, threats targeting control elements, network-based attacks, and AI/ML-related attacks are security vulnerabilities.

f. Industry 5.0

Industry 5.0 improves efficiency with automated robots and smart machines usage, where the security is must to maintain integrity, availability, audit aspects, authentication, etc. The reduced operational cost, diversity of devices, and high scalability must be considered while devising security mechanisms [8].

g. Intelligent Healthcare

To improve quality of life, AI-driven intelligent healthcare with Intelligent Wearable Devices (IWD), Intelligent Internet of Medical Things (IIoMT), Hospital-to-Home (H2H) services, etc. are utilized as Body Area Networks (BANs) for personalized health monitoring and management. The critical security challenges would be related to privacy protection, ensuring the ethical aspects of user data or electronic health records. Stringent privacy rules and regulations must be enforced by the regulation bodies.

h. Digital Twin

The digital twin interconnects virtual and physical worlds by collecting real-time data from IoT devices that are connected to the physical system, which is stored in locally decentralized servers or centralized cloud servers for real-time analysis. The biggest security challenge is that an attacker can intercept, modify, and replay all communication messages between the physical and digital domains. Hence 6G supported highly scalable secure communication channels are necessary along with blockchain technology [8].

9.3 6G NETWORK-BASED CHALLENGES

The 6G enabling technologies are incorporated at three layers i.e., physical layer, connection/network layer, and service/application layer. At each layer, the security issues do arise [9].

9.3.1 At physical layer

At physical layer security exploits noise, fading, etc. of wireless channels to enhance confidentiality and perform lightweight authentication. The various recent developments and their potential threats are described below.

9.3.1.1 Security in 6G mmWave communications

mmWave and massive MIMO can be used in 6G compatible (non-standalone) networks which have three common attacks i.e., eavesdropping, jamming, and pilot contamination attacks (PCA). Eavesdropping is carried out through inferring and wiretapping (sniffing) open (unsecured) wireless communications which can be avoided by beamforming technology. The jamming attack injects radio signals to occupy a shared wireless channel which results in Denial of Service. In PCA, the attacker intentionally transmits identical pilot signals called spoofing uplink signals to contaminate the user detection and channel estimation phase of the transmitter which causes signal leakage. The security solutions can be maximizing the secrecy rate, introducing extra randomness into the modulation, creating multiple random frequency shifts or pilot sequence or frequency hopping, using covert communication with artificial noise or friendly jamming to confuse the eavesdropper, using AI/ML techniques like channel hopping, etc [10].

9.3.1.2 Security in 6G large intelligent surface

Large Intelligent Surface (LIS) uses a planar array of low-cost reflecting elements to dynamically tune the transmission signal phase shift for enhancing communication performance. The security threat being eavesdropping even though multi-path propagation channels are used to provide different secure communication links to legitimate users.

9.3.1.3 Security in NOMA for 6G massive connectivity

In Non-Orthogonal Multiple Access (NOMA) allocates channel resources fairly for the receivers by assigning more power for "weak signal" users and subtracting power of strong signal users which supports both multicast transmission (to cluster users) and unicast transmission (to specific user). The security issues of NOMA are eavesdropping, jamming attacks, etc. which can be overcome by maximizing the transmission secrecy rate along with successive interference cancellation (SIC) process.

9.3.1.4 6G holographic radio technology
with large intelligent surface

Holographic radio or beamforming and MIMO is a disruptive radio technology for 6G indoor/outdoor communications based on software-defined antenna or photonics-defined antenna arrays i.e., it doesn't use phase shifters or active amplification in the beam-steering process. These benefit from secrecy rate optimization of base station signals by canceling out reflections to eavesdroppers. But the other threat is scattering of electromagnetic waves, hardware design.

9.3.1.5 Security in 6G terahertz communications

Terahertz (THz) enables ultra-high data rate (up to terabits per second) for 6G applications like tactile Internet, XR/AR services, etc. THz faces challenges due to low penetration power, high absorption resonance, less coverage area, etc. including dense networks of THz-enabled devices for effective communications in line-of-sight (LOS) transmissions. THz is secured against jamming and eavesdropping attacks, along with the usage of frequency hopping over a large number of sub-channels. Lightweight encryption, beam encryption, and spatial modeling can be used to overcome these attacks. Other adversaries include the collection of scattered signals during re-verification during hand-off of devices [11].

9.3.1.6 Security in 6G VLC communications

Visible light communication (VLC) uses visible light between 400 and 800 THz for high-speed communication. It is an economical alternative for indoor buildings (hospitals, personal rooms), underwater applications, or electromagnetic-sensitive areas (nuclear plants) using 6G. The advantages being super-high bandwidth, but disadvantages being incapable to penetrate a wall. Eavesdropping is a threat but maximizing the secrecy rate, utilizing beamforming signals, using spatial modulation with the zeroforcing precoding strategy, etc. can overcome threats.

9.3.1.7 Security in 6G molecular communications

Molecular communication uses chemical signals or molecules as an information carrier for communication between nano/cellscale entities like drug delivery in blood vessels in healthcare, water/fuel distribution monitoring in industry, etc. They are useful for extra covert channels in harsh environments i.e., highly absorptive channels or adversarial networks with high jamming threat. The threat is in leakage of healthcare information or bio-machines may be remotely attacked due to vulnerabilities of IoT devices. To overcome these, biochemical cryptography, which uses biological molecules like DNA/RNA information or protein structure to encode information, can protect information integrity along with strict access control system or strong firewall at the gateway portal.

9.3.1.8 Other prospective technologies for 6G physical layer security

The three key physical-layer-based technologies used to mitigate special attacks in 6G applications like spoofing messages, and tampering physical data bits that occur in network and application layers.

1. Physical layer authentication is used to crack down spoofing or impersonation attacks.
2. Physical key generation protects confidentiality for communications between UEs and stations from eavesdropping by exploiting the entropy of randomness in transmit-receive channels such as CSI and Received-Signal-Strength (RSS) to generate secrecy keys for communications.

3. Physical coding for PHY data integrity protection from tampering to maximize data transmission rate among legitimate users. E.g., Low-Density Parity-check Code (LDPC) coding, polar coding, space-time coding, and quasi-cyclic multi-user LDPC [12].

9.3.2 At connection or network layer

The connection layer is a combination of network and transport layers. Security in this layer addresses spoofing attacks and massive signaling DoS attacks. The following include key security concerns and their authentication.

9.3.2.1 6G authentication and key management for mutual authentication between the subscriber and the network

Network access control to satisfy highly personalized services and requirements of holographic telepresence use 6G authentication and key management (6G AKA) expects to solve issues like authentication of subscribers, dual authentication model for huge heterogeneous network, new subscriber identifier privacy model, etc [13].

9.3.2.2 Quantum-safe algorithms and quantum communication networks for 6G secure communication

Preventing unauthorized interceptors from accessing communications by using cipher algorithms. But quantum can break any cryptosystem built on complex integer factoring and discrete logarithms. These can be avoided by enhancing existing ciphersuites and related protocols in short term, and using post-quantum algorithms in long term, or use a public-key quantum-resistant algorithm like lattice-based cryptography, use quantum key distribution (QKD), etc [14].

9.3.2.3 Enhanced security edge protection proxy (SEPP) for securing interconnect between 6G networks

Roaming Security in 6G, the TLS-based protocols upgrade cryptographic algorithms to support high-performance TCP/IP transmissions on gigabit networks. The threats include protocol downgrade attacks [15].

9.3.2.4 Blockchain and distributed ledger technologies for a vision of 6G trust networks

Trust networks and 6G services include maintaining the worth of information sharing while preventing fake/misbehaving sources, guaranteeing low likelihood of undesirable events and avoiding the single-point-of-failure. To overcome challenges in securing all of these, use digital signatures and certificates, quantum-safe encryption, blockchain and distributed ledger technologies (DLT), etc [16].

9.3.2.5 SD-WAN security

6G network management control SDNs are Software-Defined Wide Area Network (SD-WAN) and Software-Defined Local Area Network (SDLAN). The threats include DoS/DDoS attacks, insider adversaries, etc. To protect them, use abnormal Intrusion Detection System (IDS), Moving Target Defense (MTD), ML/DL to enhance detection engines, supporting systems like FlowVisor, FlowChecker, and FlowGuard, etc [17].

9.3.2.6 Deep slicing and open RAN for 6G network security isolation

9.3.2.6.1 RAN/core network slicing

Network slicing enhances security by preventing cross-talk between slices. The challenge is to develop a security-isolation solution in spite of lack of specifications to develop a framework. The threats include large-volume DoS attack.

9.3.2.6.2 Virtualized RAN, cloud-RAN, and open RAN

Virtualized radio access networks (vRAN) and Open RAN are used to improve modularity and reduce inter-dependencies with more granular security attestation. These are vulnerable if the source code is reused as a library for developing other codes.

9.3.2.7 Next-generation firewalls/intrusion detection for 6G network endpoint and multi-access edge security

Endpoints include perimeter routers, IP core network gateways that need security gateway e.g., network firewalls, web application firewall (WAF), IDS, service-oriented architecture API protection, antivirus program, VPN, etc. to prevent unauthorized traffic. To avoid threats, use deep learning models, enable enhanced protection, etc.

9.3.3 Security in the service layer

The service layer consists of edge/fog/cloud technologies that are used to provide middleware for serving third-party value-added services. The tasks required are authentication, data encryption, application security protocols, firewalls, hardware security, service identity access management, operation/kernel systems reinforcement, data-center network protection, etc. The authentications used are:

9.3.3.1 6G application authentication: distributed
PKI and blockchain-based PKI

The public-key infrastructure (PKI) supports user and application authentication by using quantum-safe cryptography with decentralization. The threat can be Certificate Authority breach, which can be overcome by using PKI based on Ethereum (a public blockchain), a complicated model like pseudonym certificate generation, Certificate Revocation List, etc.

9.3.3.2 Using service access authentication (6G AKA)
for application authentication

To protect integrity and confidentiality, credentials like session keys will be maintained at both the UE and the application server in heterogeneous networks. To provide protection against the maintenance of AKMA, use OAuth or SSO schemes.

9.3.3.3 6G biometric authentication for 6G-enabled
IoT and implantable devices

Biometric authentication systems with wearable devices and implantable equipment are used to avoid keying complicated codes or memorizing username/passwords especially for people with disabilities to obtain unique biological characteristics for higher accuracy and flexibility.

In future, brain/heart-signal-based authentication may be able to restrict frauds. Threat could be biometric spoofing attacks by using forged or synthetic face/iris samples which may be abused. To overcome this, use secure enclave equipment, cancellable biometric models, or pseudo-biometric identities.

9.3.3.4 OAuth 3.0: new authorization protocol for 6G
applications and network function services

OAuth 3.0 is used in 6G end-to-end service-based architecture with added features like key proofing mechanisms, multi-user delegation, multidevice processing, etc. Threat imposed can be to spectrum sharing allocation/networking/security can be quickly deployed as a service through on-demand authorization.

9.3.3.5 Enhanced HTTP/3 over QUIC for secure data
exchange in 6G low-latency applications

Hypertext Transfer Protocol Version 3 (HTTP/3) is used for Web applications and mobile platforms to support secure communication on the interexchange/roaming links among the serving network and the home network.

It is built on top of Quick UDP Internet Connections (QUIC) to handle congestion. The monitoring attacks prove to be threats that are overcome by quantum-safe cryptographic algorithms.

9.3.3.6 Quantum homomorphic encryption for secure computation

Secure computation to protect confidentiality and integrity in computing nodes like edge servers use encryption methods in service computing nodes e.g., identity-based encryption, attributed-based encryption, homomorphic encryption, etc. to support quantum-safe standards. The challenging problems include searchable symmetric encryption, secure multi-keyword semantic search, secure range query, etc.

9.3.3.7 Liquid software security: a step to 6G platform-agnostic security

Liquid software allows data and applications to flow from one node to the others, using enhancing containerization architecture like Kubernetes and cloudization of edge/fog nodes to accommodate multiple applications and enable interactions through API calls. The attacks include platform-agnostic security attacks for multiple devices and in combination with hardware solutions like Trusted Platform Module (TPM) and Hardware Security Module (HSM).

9.3.3.8 AI-empowered security-as-a-service transition for 6G "service everywhere" architecture

In 6G, the usage of holographic telepresence, massive IoT applications, autonomous driving, edge computing occur at every hop of 6G communication. The threats include malware/ intrusion behaviors, data leakage, etc. in distributed computing nodes. To overcome these, use security-as-a-service (SECaaS) model to the computing nodes, super-intelligent AI models.

9.4 POTENTIAL SOLUTIONS FOR 6G SECURITY THREATS

The threat landscape and possible security solutions related to few 6G technologies are discussed here.

9.4.1 Distributed ledger technology (DLT)

Blockchain technology is one of the DLTs with advantages like disintermediation, immutability, non-repudiation, proof of provenance, integrity, pseudonymity, etc. that ensure trusted and secure 6G networks. Due to the use of AI/ML in 6G, networks became vulnerable to several new attacks that target training phase (i.e., poisoning attacks) and the testing phase (i.e., evasion attacks). The AI-driven systems in a multi-tenant/multi-domain environment include services like secure VNF management, secure slice brokering, automated Security SLA management, scalable IoT PKI management, secure roaming and offloading handling, and user privacy protection, to comply with 6G requirements.

The threats that can occur include

i. Majority attack/51% attack i.e., which happens when a group of malicious users capture ≥51% nodes can take control of the blockchain which uses majority voting consensus.
ii. Double spending attacks happen when a user spends a single token multiple times due to lack of physical notes.
iii. Re-entrancy Attack can occur when a smart contract invokes another iteratively and the secondary smart contract invoked can be malicious.
iv. Sybil attacks happen when an attacker hijacks the blockchain peer network by conceiving fake identities explicitly.
v. Broken authentication and access control foreseen in authentication and access control mechanisms.
vi. Security misconfiguration due to insecure or outdated configurations usage.
vii. Privacy leakages of user with transaction data, smart contract logic privacy, etc.
viii. Other vulnerabilities include destroyable contracts, exception disorder, call stack vulnerability, bad randomness, underflow/Overflow errors, unbounded computational power-intensive operations, etc.

The potent solutions used are ensuring the accuracy of the smart contract, proper validation of correct functionality of the smart contract by identifying semantic flaws, using security check tools, performing formal verification, etc. As blockchain/DLT support different architecture types such as public, private, consortium, hybrid blockchain, etc, hence selecting the proper blockchain/DLT with spectrum management, roaming, etc. can eliminate the impact of certain attacks.

9.4.2 Quantum computing

Quantum safe cryptography is a new approach to overcome the threats imposed due to discrete logarithmic problem for asymmetric cryptography that may become solvable in polynomial time. Quantum ML algorithms further ensures absolute randomness and security to improve the transmission quality unsupervised and supervised learning for classification and clustering tasks. The quantum key distribution (QKD) is used as communication protocol to establish a secret key between two legitimate parties.

The threats foreseen are quantum-based attacks in IoT devices, oblivious transfer (OT), quantum cloning attacks, etc. The potent solutions include a few post-quantum cryptographic primitives identified as lattice-based, code-based, hash-based, and multivariate-based cryptography with verification security in the quantum-accessible random oracle model.

9.4.3 Distributed and scalable AI/ML

With autonomous networks, self-monitoring, self-configuration, self-optimization, and self-healing are envisaged. The ZSM architecture entails intent-based interfaces, closed-loop operation, and AI/ML techniques to empower full automation of network management operations including security. The AI/ML-driven cybersecurity proves advantageous in terms of autonomy, higher accuracy, and predictive capabilities for security analytics.

The threats include AI/ML-related attacks i.e., poisoning attacks, evasion attacks, Infrastructure-targeting physical attacks that tamper communication, artefacts or traditional attack vectors towards their software, firmware, and hardware elements. The potent solutions include defensive distillation based on knowledge transfer among neurons via soft labels, usage of blockchain for improving security, moving target defense, input validation, etc.

9.4.4 Physical layer security

These mechanisms rely on random and noisy wireless channels to enhance confidentiality and perform lightweight authentication and key exchange with flexibility and adaptability in resource-constrained scenarios. The various technologies used in physical layer security include

9.4.4.1 Terahertz technology

It utilizes higher carrier frequencies in the terahertz range (1 GHz to 10 THz), to improve spectral efficiency and capacity of future wireless networks as well as provide ubiquitous high-speed Internet access. The threats include data transmission exposure, eavesdropping, access control attacks, etc. The potent solutions include the usage of intercept signals by placing an object in the path of the transmission for obtaining scattered radiations and backscatter of the channel, multipath nature to enhance the information theoretic security, distance dependent-path loss-based authentication, etc.

9.4.4.2 Visible light communication technology

VLC offers high data rates, large available spectrum, robustness against interference, and low-cost for deployment. The threats include eavesdropping attacks where the potent solutions include enhancement of the secrecy performance, usage of multiple-input multiple-output (MIMO) VLC system by using linear precoding. Spread spectrum watermarking technique, etc.

9.4.4.3 Reconfigurable intelligent surface

RIS is a software-controlled metasurface composed of a planar array of a large number of passive and low-cost reflecting elements, which are capable of adjusting dynamically their reflective coefficients, thus controlling the amplitude and/or phase shift of reflected signals to enhance the wireless propagation performance. The threat landscape includes Traditional PLS techniques, such as the deployment of active relays or friendly jammers that use artificial noise (AN) for security provisioning, which may incur increased hardware cost and energy consumption. The potent solutions include intelligent phase shift control and the use of RIS-assisted PLS for secure and low-cost 6G networks.

9.4.4.4 Molecular communication

In MC, bionanomachines communicate using chemical signals or molecules in an aqueous environment, especially for healthcare applications. The threats include several security and privacy challenges in the communication, authentication, and encryption process, which can be tackled by using biochemical cryptography.

9.4.4.5 Privacy

Digital privacy envisages threats when more and more end devices tend to share local data to the centralized entities, the storage and processing of this data pile with the added privacy protection mechanisms will be difficult. Figure 9.3 details the privacy types, privacy violation, privacy protection, and related technologies. The three key challenges encountered in protecting privacy of 6G are the privacy and regulatory difficulties for extremely large amounts of data exchange resulting in large number of small chunks of data accumulations, incorporating lightweight privacy-protecting mechanisms in resource-constrained devices and data access rights and ownership, supervision and regulations for protecting privacy by using differential privacy. It protects against certain privacy attacks such as differencing, linkage, and reconstruction attacks. The privacy risks can be addressed by solutions including, risk signatures, zero-knowledge augments and coin mixing, usage of AI-emulated human brain capability with collaborative/cooperative robots (cobots), adding quantum noise to protect quantum data will lead the security concept of DP towards quantum differential privacy, etc.

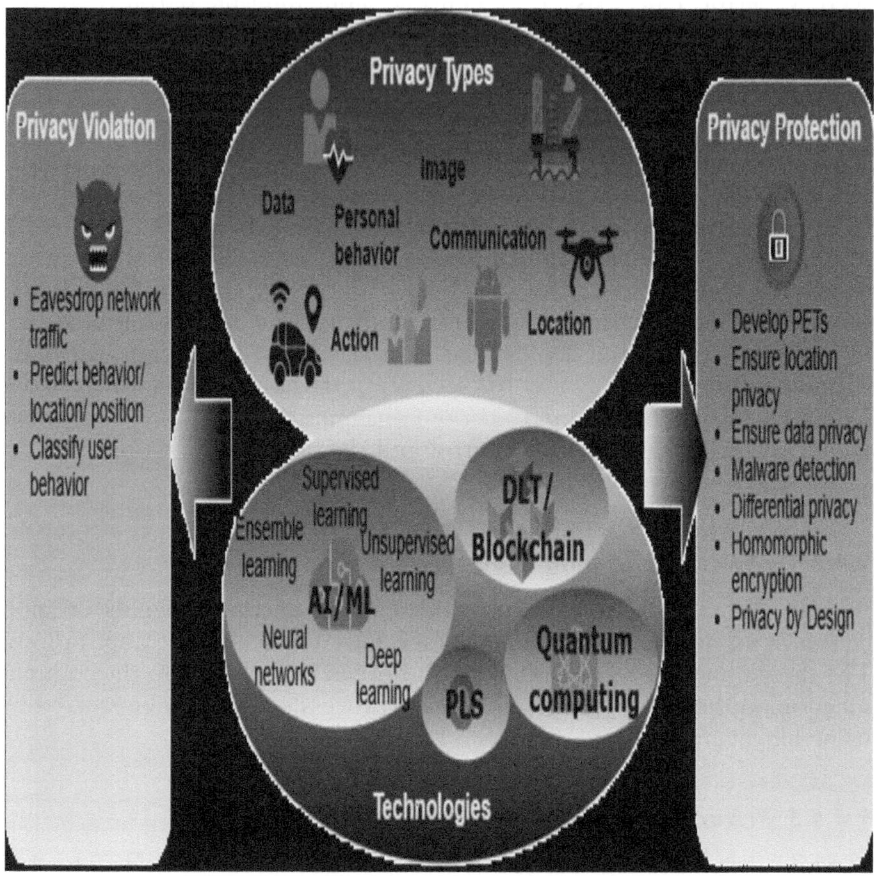

Figure 9.3 6G privacy types, issues, and solutions. (Adapted from Porambage et al. 2021).

9.5 SECURITY STANDARDIZATION

The next generation networks and digital services require active standardization. The various Standards Developing Organizations (SDOs) which are relevant to 6G security are

9.5.1 ETSI

It is launched by multiple Industry Specification Groups (ISG) to examine 5G component technologies such as NFV (ETSI NFV for group specifications and reports), AI (ETSI ISG Securing Artificial Intelligence-SAI, ETSI ISG Experiential Network Intelligence – ENI) and network automation (ETSI ISG Zero touch and service management – ZSM).

9.5.2 ITU-T

At a global level, ITU-T Focus Group on ML for Future Networks (FG-ML5G) working on technical specifications for machine learning for future networks, including interfaces, network architectures, protocols, algorithms, and data formats.

9.5.3 3GPP

3GPP addresses the use of AI/ML in the 5G Core Service Based Architecture (SBA), by introducing the Network Data Analytics Function to provide analytics and notifications to other network functions regarding the users' behavior and the network's status for solutions to slice networks and perform Network Data Analytics.

9.5.4 NIST

National Institute of Standards and Technology (NIST) standardizes post-quantum cryptographic algorithms to solicit candidates and then specify quantum-resistant algorithms each for digital signatures, public-key encryption, and cryptographic key establishment.

9.5.5 IETF

IETF Security Automation and Continuous Monitoring (SACM) Architecture RFC defines an architecture enabling a cooperative SACM ecosystem based on entities, or components, which communicate by sharing information via policy repositories, vulnerability definition data repositories, and security information repositories.

9.5.6 5G PPP

5G PPP Security Work Group tackles 5G security risks and challenges and provides insights into 5G security and how it should be addressed. It has direct implications on Beyond 5G networks such as intelligent network security, security KPIs, emerging risks, threats, and countermeasures.

9.5.7 NGMN

NGMN 5G End-to-End Architecture Framework v4.3 (2020) describes the requirements in terms of network entities and functions for the capabilities of an end-to-end framework which also includes security for the end-to-end protection of the various network features and enabling capabilities.

9.5.8 IEEE

IEEE P1915.1 Standard for Software Defined Networking and Network Function Virtualisation (SDN/NFV) Security works to provide a framework to build and operate secure SDN/NFV environments aiming at end users, network operators, service/content providers, etc. IEEE P1917.1 Standard for Software Defined Networking and Network Function Virtualisation Reliability focuses on reliability requirements and develops a framework for reliable SDN/NFV service delivery infrastructure. For quantum communications, IEEE P1913.1 (Draft) Standard for Software-Defined Quantum Communication (SDQC) defines the SDQC protocol that enables the configuration of quantum endpoints in a communication network to allow dynamic creation, modification, or removal of quantum protocols or applications in a software-defined setting.

9.6 SUMMARY OF 6G SECURITY

The 6G as an enabling technology for future too needs to address challenges for deployment around 2030. This chapter details the various newly adopted technologies for 6G with their potential security threats and possible solutions. The key 6G technologies including AI/ML, DLT, Quantum Computing, VLC, and THz communication capture the future of entire world of opportunities.

REFERENCES

[1] T. Ma, Y. Xiao, X. Lei, L. Zhang, Y. Niu and G. K. Karagiannidis, "Reconfigurable Intelligent Surface Assisted Localization: Technologies, Challenges, and the Road Ahead," in *IEEE Open Journal of the Communications Society*, doi: 10.1109/OJCOMS.2023.3292052.

[2] Joberto S. B. Martins, Tereza C. Carvalho, Rodrigo Moreira, Cristiano Bonato Both, Adnei Donatti, João H. Corrêa, José A. Suruagy, Sand L. Corrêa, Antonio J. G. Abelem, Moisés R. N. Ribeiro, José-Marcos S. Nogueira, Luiz C. S. Magalhães, Juliano Wickboldt, Tiago C. Ferreto, Ricardo Mello, Rafael Pasquini, Marcos Schwarz, Leobino N. Sampaio, Daniel F. Macedo, José F. De Rezende, Kleber V. Cardoso, Flávio De Oliveira Silva, "Enhancing Network Slicing Architectures with Machine Learning, Security, Sustainability and Experimental Networks Integration," in *IEEE Access*, doi: 10.1109/ACCESS.2023.3292788.

[3] L. Mohjazi, L. Bariah, S. Muhaidat, X. Lei and A. Shami, "Guest Editorial Special Issue on Recent Advances in Security and Privacy for 6G Networks," in *IEEE Open Journal of Vehicular Technology*, doi: 10.1109/OJVT.2023.3288457.

[4] David Soldani, Petrit Nahi, Hami Bour, Saber Jafarizadeh, Mohammed F. Soliman, Leonardo Di Giovanna, Francesco Monaco, Giuseppe Ognibene, Fulvio Risso, "eBPF: A New Approach to Cloud-Native Observability, Networking and Security for Current (5G) and Future Mobile Networks (6G and Beyond)," in *IEEE Access*, vol. 11, pp. 57174–57202, 2023, doi: 10.1109/ACCESS.2023.3281480.

[5] Y. Choi, H. Choi and S. C. Seo, "AVX512Crypto: Parallel Implementations of Korean Block Ciphers Using AVX-512," in *IEEE Access*, vol. 11, pp. 55094–55106, 2023, doi: 10.1109/ACCESS.2023.3278993.

[6] B. Pei, X. Zhou and R. Jiang, "Lattice-Based Fine-grained Data Access Control and Sharing Scheme in Fog and Cloud Computing Environments for the 6G Systems," in *2022 18th International Conference on Mobility, Sensing and Networking (MSN)*, Guangzhou, China, 2022, pp. 563–570, doi: 10.1109/MSN57253.2022.00094.

[7] R. Zhao, K. Zhang and Y. Zhang, "Energy-efficient Edge Association in Digital Twin empowered 6G Networks," in *2022 IEEE 22nd International Conference on Communication Technology (ICCT)*, Nanjing, China, 2022, pp. 869–874, doi: 10.1109/ICCT56141.2022.10073211.

[8] J. Zhang, C. Yang, A. Anpalagan, R. Dong and L. Zhang, "IDSM: Intent-Driven Slice Management and Maintenance for 6G RANs," in *2022 IEEE 22nd International Conference on Communication Technology (ICCT)*, Nanjing, China, 2022, pp. 1244–1249, doi: 10.1109/ICCT56141.2022.10073456.

[9] P. Porambage, G. Gür, D. P. Moya Osorio, M. Livanage and M. Ylianttila, "6G Security Challenges and Potential Solutions," 2021 *Joint European Conference on Networks and Communications & 6G Summit (EuCNC/6G Summit)*, Porto, Portugal, 2021, pp. 622–627.

[10] Chamitha De Alwis, Anshuman Kalla, Quoc-Viet Pham, Pardeep Kumar, Kapal Dev, Won-Joo Hwang, and Madhusanka Liyanage, "Survey on 6G Frontiers: Trends, Applications, Requirements, Technologies and Future Research," in *IEEE Open Journal of the Communications Society*, vol. 2, pp. 836–886, 2021.

[11] Xiaohu You, Cheng-Xiang Wang, Jie Huang, Xiqi Gao, Zaichen Zhang, Mao Wang, Yongming Huang, Chuan Zhang, Yanxiang Jiang, Jiaheng Wang, Min Zhu, Bin Sheng, Dongming Wang, Zhiwen Pan, Pengcheng Zhu, Yang Yang, Zening Liu, Ping Zhang, Xiaofeng Tao, Shaoqian Li, Zhi Chen, Xinying Ma, Chih-Lin I, Shuangfeng Han, Ke Li, Chengkang Pan, Zhimin Zheng, Lajos Hanzo, Xuemin (Sherman) Shen, Yingjie Jay Guo, Zhiguo Ding, Harald Haas, Wen Tong, Peiying Zhu, Ganghua Yang, Jun Wang, Erik G. Larsson, Hien Quoc Ngo, Wei Hong, Haiming Wang, Debin Hou, Jixin Chen, Zhe Chen, Zhangcheng Hao, Geoffrey Ye Li, Rahim Tafazolli, Yue Gao, H. Vincent Poor, Gerhard P. Fettweis & Ying-Chang Liang, "Towards 6G Wireless Communication Networks: Vision, Enabling Technologies, and New Paradigm Shifts," in *Science China Information Sciences*, vol. 64, no. 1, pp. 1–74, 2021.

[12] G. Gui, M. Liu, F. Tang, N. Kato and F. Adachi, "6G: Opening New Horizons for Integration of Comfort, Security, and Intelligence," in *IEEE Wireless Communications*, vol. 27, no. 5, pp. 126–132, Oct. 2020.

[13] M. Wang, T. Zhu, T. Zhang, J. Zhang, S. Yu and W. Zhou, "Security and Privacy in 6G Networks: New Areas and New Challenges," in *Digital Communications and Networks*, vol. 6, no. 3, pp. 281–291, 2020.

[14] F. Tang, Y. Kawamoto, N. Kato and J. Liu, "Future Intelligent and Secure Vehicular Network toward 6G: Machine-Learning Approaches," in *Proceeding of IEEE*, vol. 108, no. 02, pp. 292–307, 2020.

[15] S. J. Nawaz, S. K. Sharma, S. Wyne, M. N. Patwary and M. Asaduzzaman, "Quantum Machine Learning for 6G Communication Networks: State-of-the-Art and Vision for the Future," in *IEEE Access*, vol. 7, pp. 46317–46350, 2019.

[16] J. M. Hamamreh, H. M. Furqan and H. Arslan, "Classifications and Applications of Physical Layer Security Techniques for Confidentiality: A Comprehensive Survey," in *IEEE Communications Surveys & Tutorials*, vol. 21, no. 2, pp. 1773–1828, 2019.

[17] Juntti, Markku & Kantola, Raimo & Kyösti, Pekka & LaValle, Steven & Lima, Carlos & Matinmikko-Blue, Marja & Ojala, Timo & Pouttu, A. & Pärssinen, Aarno & Yrjola, Seppo & Aazhang, Behnaam & Ahokangas, Petri & Alves, Hirley & Alouini, Mohamed-Slim & Beek, Jaap & Benn, Howard & Bennis, Mehdi & Belfiore, Jean & Strinati, Emilio & Peltonen, Ella. "Key Drivers and Research Challenges for 6G Ubiquitous Wireless Intelligence", *6G First Sumit*, 2019.

Chapter 10

6G technologies

Security challenges and applications

Jain Manish

10.1 INTRODUCTION

To achieve the targets for better communication in 6th G vision, sophisticated dedicated environment including firmware, software and hardware, communication structure, network structure, workstation, database, ML/AI algo, etc. is required. Consequently, normally 6G is the hub of Information process transfer and data processing concept which is based on hardware design and software malfunction, infrastructure protection in the digital era of the recent 21st century world [2].

Most of the recent industries around the world are expecting for the 6G technology, and for that they have started research and exploration on all aspects of 6G technologies. As per observation and effective advancement in communication technology, all stakeholders will achieve the vision of 6G by around 2035 (Table 10.1).

10.2 WIRELESS COMMUNICATION FOR MACHINE LEARNING

The new era for the production of new technology has ever improved the competence by exploring the modulation bandwidth product in essence of generates abundant drape of microwave frequencies upon which the data processing is being carried out for further transmission. As technology moves to upgrades from 3G technology to 4G era its raised bandwidth range from 5.1 to 20.5 MHz, whereas the conversion from 4G technology to 5G era found the channel bandwidth upgraded from 20.1 to 100.5 MHz approximately. Now as move to emerging 6G technologies, researchers are supposed to produce the required spectral bandwidths to enhance repetitively, expecting to approach about 400.5 MHz, significantly improving the capacity of single cell generation in the channel formed [3].

Upward the bandwidth to the greatest amount with intense substantial Multi input Multi output (MIMO) channel communication. As updated band spectrum is not endowed with adequate channel strength.

DOI: 10.1201/9781003522003-12

Table 10.1 The digital world of 2030

Features of digital world	Year 2020	Tear 2030+	Analysis of constraint	Practical challenges
Range	Moderately	Omnipresent	To understand the Internet of everything for the digital outcomes and technology	Accessibility, density and required number of connections
Level	Elementary	Complicated	Proficient process of the physical digital integrated world and accomplish defined control	Delay and frequency band capability in system sensing and data storage potential
Update	Semi-stagnant	Vibrant and synchronized	Reliable and secure interconnection with updating of real-time data for efficient operation in digital scenario	Delay, bandwidth, Baud rate and memory capacity for digital system
Integration	Remote	Incorporated	To understand physical and digital interconnection of digital era, sustainable growth and effective operation	Information security concern, AI competence and privacy security, data communication capabilities and their laws and regulations

Intensifying the channel modulated bandwidth from about 10.50 to 400.5 MHz equipped with four times improving the capacity, As growing 20 times higher demand of the 6G future technology of emerging era. Researcher makes use of re-use the band spectrum in effective ways to processing and controlling AI/ML technology super flow in the required channel network.

Today's current scenario of AI has been fuelled by the accessibility of powerful computing frameworks and the big data revolt [4]. Observed two ingredients are conventionally leveraged following the centralized machine learning (CML) concept. Consequently, the training data is assembled for a processing unit, or a cluster of processing units interconnected through the wired links, and the optimization of data is encapsulated locally. Conversely, as the amount of IoT devices or smart phone, data has started being generated in a distributed manner at the edge of the latest network by significantly enhancing powerful devices [8]. Any distributed data sets can be collected at a central node and deeply processed as per the principle of the CML framework. Though the volume of the data generated at the network side, the wireless link and the privacy matters concerned with off-loading personal data are an insurmountable barrier to the application of CML in 6G network devices. The CML approach posses' limitations and motivates the search for novel communication processes and protocols to the physical layer technologies for further enhancement and enable efficient distributed machine learning (DML) at the wireless network edges of the network [5]. Machine learning is not specified absolutely. Exceptionally network makes wide-ranging uses of its application. This new approach to network organization is applicable on fundamentals law applying traditional to machine-learning techniques to further changes in the appropriate functional information to obtain the most optimized channel band parameters to settings for further advanced communication. For the moment, AI and deep learning techniques are being functional to upgrade the channel bandwidth and radio frequency management problems nowadays to further process the optimized, load examination and carrier augmentation process [6, 7].

The current uses of AI along with other intelligent technologies make possible a paradigm shift to transform the 5G technology features into a compact intelligent network. 6G communications is further augmented with lots of emerging technologies to enhance spectral efficiency. 6G communication technologies manage the amalgamation of space air and ground control sea channel network to establish the link among all the networking channels nodes in the atmosphere. Further the current issue of blockchain emerges to enhance efficiency to govern the spectrum resources which will eliminate the challenges of spectral distribution among the network from the decentralized data bases available in the surrounding (Table 10.2) [8].

Table 10.2 6G services

Specification	6G services
Service types	MBRLLC/mURLCC/HCS/MPS
Device types	Sensor/BCIdevices/DLT devices
Jitter	1 µs
Data rate	100 Gbps
Latency	0.1 ms
Mobiiity	Upto 1,000 km/h
Reliability	Upto 99.9999%
Frequency	Sub THZ band, optical VLC etc.
Multiplexing	Smart OFDM
Power consumtion	Ultra low
Processing delay	≤10 ns
Security and privacy	Very high
Wireless charging	Support BS to device transfer

10.3 THE 6G SYSTEM ARCHITECTURE

To structure the 6G channel network which provides various types of patterns and system-specific targets will attain diverse types of architectural innovation that extend various fields for channel communication. A future 6G technological network are new era and can spread across multiple heterogeneous structures to provide scattered open access and cloud configuration from various providers to attach and to provide appropriate cloud platform which may vary the competence like hardware configuration. Acquiring such kind of architecture for effective presentation would achieve a high integrated level of software configuration implementation to achieve the latest demands of predefined and different kinds of proficient outcome [7, 12]. The proposed type of channel structural design will provide an exceptional form of selection of outputs and the highly intense types of program structure to run it to set up in large-scale scattered networks. Further in particular surrounding local and wide area channel profound. This new era of 6G will profound that degree of customization in the network that to the generally rough degree of development to intend to deploy in the today's market needs in technological form.

10.3.1 Security and privacy for 6G

For cellular mobile communication, the 6G plays an advanced and new era in communication technology as compared to 5G hierarchical structure. As per the vision of 2030, the 6G technology provides autonomy and an advanced network for future vision [1]. The new era will lead to enhancement of speed up to 1 Tbps which is 50× more than what is 5G communication. The industrial and researcher had shown the vision to expand the channel convergence up to ground level sea, space level air for various kinds of application in near future technological development in defence. It further expands the capabilities to telepresence communication, wearable display devices, implantable products, mixed signal processing, automatic vehicle with the advancement and sustainability in the system (Figure 10.1) [2, 4].

Security challenges	Attacks and threats
Visibility	API based attacks
Trustiness	Poisoning attacks
Ethical and legal aspects	Evasion attacks
Extendibility and viability	Physical Layer attack
Controlled Security task	Model Inversion attack
Adaptability of modes	AI middle layer attack

Figure 10.1 AI/ML threat scope and security issues in 6G network.

10.4 SECURITY CHALLENGES IN 6G AND QUANTUM COMMUNICATION

Many technologies explained above are proven for providing better security services in 6G communication. 6G presents different types of new security challenges due to the ultra-dense heterogeneous networks and many different types of connected devices. The new technology must ensure security over multiple trust domains [9, 10].

Quantum computing has several applications in 6G communication. Security and reliability are the two major interests of quantum computing. In quantum communication, if an attacker changes something the quantum status will be changed. Quantum communication assisted 6G communication is for replacing the traditional communication channel to achieve extremely high reliability. Classical cryptography with traditional encryption and authentication may not be suitable for 6G. To provide better security the public-private key cryptosystem can be created by using quantum computing methods [5].

Quantum computers currently cannot solve the existing cryptographic problems with the available number of qubits. Solve these problems, it needs millions of qubits. Only with quantum cryptographic algorithms, we can fix this problem. Combining post-quantum computing security scenarios with the security mechanisms applicable may ensure privacy and security in the 6G communication channel with the machine learning technique. Many 6G applications use quantum security mechanism with quantum key distribution (QKD). For the ease of implementation quantum cryptography in 6G, NIST has proposed numerous new standards. This enables 6G to deliver secure data communication including QKD, quantum cryptography-enabled communication, and key sharing features such as quantum teleportation, super dense coding, and entanglement distribution [11].

The development of deep learning and ML-based air communication yields intelligence in the network which provides the real-time mixed signal processing application in the source and destination of various paths. Mathematical modelling plays an important role in designing algorithms to make new techniques as par the generation gap improving staidly over time in the application of mobile and wireless communication systems, signal transmission and reception at various transmitters and receivers. Further advancement in ML/AI may possess self-optimized interfaces that are broadly adapted to acquire ant channel transmission, their software and hardware, further for merely sophisticated applications [8, 12]. Investigation on the applications of deep learning and ML proposed to enhance various network layers like physical and medium access layer for data transmission and reception. Researchers had developed receiver blocks for 5G which perform perfect baselines for communication. By optimization algorithm produced high throughputs in the required gain. Further advancement in this cause energy efficient practical outcomes for the channels.

10.5 SIXTH-SENSE APPLICATIONS FOR NEW ERA

The sensing network provides new services and applications for superfluity. In the proposed channel network could sense the place, speed and route of all the vehicle available in the area, further the technology detect the appropriate path for pedestrian and proper working area for vehicle transportation in the automated route to access the obstacle with the sensor technology embedded [9].

10.5.1 New mobile revolution mobile networks

The future new mobile communication offers the transformation of technical equipment effectively on planets and satellite systems for effective implementation anywhere on the system. This sensor-based mobile structure creates revolutionary changes in the human tendency for superficies needs of the communication among the surrounding sensing network and devices [11].

10.5.1.1 Where 6G performs better

In various applications for future technologies, the 6G supports to control autonomous four-wheeler vehicles, robots, space, drones control and so on 6G is also able to work with Virtual reality and augmented reality (AR) which is becoming the most popular entertainment in the recent days of communication. However, 6G is still in the state of further research and hence the cope up of artificial intelligence, and machine learning and other methods are in testing to make it for versatility of daily uses in the new era.

REFERENCES

1. V. Ziegler, H. Viswanathan, H. Flinck, M. Hoffmann, V. Räisänen, and K. Hätönen, "6G Architecture to Connect the Worlds," *IEEE Access*, vol. 8, pp. 173508–173520, 2020.
2. P. Ranaweera, A. D. Jurcut, and M. Liyanage, "Survey on Multi-Access Edge Computing Security and Privacy," *IEEE Communications Surveys Tutorials*, pp. 1–1, Volume: 23, Issue: 2, second quarter 2021.
3. S. Wijethilaka and M. Liyanage, "Survey on Network Slicing for Internet of Things Realization in 5G Networks," *IEEE Communications Surveys & Tutorials*, 2021.
4. Y. Sun, J. Liu, J. Wang, Y. Cao, and N. Kato, "When Machine Learning Meets Privacy in 6G: A Survey," *IEEE Communications Surveys & Tutorials*, vol. 22, no. 4, pp. 2694–2724, 2020.
5. M. Pawlicki, M. Choras, and R. Kozik, "Defending Network Intrusion Detection Systems against Adversarial Evasion Attacks," *Future Generation Computer Systems*, vol. 110, pp. 148–154, 2020.

6. C. de Alwis, A. Kalla, Q. V. Pham, P. Kumar, K. Dev, W. J. Hwang, and M. Liyanage, "Survey on 6G Frontiers: Trends, Applications, Requirements, Technologies and Future Research," *IEEE Open Journal of the Communications Society*, Vol. 2, pp. 836–886, 2021.

7. N. Khurana, S. Mittal, A. Piplai, and A. Joshi, "Preventing Poisoning Attacks on AI Based Threat Intelligence Systems," in *2019 IEEE 29th International Workshop on Machine Learning for Signal Processing (MLSP)*. IEEE, 2019, pp. 1–6.

8. G. Gui, M. Liu, F. Tang, N. Kato, and F. Adachi, "6G: Opening New Horizons for Integration of Comfort, Security and Intelligence," *IEEE Wireless Communications*, 2020.

9. You, C.-X. Wang, J. Huang, X. Gao, Z. Zhang, M. Wang, Y. Huang, C. Zhang, Y. Jiang, J. Wang et al., "Towards 6G Wireless Communication Networks: Vision, Enabling Technologies, and New Paradigm Shifts," *Science China Information Sciences*, vol. 64, no. 1, pp. 1–74, 2021.

10. M. Wang, T. Zhu, T. Zhang, J. Zhang, S. Yu, and W. Zhou, "Security and Privacy in 6G Networks: New Areas and New Challenges," *Digital Communications and Networks*, vol. 6, no. 3, pp. 281–291, 2020.

11. Mika Ylianttila, Raimo Kantola, Andrei Gurtov, Lozenzo Mucchi, Ian Oppermann, Zheng Yan, Tri Hong Nguyen, Fei Liu, Tharaka Hewa, Madhusanka Liyanage, Ahmad Ijaz, Juha Partala, Robert Abbas, Artur Hecker, Sara Jayousi, Alessio Martinelli, Stefano Caputo, Jonathan Bechtold, Ivan Morales, Andrei Stoica, Giuseppe Abreu, Shahriar Shahabuddin, Erdal Panayirci, Harald Haas, Tanesh Kumar, Basak Ozan Ozparlak, Juha Röning. "6G White Paper: Research Challenges for Trust, Security and Privacy," arXiv preprint arXiv:2004.11665, 2020.

12. M. Giordani, M. Polese, M. Mezzavilla, S. Rangan, and M. Zorzi, "Toward 6G Networks: Use Cases and Technologies," *IEEE Communications Magazine*, vol. 58, no. 3, pp. 55–61, 2020.

Chapter 11

6G communication challenges

Infrastructure, security and standard

Palakshi Sinha, Manivel Kandasamy,
Raju Shanmugam, and Nageswari P.

11.1 INTRODUCTION TO 6G COMMUNICATION

The world has seen groundbreaking advances in communication technologies that have revolutionized the way we connect, communicate, and engage in an era characterized by rapid technical change. As we continue to tap into the potential of 5G networks, more and more connectivity opens up, providing a preview of the exciting prospects for 6G communication.

It is anticipated that the sixth-generation (6G) wireless communication network will combine marine, aerial, and terrestrial communications into a dependable, quick, and scalable network that can accommodate a large number of devices with extremely low latency needs [1].

11.1.1 Overview of 6G communication

There is much more to 6G than just faster speeds. It includes a comprehensive shift in perception with the introduction of innovative architectures and revolutionary technologies that go beyond traditional boundaries of communication.

6G aims to create a seamlessly connected world where information exchange happens at an unprecedented scale and pace, empowering a variety of sectors including healthcare, transportation, manufacturing, entertainment, and beyond. It does this by utilizing advancements in artificial intelligence (AI), terahertz frequency bands, quantum computing, nanotechnology, and beyond.

The Internet of Things (IoT) will be fully developed and personalized communication will be surpassed by the sixth-generation (6G) mobile communication technology. In addition to people, it will link computers, devices, wearable technology, sensors, cars, and even robots. Only with a 6G network will we be able to deliver sophisticated network management and adaptation as well as cutting-edge services [2]. In the course of the process, we explore the unexplored domains of 6G communication, revealing its complex nature and imagining the potential it has to transform the technological environment in our modern world.

DOI: 10.1201/9781003522003-13

11.1.2 Challenges and features

Many important challenges and possibilities emerge as scientists, engineers, and entrepreneurs come together to design the architecture of 6G networks. This transformative path towards the sixth generation of wireless communication presents complex problems to be solved, including issues with spectrum allocation, energy efficiency, security, standardization, and ethical consequences.

Several significant challenges must be overcome for 6G communication to be implemented, necessitating substantial study, creativity, and cooperation across numerous fields. Let's examine these issues in more detail:

i. Terahertz Band Challenges:

Using frequencies in the Terahertz range (0.1–10 THz) for data transmission is the difficulty associated with the Terahertz (THz) band in 6G communication. Since Tbps communications cannot be supported by technologies utilizing frequency bands smaller than 0.1 THz, 6G will be the first wireless communication system to support Tbps for high-speed communication [3]. Although these frequencies offer noticeably faster data rates, there are technological obstacles:

- Transmission Losses: The strong absorption of terahertz waves by solids and gasses in the atmosphere results in significant signal loss over extended distances. Their range and dependability are restricted by this absorption, necessitating creative fixes to keep connectivity steady.
- Antenna Design: Specialized antennas are required for the efficient transmission and reception of signals in the Terahertz band. One of the biggest engineering challenges is creating compact, effective, and Terahertz-capable antennas.

ii. Security and Privacy Concerns:

- Data Security: As the number of connected devices and data transmission increases exponentially, it is imperative to take strong security precautions to guard against cyberattacks, illegal access, and data breaches. It is essential to develop secure communication protocols and encryption techniques that are resistant to advanced attacks.
- Preservation of Privacy: In an era of ubiquitous connection, maintaining user privacy necessitates striking a balance between the collection of data for network optimization and the defense of individual privacy rights. It's difficult to design systems that preserve functionality while protecting or anonymizing personal data.

iii. Standardization and Compatibility:

- Global Standards: To guarantee smooth connectivity and compatibility, it is crucial to establish global standards for 6G technology across many industries and geographical areas. It is very difficult to coordinate the activities of many organizations, governments, and stakeholders to establish common standards.

- Integration with Current Technologies: One of the interoperability issues in 6G networks is making sure that new features and capabilities don't conflict with older wireless technology generations. It is essential to harmonize several technologies for them to work together in the 6G framework.

iv. Infrastructure and Investment:

- Network Infrastructure Deployment: Massive infrastructure upgrades are necessary for the implementation of 6G, including the deployment of a dense network of tiny cells, backhaul links with high capacity, and base stations. This infrastructure rollout comes with a significant financial and logistical cost.
- Research and Development Investment: To investigate and develop the technologies—such as AI, nanotechnology, sophisticated materials, and more—that will power 6G, substantial investment in research and development is necessary.

v. Implications for Ethics:

- AI Governance and Ethics: Given the prominence of AI in 6G networks, it is critical to address ethical issues with algorithmic biases, AI governance, and responsible AI deployment. There are several ethical issues in ensuring accountability, justice, and openness in AI-driven decision-making systems.
- Data Usage and Ownership: Using the massive volumes of data in 6G networks while resolving concerns about data ownership, permission, and fair usage is still a difficult ethical task.

vi. Effect on the Environment:

- Energy Usage: Energy consumption optimization becomes increasingly important as network complexity rises. There are several obstacles in the way of minimizing carbon footprints and decreasing e-waste to reduce the environmental impact of 6G networks through the design of energy-efficient hardware and the implementation of sustainable practices.

To overcome technical, regulatory, ethical, and environmental constraints, researchers, industry leaders, legislators, and stakeholders must work together to address these complex difficulties in the implementation of 6G communication. By overcoming these obstacles, a revolutionary and inclusive 6G ecosystem that enhances society while reducing hazards will be possible.

The possibilities of 6G communication include a variety of game-changing technologies that are anticipated to completely reshape the wireless connectivity and technological landscape. These characteristics, which go beyond what current networks like 5G can offer, have the potential to transform several industries completely. Here is a thorough examination of the salient characteristics of 6G communication:

i. Hyperconnectivity and Bandwidth Enhancement:

6G is expected to provide previously unheard-of levels of hyperconnectivity, allowing for the smooth integration of a vast array of sensors, devices, and systems. More available bandwidth will be made available to support this improved connectivity, even exceeding the capabilities of present 5G networks. To accommodate a wide range of simultaneous connections and help the IoT realize its full potential, higher data transmission rates are the goal.

ii. Ultra-Low Latency:

6G's potential to achieve ultra-low latency, which cuts the amount of time it takes for data to travel between devices to a fraction of what is possible with current technologies, is one of its most important aspects. This incredibly low latency—which could be as low as microseconds—will transform real-time applications. This almost instantaneous response time will be very helpful to gaming, immersive experiences like AR and VR, healthcare (remote surgery), autonomous vehicles, and gaming industries.

iii. Terahertz Frequency Bands:

The goal of 6G networks is to take advantage of the mainly untapped Terahertz frequency bands, which have the potential to transmit data far faster than those that are already in use. However, because of their vulnerability to signal degradation and atmospheric absorption, these frequencies pose significant technological hurdles. Investigating solutions for these obstacles is essential to making efficient use of Terahertz frequencies in 6G networks.

iv. AI Integration:

The foundation of 6G communication is AI integration. It will be essential for predictive analytics, decision-making process automation, resource management and optimization, and overall network intelligence enhancement. By dynamically adjusting to shifting network circumstances and user demands, AI algorithms will help 6G networks operate at peak efficiency.

v. Energy Efficiency and Sustainability:

Making sure 6G networks are not just cutting edge technologically but also environmentally sustainable is a goal now in pursuit. This entails creating systems that efficiently handle enormous data loads while consuming the least amount of energy. The advancement of 6G technology is centered on implementing energy-efficient hardware, sustainable practices, and tackling issues with e-waste management.

vi. Global Connectivity and Ubiquitous Networking:

The goal of 6G networks is to offer worldwide connectivity, guaranteeing smooth networking experiences regardless of location. Global users will be able to access high-speed, low-latency networks thanks to ubiquitous connectivity, which will improve information sharing, cooperation, and communication among various demographic groups and geographical areas.

11.1.3 Development of cellular communication

People's ability to interact and communicate has changed dramatically as a result of the revolutionary developments in cellular communication from the first generation (1G) to the highly anticipated sixth generation (6G). Significant technological advancements have characterized this development, with each generation adding new features and capabilities while improving upon the work of its predecessor.

Public voice service was the very first generation of providing cellular services [4]. The first generation (1G) of cellular communication signaled the beginning of the wireless technology revolution and laid the groundwork for later developments. 1G was the first commercial wireless telephone technology that enabled wireless networks to carry analog voice calls. These networks sent signals using analog modulation techniques. In comparison to networks of succeeding generations, 1G networks have a smaller coverage area and a lower capacity to handle several simultaneous calls. However, 1G was primarily concerned with delivering voice communication services; it did not have any sophisticated features or data capability.

An important turning point from analog to digital communication technologies was the introduction of 2G. With the advent of digital voice calls, there has been an improvement in speech quality, security, and efficient use of the spectrum. GSM (Global System for Mobile Communication) technology serves as the foundation for 2G. A combination of Time Division Multiple Access (TDMA) and Frequency Division Multiple Access (FDMA) was utilized in 2G systems. As a result, more people might connect at once in a particular frequency range [5].

The development of 3G networks resulted in a major advancement in mobile communication. 3G technology allowed for higher-quality phone calls, video calling, and the first steps towards mobile internet because it had quicker data transfer rates than 2G. By enabling customers to access services like mobile TV, video streaming, and internet surfing on their mobile devices, this generation set the foundation for mobile internet access. The widespread use of cell phones further spurred the development of data-centric usage behaviors. WCDMA (WideBand Code Division Multiple Access) and CDMA (Code Division Multiple Access) were used by 3G systems [5].

Data speed and network capacity have increased with the transition to 4G networks. Long-Term Evolution (LTE) technology was widely adopted, which greatly improved data rates and allowed for faster internet connectivity, more fluid video streaming, and better mobile gaming. With the availability of high-speed mobile broadband, 4G networks enabled customers to expand their use of bandwidth-intensive applications and services.

A new era of connectivity marked by extremely high speeds, extremely low latency, and huge interconnectedness was brought about by the introduction of 5G. 5G networks, which provide speeds many times faster than 4G, allow almost instantaneous data transfer, supporting real-time applications like remote operations and autonomous vehicles as well as technologies like the IoT, augmented reality (AR), and virtual reality (VR). 5G networks have capabilities that go beyond standard communication, enabling revolutionary applications and services for a range of industries.

6G is expected to be the next big thing in wireless communication, with much greater capabilities than 5G. It offers hyperconnectivity, dramatically increased bandwidth, and the ability to support an unparalleled number of simultaneous connections. With the potential to reach microseconds, the ultra-low latency will transform real-time applications in a variety of industries by facilitating instantaneous data transfer and response times. Furthermore, Terahertz frequency utilization is anticipated to be explored by 6G, potentially resulting in quicker data transmission rates. The capabilities of 6G networks will be further improved by deeply integrating AI into resource optimization, predictive analytics, and network management (Figure 11.1).

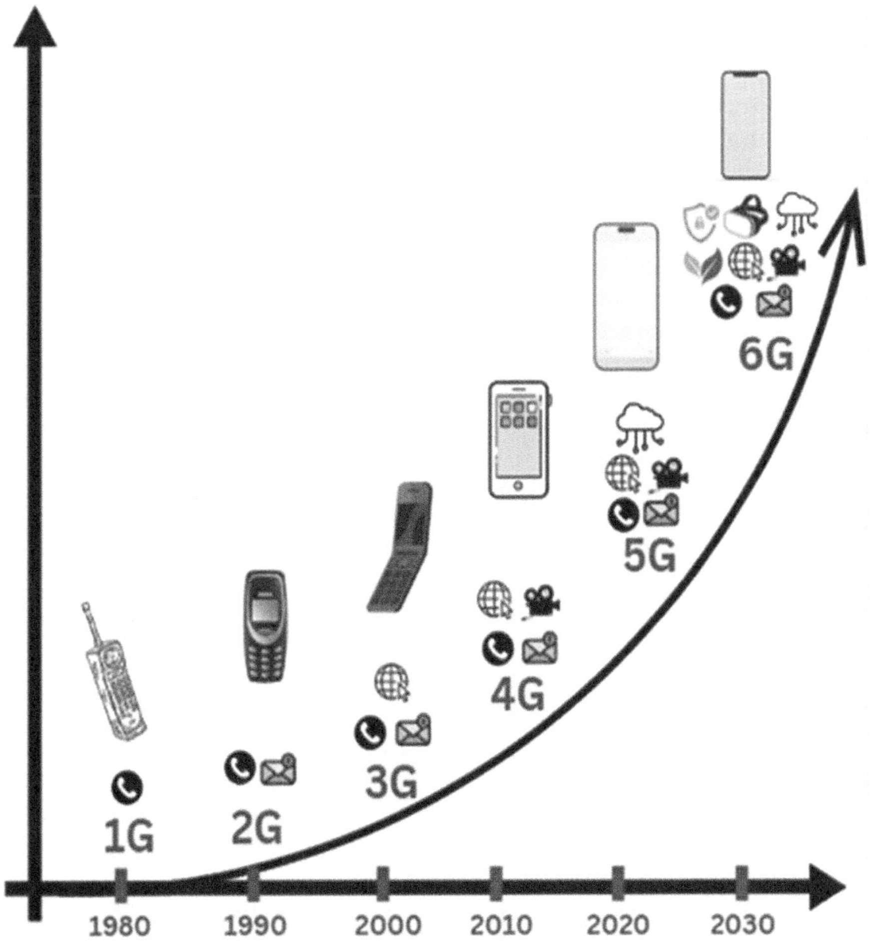

Figure 11.1 Development of 6G communications.

It has been a radical leap from voice-centric communication to a highly networked, data-driven ecology as 2G and the upcoming 6G evolve. Every generational shift has increased communication options while also laying the foundation for cutting-edge services and applications that are redefining how people engage and develop in the digital age as a society, industry, and individual. The goal of moving towards 6G is to redefine the limits of wireless communication technology and build a more intelligent, efficient, and connected future.

11.1.4 Importance of 6G communication

The advent and advancement of 6G communication technology represent a turning point in the history of wireless networks, with far-reaching implications in a variety of fields. 6G is significant because it can surpass the capabilities of its predecessors and redefine the core tenets of connectivity, speed, latency, and transformational applications. The promise of hyperconnectivity, which provides the ability to seamlessly integrate an unprecedented number of devices, sensors, and systems, lies at the heart of this relevance. With far more bandwidth available, 6G promises to enable quicker data rates and support a large number of simultaneous connections, allowing the IoT to be fully realized.

In addition, the ultra-low latency that 6G networks are expected to achieve—possibly as low as microseconds—signifies a revolutionary change that will revolutionize real-time applications that are essential to several industries, including manufacturing, entertainment, healthcare, and transportation. The use of terahertz frequency bands and the widespread integration of AI are essential components that enable increased data transmission speeds and enable predictive analytics, intelligent network management, and self-governing decision-making. Beyond merely advancing technology, 6G envisions a world in which information is exchanged effectively and smoothly, stimulating innovation, economic expansion, and societal progress.

11.1.5 Capacity and data rates for 6G

It is projected that 6G communication technology would transform wireless network capacity and data rates, to greatly outpace current generations. With the capacity and data rates predicted for 6G networks, throughput can be achieved at a quantum jump, allowing an ecosystem to sustain massive volumes of data and support the exponential expansion of connected devices and applications (Table 11.1).

The above table depicts the data rate according to the network generation and with this comparative analysis, we may forecast the advancement for 6G. It is reasonable to expect that 6G will achieve data speeds several times faster, possibly in the gigabits per second (Gbps) level.

Table 11.1 Data rate vs. to the
network generation

Network generation	Data rate
1G	1 Kb/s
2G	5 Kb/s
3G	3.1 Mb/s
4G	11.4 Mb/s
5G	400 Mb/s

11.1.5.1 Increasing capacity

- Hyperconnectivity: Hyperconnectivity, which will be made possible by 6G, will allow for a vast network that can manage an unprecedented amount of devices, sensors, and systems at once. A network infrastructure that can handle the growing demands of the IoT and other data-intensive applications across industries will be produced by this improvement in connectivity.
- Enhanced Bandwidth: When compared to current technologies, the predicted increase in capacity in 6G networks will enable far higher data transfer rates. It is anticipated to offer improved spectral efficiency, allowing higher data volumes to be sent at quicker rates and satisfying the changing needs of high-capacity networks.

Data rates on 6G networks are expected to be several times quicker than those on 5G networks. Although 5G networks have shown remarkable data rates thus far, 6G is anticipated to push these speeds even higher, possibly to terabit-per-second (Tbps) levels. With unmatched efficiency, this lightning-fast data transfer will enable almost immediate downloads, ultra-high-definition streaming, and real-time communication experiences.

11.1.6 Infrastructure challenges and development

To fully utilize this next-generation communication technology, infrastructural issues must be resolved and strong foundations for 6G networks must be developed. In terms of infrastructure development, the move to 6G brings both benefits and challenges:

i. Network Layout:
- Dense Infrastructure Implementation: Constructing a dense network of antennas, small cells, and base stations is necessary to meet 6G's increased coverage and connection requirements. This necessitates significant infrastructure investment and expansion.

- Combining Various Technologies: Harmonizing these components into a coherent and effective network design presents issues when integrating many technologies, such as Terahertz frequencies, sophisticated antenna systems, and AI-driven.

ii. Using the Spectrum:
  - Using Terahertz Frequencies: To fully utilize Terahertz frequencies for data transmission, new strategies are needed to address issues including propagation loss and signal loss. One major technical challenge is to maximize the utilization of these frequencies without sacrificing reliability.
  - Spectrum Management: For 6G networks, it's critical to allocate and manage spectrum resources effectively. Regulatory and logistical issues arise in coordinating spectrum allocation globally and guaranteeing interference-free utilization.

iii. Investment in Infrastructure:
  - Research and Development: To create the technologies that enable 6G, including AI, Terahertz communications, new materials, and network architecture, a significant investment in research and development is required.
  - Financial Aspects: The costs of deploying 6G infrastructure are high. It is crucial to address the funding models, ROI (Return on Investment), and economic viability for all parties involved, including governments, telecom carriers, and technological businesses.

11.2 THE ARCHITECTURE OF THE 6G MODEL

To fulfill the demands of a hyperconnected and data-driven future, 6G networks aspire to establish a highly sophisticated, flexible, and intelligent framework. By emphasizing energy-efficient designs and environmentally responsible infrastructure, the architecture should highlight sustainability. Important components of this sustainable strategy include minimizing electricity usage, implementing eco-friendly methods in infrastructure development, and resolving issues with e-waste disposal should be considered.

11.2.1 Mobile network density and coverage

Mobile network coverage and density in the context of 6G networks are important factors influencing how future wireless communications will be connected. To meet the expanding demands of a world that is becoming more and more connected, 6G networks are expected to reach previously unheard-of levels of network density and coverage.

It is expected that 6G networks would develop an infrastructure that is far denser than that of its predecessors. The dispersion of network nodes, antennas, and base stations throughout geographic regions is referred to as this density. To accommodate the growing number of connected devices and users, the goal is to build a highly interconnected network with a large number of access points. To achieve this density and guarantee flawless connectivity everywhere, a large number of small cells, antennas, and base stations must be installed in urban, suburban, and rural locations. 6G intends to alleviate capacity limitations, lessen congestion, and enhance network performance by densifying the network.

Improving coverage is a key goal of 6G networks, which seek to provide ubiquitous access, even in hard-to-reach areas. The goal is to bring dependable communication and high-speed data services to previously unserved or underserved areas by expanding network coverage. To increase the network's reach, this expansion makes use of cutting-edge technology including beamforming, huge MIMO (Multiple Input Multiple Output), and creative antenna designs. Higher frequency bands, such as Terahertz frequencies, may also be used to provide more inclusive network coverage by filling in connectivity gaps and enabling wider coverage.

In 6G networks, the focus on higher network densities and wider coverage has major implications. It promises to provide incredibly dependable, low-latency communication, promoting the creation of a wide range of applications in a variety of industries, such as smart cities, healthcare, and transportation. Robust connectivity in diverse settings guarantees uninterrupted availability of fast data services, facilitating inventive uses like virtual reality experiences in real-time, self-driving cars, and remote surgery.

11.2.2 Challenges in building network and density maintenance

The development and upkeep of 6G communication networks provide substantial hurdles in terms of creating and sustaining network density. There are several obstacles and complications involved in building a strong and highly integrated network infrastructure to handle the expected demands of 6G technology.

It takes a significant infrastructure rollout, including more small cells, base stations, and antennas, to deploy a dense 6G network. Finding appropriate placements for these network components can be difficult, particularly in crowded urban settings. For infrastructure deployment, getting permissions, overcoming zoning restrictions, and gaining rights-of-way become difficult jobs that frequently involve drawn-out discussions and bureaucratic procedures.

To guarantee smooth data transfer between the distributed access points and the core network, maintaining high-density networks requires effective backhaul connections. Network performance may be hampered by difficulties in establishing dependable backhaul connections, particularly in places with high population density. Technical difficulties in controlling interference between densely populated cells can affect network dependability and signal quality.

When it comes to growing network density, navigating regulatory frameworks and administrative procedures provides significant challenges. Infrastructure rollout is further complicated by addressing environmental concerns, obtaining spectrum licenses, and adhering to local legislation. The complexity is increased by the differences in rules across different countries and areas, necessitating the development of comprehensive techniques to maintain compliance while maximizing network density.

Sustainability issues ought to be taken into consideration when creating and sustaining dense networks. It is imperative to take into account the environmental impact of infrastructure deployment, manage e-waste, and implement eco-friendly methods. Network density must be achieved while maintaining sustainability, which calls for creative thinking and a proactive approach to ecologically responsible infrastructure construction.

In summary, complex regulatory, technological, administrative, and environmental factors all play a role in the difficulties associated with creating and sustaining network density for 6G communication. To overcome these obstacles, stakeholders must work together, find creative ways to deploy infrastructure, and implement adaptable tactics to maintain dependable and sustainable network density across a range of environments. Achieving the goal of a highly integrated and productive 6G communication environment requires successfully addressing these obstacles.

11.2.3 Energy challenges in 6G infrastructure

Significant energy difficulties arise throughout the development and implementation of 6G infrastructure, requiring creative solutions to maintain efficient and sustainable network operations. To minimize the environmental effect and ensure long-term viability, 6G networks must manage energy usage while they strive for unparalleled performance and connectivity.

Energy usage may rise as a result of the densification and development of 6G networks to meet the growing demand for high-speed connectivity. It takes a lot of energy to deploy a large number of tiny cells, base stations, and network equipment—potentially more energy than earlier generations. Higher energy consumption may also result from the usage of cutting-edge technology like AI-driven network optimization and Terahertz frequencies.

11.2.3.1 Mitigating energy difficulties

Various strategies are required to address energy difficulties in 6G infrastructure.

1. Energy-Efficient Hardware and Technologies: It's critical to develop and put into practice cutting-edge technologies and hardware that can achieve great performance while consuming the least amount of electricity. This entails making base stations, antennas, and other network components perform as effectively as possible.
2. Network Optimisation: Resource allocation can be more effectively handled, energy waste can be minimized, and demand-based power consumption can be optimized by utilizing AI and machine learning algorithms for intelligent network management and optimization.
3. Renewable Energy Integration: The carbon footprint of 6G infrastructure can be considerably decreased by investigating renewable energy sources, such as solar, wind, or fuel cells, to power network components. Sustainability may be improved by putting energy harvesting and hybrid energy solutions into practice.

The coordinated efforts of industry players, governments, researchers, and technological developers are necessary to address energy concerns in 6G infrastructure. Overcoming these obstacles will need funding research and development to investigate novel energy-efficient technology and sustainable practices.

11.2.4 Network design and sustainability

A key component of 6G network design is sustainability, with an emphasis on eco-friendly operations, economical resource use, and minimal environmental impact over the network's lifetime. In addition to offering previously unheard-of capability, 6G, the next generation of wireless communication technology, seeks to build an environmentally responsible communication infrastructure.

11.2.4.1 Eco-friendly infrastructure

Throughout the network lifetime, the adoption of sustainable practices, energy-efficient components, and eco-friendly materials must be prioritized in 6G network designs. This entails reducing the environmental impact of infrastructure installation, use, and eventual decommissioning. Recyclable materials and energy-saving hardware are included to help reduce e-waste and encourage sustainable habits.

11.2.4.2 Energy-efficient operations

A key component of sustainable network design is energy efficiency. Intelligent network management techniques, like power-saving protocols and AI-driven optimization algorithms, can be used to optimize energy use without degrading network performance. Reducing carbon emissions and total energy consumption is facilitated by the use of renewable energy sources in conjunction with energy-efficient technology and practices.

11.2.4.3 Life cycle assessment and environmental impact

Understanding the environmental impact at different phases of deployment and operation is made easier by conducting thorough lifecycle evaluations for network equipment. In this study, the environmental impact of network component production, use, and disposal at end of life are all evaluated. Network designers are better able to minimize ecological effects by taking into account the environmental impact of their decisions, from procurement to decommissioning.

11.2.4.4 Deployment in harmony with nature

Coexistence with the natural world should be given top priority in the design of 6G networks. Mitigating negative effects on biodiversity and ecosystems involves the strategic deployment of infrastructure, such as minimizing disturbance to ecosystems, optimizing antenna location, and abiding by environmental standards.

11.2.4.5 Collaborative industry initiatives

Promoting sustainable network design requires industry cooperation, which is backed by regulators, legislators, and technology developers. A sustainable 6G landscape is shaped in large part by initiatives that support the standardization of environmentally friendly practices, provide incentives for green technologies, and establish rules for the deployment of sustainable infrastructure.

11.2.4.6 Extended environmental care

Upholding extended environmental care in network architecture necessitates ongoing innovation, investigation, and advancement of eco-friendly technologies. Investing in R&D for sustainable materials and energy-efficient technologies, adopting circular economy ideas, and investigating creative solutions all help to build a network infrastructure that is both environmentally responsible and future-proof.

11.2.5 Channel capacity

The optimum data rate or throughput that may be successfully sent via a communication channel is known as channel capacity in 6G networks.

Intending to transform wireless communication, 6G technology places a premium on channel capacity to support the increasing data needs and wide range of applications that are anticipated in the future.

Terahertz frequencies, which offer far higher frequencies than those utilized in earlier generations, are expected to be utilized by 6G networks. Because these frequencies have bigger bandwidths, more data may be transmitted more quickly and in greater volumes. The goal of Terahertz band investigation and use is to maximize channel capacity through more effective use of the available spectrum.

Modern multiple access schemes and advanced modulation techniques are essential for increasing channel capacity in 6G networks. Combining cutting-edge multiple access strategies like NOMA (Non-Orthogonal Multiple Access) with creative modulation methods that can process large data rates allows for enhanced spectral efficiency and supports more simultaneous connections, which increases channel capacity.

Optimizing the use of spectrum is essential to increasing channel capacity. To make better use of the available frequency bands, 6G networks plan to implement spectrum aggregation, cognitive radio systems, and dynamic spectrum-sharing techniques. Putting these tactics into practice increases channel capacity and improves spectral efficiency.

11.2.6 Upgrading challenges and expanding capacity

For communication systems to scale and evolve seamlessly in the context of 6G networks, it is imperative that the issues surrounding capacity expansion and upgrade are resolved. Upgrading to 6G entails a plethora of technical, operational, and strategic issues that need to be addressed successfully for successful deployment and development of network capacity.

11.2.6.1 Technical difficulties with updating

  i. Backward Compatibility: Ensuring backward compatibility with existing network infrastructures (like 5G, 4G, etc.) while introducing new features and technologies in 6G networks is a significant challenge. This needs shifting smoothly without compromising ongoing services.
  ii. Technology Integration: Integrating and standardizing a plethora of modern technologies such as Terahertz frequencies, AI-driven systems, massive MIMO, and others poses hurdles in harmonizing these pieces into a cohesive network design.
  iii. Infrastructure Requirements: Investing heavily in and deploying new hardware, such as small cells, antennas, and network nodes, is necessary to upgrade infrastructure to meet the growing demand for data rates and connectivity.

11.2.6.2 Problems in increasing network capacity

i. Spectrum Allocation: Obtaining and overseeing spectrum resources is essential to increasing network capacity. Global spectrum distribution is difficult to coordinate while preventing interference, necessitating collaboration between industry players and regulatory agencies.

ii. Infrastructure Densification: Expanding network capacity entails densifying infrastructure by deploying more small cells and base stations. Challenges occur in selecting suitable places, managing zoning rules, and assuring constant density throughout regions.

iii. Managing Data Traffic: Increasing network capacity has become more difficult due to the rise in data consumption and the number of connected devices. Effective data routing and traffic management are necessary to avoid congestion.

11.2.7 Security challenges in 6G communications

For 6G communication technologies to be developed and implemented, security issues must be resolved. As 6G technology develops, a number of intricate security issues surface because of the extensive connectivity, faster data speeds, and wide range of applications that are anticipated in the future.

Significant Security Challenges for 6G Communications:

i. Big Data Streams and Privacy Issues: Huge data feeds from many sources are anticipated for 6G networks, which raises privacy problems for users. It becomes increasingly difficult to manage and secure sensitive data while maintaining user anonymity.

ii. Potential Vulnerabilities in New Technologies: New attack vectors are introduced by incorporating newer technologies like edge computing, AI-driven systems, and terahertz frequencies. It is now crucial to protect these technologies against cyber threats and vulnerabilities.

iii. Authentication and Access Control: It is becoming more difficult to maintain reliable authentication methods and access control with the growth of IoT devices and a variety of access points. Ensuring the safe authentication of devices and users across a highly diverse network is crucial.

iv. Securing Edge Computing Environments: Since edge computing relies on decentralized processing, integrating it with 6G networks raises security issues. It is very difficult to prevent unwanted access or alteration of sensitive data at the network edge.

v. Resilience against Advanced Threats: 6G networks need to be able to resist advanced cyberattacks, such as malware, zero-day exploits, and attacks powered by AI. It is essential to have strong cybersecurity defenses against these cutting-edge attacks.

11.3 INTRODUCTION TO 6G NETWORK SECURITY

The 6G network security paradigm goes beyond conventional methods. It requires an all-encompassing, multi-layered approach that includes authentication, encryption, and cutting-edge security procedures designed to lessen the impact of the advanced threats that will unavoidably surface. 6G's confluence of cutting-edge technologies necessitates a proactive, flexible security architecture that can counter known as well as unknown threats.

The interconnection of 6G is fundamental to its capabilities; it is a large network that includes not just individual devices but entire ecosystems, such as autonomous cars and smart cities. Due to this complex web's increased attack surface, 6G networks are vulnerable to a variety of risks, from established cyberattacks to cutting-edge dangers like quantum hacking.

11.3.1 Privacy concerns in 6G communication

Following the widespread use of 6G technology, there will be unprecedented levels of connection and a pervasive network connecting users, devices, and sensors on a never-before-seen scale. But the security of personal data and individual privacy rights are seriously jeopardized by this widespread connectedness.

- Data Proliferation and Granularity: Massive amounts of data, frequently ranging from unique IDs to complex behavioral patterns, can be exchanged across 6G networks. Concerns concerning the granularity of data being collected are raised by this growth, since it may expose private information about people's preferences, actions, and even whereabouts.
- Surveillance and Tracking: Concerns over widespread surveillance and tracking have arisen as a result of 6G connectivity's ability to facilitate extensive data collection. The capacity to combine data from various sources—such as wearables, smart infrastructure, and IoT devices—raises concerns about the possibility of unauthorized monitoring that could jeopardize personal privacy.
- Consent and Control: It is critical to provide informed consent and give people control over their data. Strong mechanisms that let consumers comprehend and control how their data is gathered, processed, and used throughout this hyper-connected ecosystem are essential given 6G's advanced capabilities.
- Security Risks and Unauthorized Access: Because 6G networks interchange a lot of data, they have vulnerabilities that make them attractive targets for hackers. Unauthorized access to this vast amount of personal data carries serious concerns since it can result in financial fraud, identity theft, or even the manipulation of vital systems.

- Regulatory and Ethical Difficulties: With 6G's worldwide connectivity, navigating the regulatory environment becomes complex. It is difficult to harmonize privacy laws across different jurisdictions; therefore, cooperative efforts are needed to create common standards that safeguard personal privacy and promote innovation.

A multifaceted approach is necessary to address these challenges. By putting strong encryption methods, decentralized identity management systems, and privacy-by-design principles into practice, users can have more control over their data while reducing the possibility of unwanted access. Furthermore, to develop confidence among users and stakeholders, ethical standards around data collection, usage, and retention in 6G networks must be established along with public discourse.

To put it simply, maintaining privacy in the context of 6G communication necessitates striking a careful balance between innovation and protection so that the advantages of connectivity are realized without jeopardizing people's basic right to privacy.

11.3.2 User privacy management

Strong user privacy management is vital, as seen by the growth of linked devices, IoT ecosystems, and hyper-connectivity in 6G networks. Safeguarding individual data in this vast network architecture necessitates a proactive strategy that gives users control over their personal data while maintaining uninterrupted connectivity.

- Encryption and Anonymization: The cornerstone to protecting user data in 6G networks is the deployment of strong encryption methods. Sophisticated encryption techniques guarantee the confidentiality of data transferred across a network and prevent unwanted access. Furthermore, anonymization methods are essential for safeguarding user identities while enabling data use.
- Decentralized Identity Management: 6G networks require decentralized identity management solutions since they go beyond centralized identity models. Users can keep control over their identities thanks to distributed ledger technologies like blockchain, which allow for selective access to personal data while maintaining anonymity and lowering the danger of identity-related cyber threats.
- Granular Consent and User-Centric Controls: It is critical to provide users with precise control over their data. Incorporating user-centric controls, such as permission methods for data sharing and processing, that enable individuals to monitor and personalize their privacy preferences, promotes a sense of agency and trust within the network.

- Accountability and Transparency: It's critical to provide clear guidelines for the gathering, storing, and use of data in 6G networks. Transparency is ensured and trust is developed by giving users clear and straightforward information about how their data is used, together with accountability measures for the entities handling the data.
- Ongoing Education and Awareness: It's critical to keep users informed about privacy and security procedures in 6G networks. Users can take an active role in protecting their personal data by learning about potential threats, best practices for data protection, and how to use features that enhance privacy.
- Adaptive Privacy Frameworks: Because 6G networks are dynamic, it is essential to have adaptive privacy frameworks in place. These frameworks are always developing to incorporate cutting-edge privacy-enhancing technologies, respond to new threats, and adjust to shifting legislative environments.

In conclusion, user privacy management in 6G networks requires a multipronged strategy that includes user empowerment, transparency, education, and flexible frameworks along with reliable technological solutions. Achieving this balance guarantees the realization of connectivity's benefits while upholding and safeguarding peoples' right to privacy.

11.3.3 Security for 6G environment

Safeguarding the 6G environment necessitates a thorough plan that combines state-of-the-art technologies, strict authentication procedures, proactive threat detection, and international cooperation to strengthen the network against a variety of cyberthreats.

- Secure Hardware Design: This is the first step towards establishing 6G security. This includes hardware-level security mechanisms that guard against logical and physical intrusions, guaranteeing the dependability and integrity of hardware and network components.
- Sturdy Authentication Mechanisms: Using sophisticated authentication techniques is essential for the 6G environment. By authenticating the identities of users and devices interacting within the network, multi-factor authentication, biometrics, and continuous authentication techniques strengthen security.
- Data Protection and Encryption: The foundation of 6G security is still end-to-end encryption. Sensitive data is kept private and secure during transmission, storing, and processing thanks to sophisticated cryptographic mechanisms that guard against unwanted access.
- AI-Powered Threat Detection: In the context of 6G, it is essential to use AI and machine learning to detect threats in real time. Anomaly detection systems powered by AI constantly observe network activity, spotting and averting possible security risks before they become serious.

- Zero-Trust Architecture: In 6G networks, implementing a zero-trust security model is essential. By demanding constant authentication and authorization for all entities attempting to access network resources, this method minimizes the possibility of lateral movement by attackers and does not presume any implicit confidence.
- Secure-by-Design Protocols: Using a secure-by-design methodology, it is crucial to incorporate security into each tier of the network architecture. This entails implementing security procedures at the development stage to guarantee resistance to changing cyberthreats.
- Immutable Ledger Technologies: By offering visible and impenetrable transaction records, immutable ledger technologies—like blockchain—improve security. Blockchain-based solutions strengthen the transactional and essential data integrity in 6G networks.
- Resilience against Quantum Threats: Post-quantum cryptography methods are essential as we anticipate the development of quantum computers. These methods provide long-term security for 6G networks by withstanding future threats from quantum computers.
- Worldwide Cooperation and Standards: It is crucial to promote worldwide cooperation in order to create uniform security standards. Global stakeholders working together to ensure universal protection, consistency, and interoperability across various 6G networks.
- Constant Monitoring and Adaptation: It is essential to put in place strong monitoring systems that keep an eye on network activity and react to new threats. Adapting tactics to quickly reduce changing hazards is part of a proactive security approach.

11.3.4 Security and privacy protection mechanisms

The implementation of a variety of sophisticated security and privacy safeguards in 6G networks includes the use of encryption methods, decentralized systems, AI-powered solutions, and privacy-preserving protocols. By combining these tactics, the network is strengthened against external attacks, and personal privacy is protected.

- Homomorphic Encryption: By enabling calculations on encrypted data without first decrypting it, homomorphic encryption is essential to 6G security. This method maintains confidentiality in data analytics and calculations while guaranteeing data privacy and facilitating processing.
- Zero-Trust Architecture: Reducing the number of potential attack surfaces in 6G networks by ensuring that trust is never taken for granted and is constantly confirmed. This method strengthens security and privacy by limiting access privileges and keeping an eye on activities.
- Blockchain for Decentralized Security: 6G delivers decentralized security features by utilizing blockchain technology. By using distributed consensus techniques to preserve user privacy, immutable

ledgers provide safe transactions and data integrity. They also guarantee clear, unchangeable records.

- AI-Driven Anomaly Detection: Using machine learning and AI to detect anomalies improves security and privacy in 6G networks. AI systems monitor network behavior continually, quickly spotting departures from the norm and thwarting possible attacks in real time.
- Differential Privacy Approaches: In 6G networks, individual data is further safeguarded by the use of differential privacy approaches. These methods maintain the accuracy of aggregated results while preventing the identification of specific user information by adding controlled noise to data sets.
- Secure Multi-Party Computation: Multiple parties can work together to calculate over their inputs while maintaining their privacy thanks to secure multi-party computation protocols. By protecting confidential data, this method promotes cooperation without jeopardizing personal privacy.

11.3.5 Cybersecurity risks and challenges

Cybersecurity threats and issues pose a serious threat to 6G networks, necessitating careful consideration and creative solutions to protect the confidentiality, integrity, and dependability of communication networks. With the integration of cutting-edge technologies, a wide range of devices, and extensive connection, the shift towards 6G technology presents a plethora of cybersecurity vulnerabilities.

11.3.5.1 Cybersecurity risks in 6G

i. New Technology Vulnerabilities: The combination of cutting-edge technology such as edge computing, AI-driven systems, and terahertz frequencies creates new attack surfaces that could be exploited by hostile actors.

ii. Increasing Attack Vectors: The attack surface is widened by the growth of linked devices, which includes a huge variety of IoT sensors and endpoints. Every device becomes a possible point of entry for online threats, making network security more difficult to achieve.

iii. Privacy and Data Protection Issues: Handling enormous amounts of data streams gives rise to worries about data breaches and privacy violations. A crucial difficulty arises when trying to protect sensitive user data while maintaining data integrity within a highly connected network.

iv. Sophisticated Cyber Threats: Defending 6G networks requires sophisticated threat detection and mitigation techniques due to the rise of cyber threats including ransomware, zero-day exploits, and AI-driven attacks.

v. Open APIs: Open APIs in 6G networks are susceptible to various security threats, including parameter attacks, identity attacks, man-in-the-middle attacks, and DoS/DDoS attacks. Parameter attacks exploit unauthorized data transfer and cross-domain data services, while identity attacks exploit authentication and authorization flaws. Unencrypted transmission of API messages can lead to man-in-the-middle attacks. Additionally, open APIs are vulnerable to DoS/DDoS attacks through overwhelming requests [6].

11.3.6 Strategies for securing 6G networks against attacks

Protecting 6G networks from a wide range of potential cyberattacks requires a complex approach that combines cutting-edge technology, strict standards, and preventative measures. These tactics seek to strengthen 6G communication systems' resilience, integrity, and confidentiality in the face of changing attack vectors. Network security is based on advanced encryption protocols, multi-factor authentication systems, and the implementation of a zero-trust architecture, which guarantees safe data transfer and access control. Utilizing blockchain technology and integrating AI-driven threat detection systems improves the network's capacity to identify abnormalities, safeguard transactions, and create transparent, unchangeable records.

A multi-layered defense strategy that strengthens the network against several attack vectors includes intrusion prevention systems, firewalls, and constant monitoring. Comprehensive security strategies should prioritize secure software development techniques, secure supply chain management, and strong incident response and recovery procedures. Fostering a cohesive and robust security posture throughout the vast 6G ecosystem depends critically on cooperative threat intelligence sharing and regulatory standard compliance. By combining these tactics, 6G networks can protect against cyberattacks in advance and guarantee the security, dependability, and trustworthiness of communication systems in the future.

11.4 INTRODUCTION TO 6G STANDARDS

6G, the groundbreaking 5G's successor, promises to be an enormous shift in the world of wireless communication. While this next-generation network is still in its early phases of development, research and conversations are defining its vision. The International Telecommunications Union (ITU) and the Third Generation Partnership Project (3GPP) have been crucial in creating the technical specifications and standards for 5G. The term "standards" refers to the specifications and technical protocols that specify how the various components of a 6G network should function.

These standards are critical for maintaining interoperability and compatibility among the equipment and devices of various suppliers, allowing for a seamless and effective communication network. These standards guarantee that devices from many manufacturers may interoperate and work consistently.

6G promises a quantum improvement in performance and functionality beyond 5G's present capabilities. Its 1 Tbps data rates dwarf 5G's 20 Gbps, and its near-instantaneous microsecond latency enables real-time engagement. Beyond speed, 6G explores uncharted terahertz frequency regions, opening up tremendous capacity for new applications such as holographic communication and sophisticated robotics. AI becomes fully integrated, optimizing network resources and constantly reacting to user demands.

The reach of 6G expands beyond metropolitan hubs, with the goal of providing seamless connection in rural and isolated places, as well as allowing linked drones and aircraft. While 5G largely focuses on augmented reality, virtual reality, and the IoT, 6G ushers in the metaverse and enables human-machine cooperation through augmented reality. And, while we weave an ever-increasing web of interlinked technology, security and sustainability continue top priorities, with 6G combining the most recent improvements in both. In essence, 6G represents a fundamental revolution in how we communicate and interact with the world, altering everything through urban planning to human-machine collaborations (Table 11.2).

Table 11.2 Feature comparisons between 5G &6G communications

Feature	5G	6G
Peak data rate	Up to 20 Gbps	Up to 1 Tbps
Latency	Milliseconds	Microseconds
Frequency bands	Low, mid-band	Low, mid-band, THz
AI integration	Limited	Extensive
Coverage	Urban focus	Broader, including rural and remote areas
Applications	AR, VR, IoT	Holographic communication, advanced robotics, metaverse
Network slicing	Virtual slices	Dynamic slicing based on real-time demand
Energy efficiency	Significant improvements over 4G	Further improvements over 5G
Security	Enhanced 5G security	Potentially quantum-resistant cryptography
Deployment timeline	Currently rolling out	Expected in the late 2030s

11.4.1 Understanding the interconnected nature of infrastructure

As we stand on the verge of the 6G revolution, we can expect a profound transformation – not just in the scorching speeds and near-real-time responsiveness, but also in the very essence of how our technological infrastructures will interact and work. 6G is more than just a speedier version of 5G; it symbolizes a paradigm change, weaving a complex web of interconnection that will reshape how we live, work, and interact with our surroundings.

Consider a cityscape where sensors implanted in buildings whisper real-time data to self-driving cars, drones flit through the air bringing essential medical supplies, and holographic avatars come to life during remote meetings. This isn't science fiction; it's a peek of the interconnected future that 6G promises.

The notion of infrastructural interdependence is central to this approach. From the THz spectrum to the AI-powered network core, every component of the 6G ecosystem will be closely integrated, providing information and adjusting to each other in real-time. Let us unravel the tapestry's complex threads:

6G networks will not function in isolation. Edge computing will be tightly linked with them, bringing intelligence and processing capacity closer to the source of data. This distributed design will not only minimize latency, but will also enable dynamic resource allocation, with network resources responding to the fluctuating demands of various applications – envision AI-powered traffic signals regulating flow based on real-time sensor data from connected automobiles.

Sensors will be implanted in everything from clothing to streetlights. This massive network of eyes will provide a steady stream of real-time data to the 6G ecosystem – air quality, traffic patterns, energy usage – enabling AI-powered optimization and automation. This interwoven tapestry of sensing and actuation will convert our cities into living, breathing creatures, from changing building ventilation depending on air quality to dynamically regulating energy networks based on demand.

The heightened danger of cyberattacks comes with such seamless interconnectedness. 6G security will need a comprehensive strategy that incorporates trust into the network's basic fabric. Blockchain technology, quantum-resistant encryption, and AI-powered threat detection will combine to form a dynamic and adaptable security barrier that will continually evolve to outwit bad actors.

6G will not replace humans; rather, it will strengthen them. Augmented reality will effortlessly layer digital information on top of our actual environment, helping doctors through intricate surgeries or supporting firemen in disaster zones. Consider technicians remotely managing hazardous robotics or instructors utilizing holograms to bring history to life. 6G will operate as

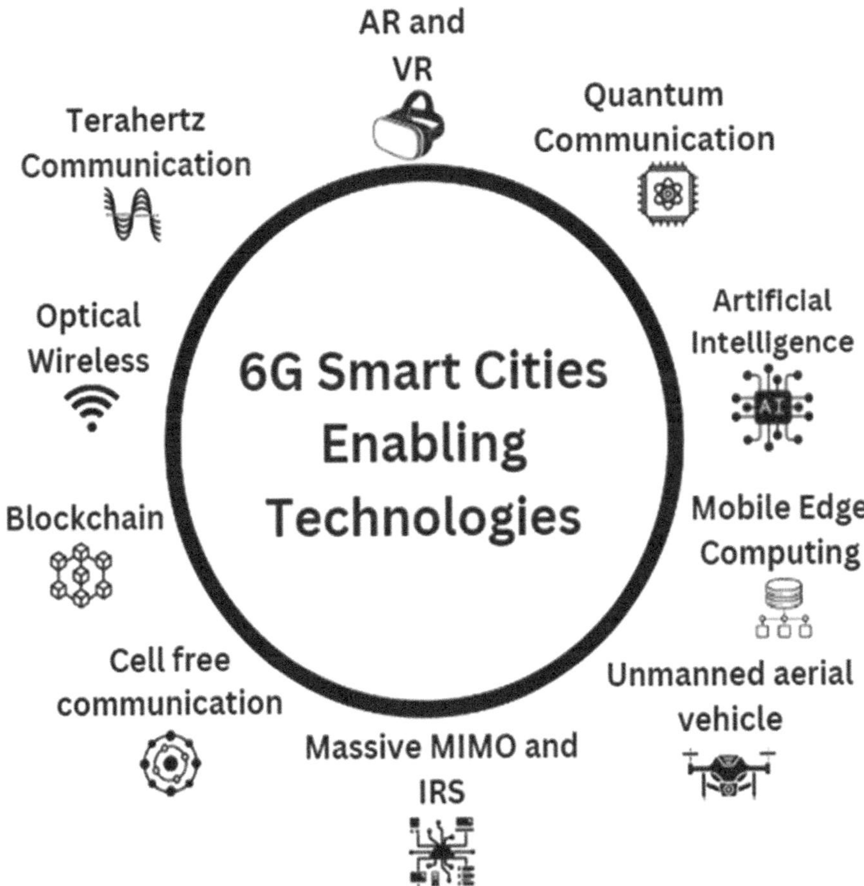

Figure 11.2 Smart cities enabling technologies.

the conductor in this human-machine dance, controlling the flow of information and enabling unprecedented cooperation (Figure 11.2).

The exponential increase of data and linked devices in the 6G era poses energy consumption problems. Interdependence, on the other hand, can pave the path for sustainability. AI-powered network optimization may reduce energy waste, while dynamic resource allocation can guarantee infrastructure use is optimal.

Understanding the interrelated nature of infrastructure in 6G is more than just a technical exercise; it is a vital step toward realizing the revolutionary promise of this technology. We must explore the ethical implications of weaving these varied threads together, guaranteeing fair access and responsible growth. The 6G future beckons as a complex ecosystem, a web of threads where technology, sensing, security, and human inventiveness entwine to create a world of inconceivable possibilities.

11.4.2 Future planning and conclusions

The full standards and specifications for 6G are still being worked on, and commercial deployment is not planned until the late 2030s. However, 6G has the potential to transform the way we connect and interact with our surroundings. It is expected that ITU will finish 6G standardization (ITU-R IMT-2030) by the end of 2030, whereas 3GPP will complete 6G standardization in R23.

It follows that the 6G wireless network standard should be expected to be up and running just under 10 years from now [7].

11.5 FUTURE TRENDS AND TECHNOLOGIES IN 6G COMMUNICATION

6G, the next stage in wireless communications, promises to transform our digital world. While research and development are still in their early phases, they are painting a vision of a future in which ultra-fast speeds, near-zero latency, and significantly greater capacity open previously inconceivable possibilities.

In 6G networks, latency, or the time required for data to transit between equipment, will be nearly undetectable. This will be critical for real-time applications such as remote surgery, self-driving cars, and realistic virtual reality experiences. Consider haptic feedback with real-time data feeds to help surgeons do difficult surgeries from afar, or self-driving automobiles negotiating complex surroundings with lightning-fast decision-making.

6G intends to deliver seamless, high-speed connections anywhere on the planet, from heavily crowded cities to distant corners of the world. This will be accomplished through the use of both terrestrial and non-terrestrial networks, such as satellites that are high-altitude platforms, and drones. Consider farmers in rural areas using linked sensors to check their crops in real time, or students in faraway communities easily accessing educational materials.

In 6G, AI will play a critical role in improving network performance, distributing resources dynamically, and assuring security. Machine learning algorithms will evaluate massive quantities of data in order to forecast traffic patterns, identify possible risks, and automate network management duties, resulting in a more efficient and robust infrastructure.

6G will serve as the connection point for a truly pervasive IoT, connecting billions of gadgets across several sectors. From smart cities and smart homes to linked wearables and commercial sensors, 6G's increased capacity and reduced latency will enable immediate exchange of information and seamless interaction between devices, enabling automated and intelligent systems.

Emphasis on Sustainability: As data traffic and device connectivity increase, sustainability becomes an increasingly important issue in 6G development. To reduce the environmental effect of this next-generation network, researchers are investigating energy-efficient technology and network architectures. This involves the use of renewable energy sources, the optimization of power usage in devices, and the use of intelligent network management solutions.

11.5.1 Key challenges and potential

The paucity of accessible spectrum for wireless communication is one of the key obstacles of implementing 6G networks. The extremely high frequency (EHF) bands, also known as the terahertz (THz) bands, which span from 300 GHz to 3 THz, are projected to be used by 6G networks. These bands provide a lot of capacity and have minimal latency, but they also have a lot of propagation loss, atmospheric absorption, and interference. To address these issues, 6G networks will need to use sophisticated technology such as beamforming, massive MIMO, and intelligent reflecting surfaces to improve signal quality and coverage.

Another issue that must be addressed while establishing 6G networks is the energy efficiency of the devices and infrastructure. Massive volumes of data transmission and processing will be required for 6G networks, which will consume a lot of power and produce a lot of heat. Furthermore, 6G networks will need to serve a wide range of applications and services, including augmented reality, virtual reality, holographic communication, and AI, all of which will need varying degrees of performance and service quality. 6G networks will need to embrace green and sustainable solutions to meet these difficulties, such as energy harvesting, wireless power transmission, edge computing, and network slicing.

The security and privacy of users and data is a third barrier in deploying 6G networks. Sixth-generation networks will enable pervasive connection and sensing, generating vast volumes of data and metadata that might disclose sensitive information about users and their surroundings. Furthermore, 6G networks will rely on complex and diverse architectures and protocols, increasing the system's susceptibility and attack surface. 6G networks will need to integrate strong and adaptive security and privacy measures, such as encryption, authentication, blockchain, and federated learning, to meet these problems.

Security technology enablers are critically required to address the 6G threat vector, which includes architectural disaggregation, open interfaces, and multiple stakeholders. This poses challenges in ensuring cyber-resilience, privacy, and trust across IoT, heterogeneous cloud and networks, devices, sub-networks, and applications [8].

The standardization and governance of the technology itself and the marketplace is a fourth obstacle of implementing 6G networks. 6G networks will involve a diverse range of parties and domains, including telecom operators, device makers, service providers, authorities, and consumers, each with its own set of interests and expectations. Furthermore, 6G networks will be required to adhere to a variety of technical, moral, and social standards and norms, including spectrum allocations, interoperability, quality of assistance, data protection, and rights for humans. To address these obstacles, 6G networks will need to enhance stakeholder engagement and coordination, as well as develop a clear and flexible structure for standardization and governance.

11.5.2 Future of 6G technology

The future of 6G wireless communication holds the potential of dramatic improvements, ushering in an era of unprecedented connectedness and technical possibilities. 6G is predicted to outperform the already revolutionary 5G in terms of speed, latency, and capacity, attaining terabit-per-second data rates and lowering latency to microseconds. Beyond mere speed increases, 6G is expected to usher in ground-breaking applications across a variety of industries. AI integration is intended to be a keystone, enabling dynamic network optimization, tailored services, and intelligent decision-making. With immersive augmented and virtual reality experiences, seamless integration of the physical and digital worlds, and extremely responsive apps, the sixth generation is ready to transform how we engage with technology.

6G wireless sensing systems, such as identification of threats, health monitoring, and environment assessments, will have an influence on government and industrial approaches to safety for the public and critical asset security. We should expect increased decision-making skills utilizing real-time information, which will improve the receptivity of law enforcement and first responders.

In healthcare, 6G may enable remote procedures and real-time monitoring, and in education, immersive, interactive experiences may transform remote learning. Manufacturing and transportation industries, for example, stand to profit from increased automation and connection, promoting the expansion of Industry 4.0 and intelligent infrastructure. As 6G technology matures, it is expected to contribute to the establishment of smart cities, monitoring of the environment, and sustainable practices by utilizing its abilities to create improved, integrated, and intelligent systems. However, as technology advances, difficulties like global standards, security concerns, and moral dilemmas will have to be addressed.

Further research is needed to develop advanced encryption algorithms and protocols that can effectively address the security challenges in 6G networks. Research on quantum-resistant cryptography should be continued to develop robust security mechanisms that can protect against future quantum computing attacks in 6G networks [9].

Future studies should focus on developing physical layer protection mechanisms, deep network slicing techniques, and quantum-safe communications to mitigate the attack magnitude and protect personal data in 6G networks [10].

The future of 6G has the ability to change the digital environment, reinvent customer experiences, and drive creativity across a wide range of industries, eventually influencing how we live, work, and interact with one another in the next decades.

11.5.3 Application and advantages of 6G technologies

One of the primary applications in which 6G is projected to play a crucial role is autonomous driving, which will enable higher precision and dependability. The newly issued IEEE 2846, a new standard for autonomous vehicle (AV) safety, is a critical step toward broad testing of AVs in the United States. In the future, 6G and subsequent networks will be required to power an AV civilization. It's obvious, for example, that data speed with total coverage will be necessary to allow thousands of autonomous vehicles to navigate the traffic in a geographical region. It will, however, be required for communication with a network of sensors that will steer the AV to an open parking place at the intended conclusion of the journey.

Immersive communication experiences will also be possible with 6G thanks to location and context-aware internet services, sensory experiences such as completely realistic extended reality (XR), including high-fidelity holograms. Look for augmented reality to replace virtual reality, which often requires a bulky headgear. Numerous fields, including communication, healthcare, interior design, construction, and gaming, will include holographic technology. Instead of today's video conferences, users will be able to communicate with others in real time via virtual reality (VR), thanks to wearable sensors that simulate being in the same room.

Some of the advantages of 6G in various fields is listed below:

1. Healthcare: 6G is predicted to change healthcare by providing ultra-low latency enhanced telemedicine and remote operations. Surgeons might do delicate surgeries from a distance, and real-time patient monitoring could be improved, resulting in faster reaction times in crises. Furthermore, the large data capacity of 6G might make it easier to transport high-resolution medical imaging for more accurate diagnosis.
2. Education: 6G might facilitate immersive augmented and virtual reality experiences in education, making distant learning more interesting and participatory. High-quality, real-time collaboration amongst learners and educators would be enabled by ultra-fast data rates and minimal latency, enabling a more flexible and effective learning environment.
3. Transportation: 6G has the potential to dramatically enhance autonomous cars and intelligent transportation systems. Low-latency

communication can improve vehicle coordination, resulting in a more secure and effective traffic flow. Furthermore, the capabilities of 6G might provide real-time navigation system upgrades and enable smooth communication between cars and traffic infrastructure.

4. Manufacturing and Industry 4.0: In manufacturing, 6G might be critical to the advancement of Industry 4.0. High communication throughput and low latency would enable the development of more advanced and adaptive robotic systems, increasing automation and enhancing overall efficiency. Improved connectivity and interaction among smart devices may result in more efficient and adaptable manufacturing processes.

5. Smart Cities: 6G's improved capabilities have the potential to convert cities into smarter, more efficient entities. 6G might provide extensive data gathering and analysis for improved urban design, management of resources, and public services by connecting a large number of linked devices, sensors, and actuators. This has the potential to increase energy efficiency, waste management, and general quality of life.

6. Entertainment and Media: 6G is expected to revolutionize the entertainment sector by providing complete immersion with augmented reality (AR) or virtual reality (VR) content. High-quality, immediate streaming of 3D and holographic material might become widespread, improving people's media consumption and interaction.

7. Environmental Monitoring: 6G has the potential to be extremely useful in tracking ecological and sustainability activities. Because of the rapid transmission of data and connection, an enormous number of sensors to monitor air and water quality, weather conditions, and habitats for wildlife might be supported. This real-time data may help to improve environment conservation and management efforts.

8. Energy Industry: 6G might aid the energy industry by improving power grid monitoring and control. Improved connection might allow smart grids to respond more quickly to swings in energy demand, improve energy distribution, and more efficiently integrate renewable energy sources into the system.

9. Disaster Services: The low latency and high data speeds of 6G might help improve communication and coordination among first responders in disaster circumstances. During an emergency, real-time data sharing, including high-resolution imagery and sensor data, might help in enhanced decision-making and response operations.

10. Finance and Banking: 6G has the potential to boost the finance sector by enabling quicker and more secure connectivity for financial activities. Improved connection and decreased latency might boost financial application speed, provide real-time analytics, and increase the overall protection of digital financial products and services.

REFERENCES

[1] Akhtar, M.W., Hassan, S.A., Ghaffar, R. et al. (2020). The Shift to 6G Communications: Vision and Requirements. *Human-centric Computing and Information Sciences*, 10:53. doi:10.1186/s13673-020-00258-2.

[2] Xu, G. (2020). Research on 6G Mobile Communication System. *Journal of Physics: Conference Series*. doi:10.1088/1742-6596/1693/1/012101.

[3] Elmeadawy, S. et al. (2019). 6G Wireless Communications: Future Technologies and Research Challenges, IEEE International Conference on Electrical and Computing Technologies and Applications (ICECTA 2019), doi:10.1109/icecta48151.2019.8959607.

[4] Bhandari, N., Devra, S., and Singh, K. (2021). Evolution of Cellular Network: From 1G to 5G. *International Journal of Engineering and Techniques*, 3(5), pp. 98–105.

[5] Patel, S., Shah, V., and Kansara, M. (2018). Comparative Study of 2G, 3G and 4G. *International Journal of Scientific Research in Computer Science, Engineering and Information Technology*, 3(3), pp. 1–16.

[6] Pawani, P., Gurkan, G., Diana, P. et al. (2021). The Roadmap to 6G Security and Privacy, International Conference on Multimedia Information Technology and Applications (MITA 2021). 2:1094–1122. doi:10.1109/OJCOMS.2021.3078081.

[7] Anoh, K., See, C.H., and Dama, Y.A.S. et al. 6G Wireless Communication Systems: Applications, Opportunities and Challenges. *Future Internet*. doi:10.3390/fi14120379.

[8] Volker, Z., Peter, S., Harish, V. et al. (2021). Security and Trust in the 6G Era. *IEEE Access*, 9:142314–142327. doi:10.1109/ACCESS.2021.3120143.

[9] Yang, L. (2020). Security in 6G: The Prospects and the Relevant Technologies, *Journal of Industrial Integration and Management*, 5(3):271–289. doi:10.1142/S2424862220500165.

[10] Van-Linh, N., Po-Ching, L., Bo-Chao, C. et al. (2021). Security and Privacy for 6G: A Survey on Prospective Technologies and Challenges. *IEEE Communications Surveys and Tutorials*, 23(4):2384–2428. doi:10.1109/COMST.2021.3108618.

Chapter 12

Racing against the machine
Data protection in the 6G era

Ajay Sudhir Bale

12.1 INTRODUCTION

A new era of hyperconnectivity and improved digital experiences made possible by next-gen communication technology has dawned with the advent of 6G networks. The goal of 6G is to bring about significant advancements in areas like as capacity, speed, use cases, and sensory interface richness. In particular, 6G aims to achieve data speeds ranging from terabits to petabits, which would enable the transmission of immersive material to occur almost instantly. With delays down into the millisecond region, real-time interaction will be possible. With the extension of spectrum, prolific device connection scales, and node densification, network capacity is expected to become almost endless. To promote digitally mediated perceptions and seamless environmental integration, these hyperconnected infrastructure developments aim to catalyse large internet-of-senses ecosystems. In these ecosystems, countless gadgets cooperate to provide a more immersive and immersive digital experience. Paradigm has several potential uses, including as digital twins, holographic telepresence, networked robots, streaming media for many senses, and expanded worlds. Additionally, 6G will be the first of its kind to combine sensing, computation, storage, and communication as it faces the trifecta of the digital, biological, and physical worlds. 6G systems will be distinguished by characteristics such as tetra-hertz operating frequencies, cell-free topologies, awareness-driven resource allocation, and 3D linked networks.

12.1.1 Overview of 6G capabilities and advancements

In light of the symmetrical technologies underlying the Internet of Things and the lightning-fast technical advancements over the last several decades, wireless communication systems are the modern-day counterparts of the Eureka! moment. The most current iteration of mobile wireless cellular communications technologies is the fifth generation (5G) wireless network, the

DOI: 10.1201/9781003522003-14

fifth generation overall. There has been a new generation of wireless cellular communication systems introduced around every ten years since 1980. The first generation, known as analogue FM cellular systems, came out in 1981. The second generation, 3G, was introduced in 2001. The fourth generation, 4G, also called the long-term evolution [LTE], was released in 2011 [1]. The ever-changing wireless technologies are summarised in Figure 12.1. In general, wireless communications have come a long way in the last ten years, paving the way for data-hungry applications like online gaming, multimedia, and HD video streaming to thrive. Many cutting-edge user-defined services, such as mobile gaming, smart homes/cities, and mobile retail and payment, are being made possible and spread by the rapidly expanding mobile Internet infrastructure [1,2].

5G communications standards have been finalised, and the technology is already being implemented all across the globe. By April 2019, South Korea has deployed 5G extensively throughout 85 cities and 86,000 base stations, making it the first country to do so. Seoul, Busan, and Daegu were among the six cities where 85% of the 5G base stations were situated. In these cities, speeds ranging from 193 to 430 Mbit/s were tested using a distributed

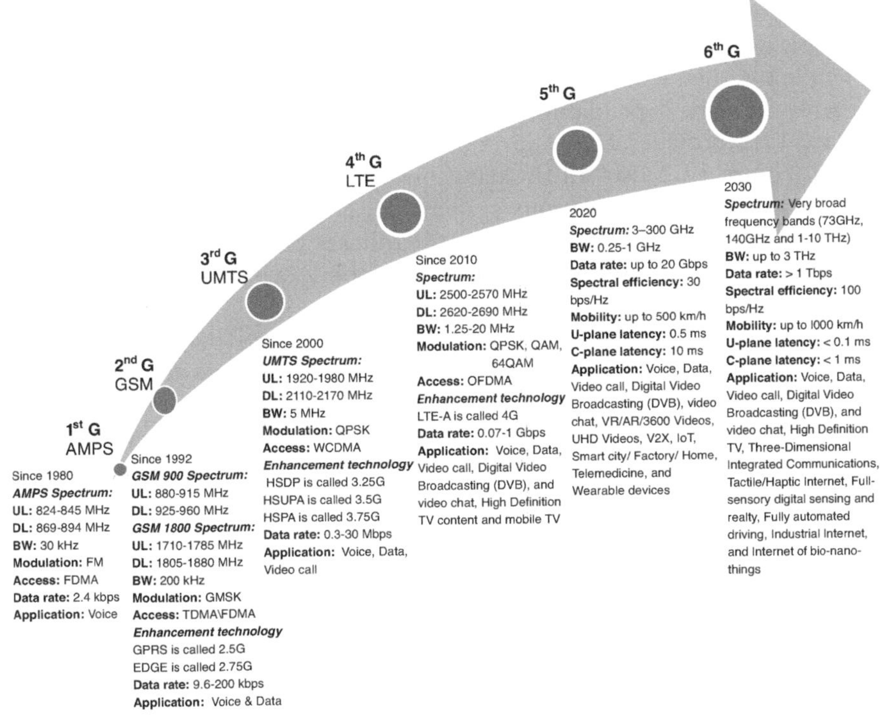

Figure 12.1 Important turning points for several telecommunication generation (1–6G) [1]. Open Access.

architecture using a 3.5 GHz (sub-6) spectrum. By the end of 2025, it is projected that over 65% of the global population will have access to 5G superfast Internet coverage [3,4].

There was a single overarching vision statement for pervasive wireless intelligence presented at the first 6G wireless cellular mobile communications conference in March 2019. An unprecedented revolution is predicted to occur in the 6G system, setting it apart from previous generations and radically altering the wireless progression from "connected things" to "connected intelligence" [5]. To be more precise, 6G will go above and beyond mobile Internet and be necessary to enable pervasive AI services throughout the whole network, from its core to its end devices. At its core, artificial intelligence will be the engine that propels the development and optimisation of 6G protocols, infrastructures, and operations.

So, in terms of network data availability, mobile data rate, and smooth ubiquitous connectivity, 6G communications are anticipated to provide better services than 5G communications. Furthermore, 6G communications will use a novel method of communication to acquire the acceptance of different types of mobile data, which will then be sent via conventional improved radio-frequency networks. This method will pave the way for a new kind of wireless emotion transmission that incorporates virtual presence and engagement. In the 2030s, there will be a plethora of new forms of wireless communication that do not yet exist, such as holographic calls and a tactile Internet. Considering the applications that will be enabled, 6G will provide the same dependability as wired networks with a low bit-error-rate. Figure 12.2 provides a high-level overview of the upcoming 6G standard [6–8]. Figure 12.2 lists all the data, but the most noticeable ones are the THz wireless communications system, AI, and programmable intelligent surfaces. These advancements mark a significant break with the established standards of design and execution in the mobile wireless telecommunications industry.

As 5G networks are being deployed globally, collaboration between industry and academics has begun to envision 6G wireless communication technologies, which will tackle the upcoming problems caused by the massive surge in wireless data traffic. In addition to a slew of new services, 6G technology enables bitrates of up to Tbps with latency below 1 ms [9–13].

12.2 PRIVACY AND SECURITY ISSUES

12.2.1 Critical privacy and security issues arising from 6G

The end-to-end latency in 6G ought to be reduced to a few µs for the Enhanced Ultra-Reliable and Low Latency Communication (ERLLC) services. System energy efficiency must also grow by a factor of 100 over 4G and

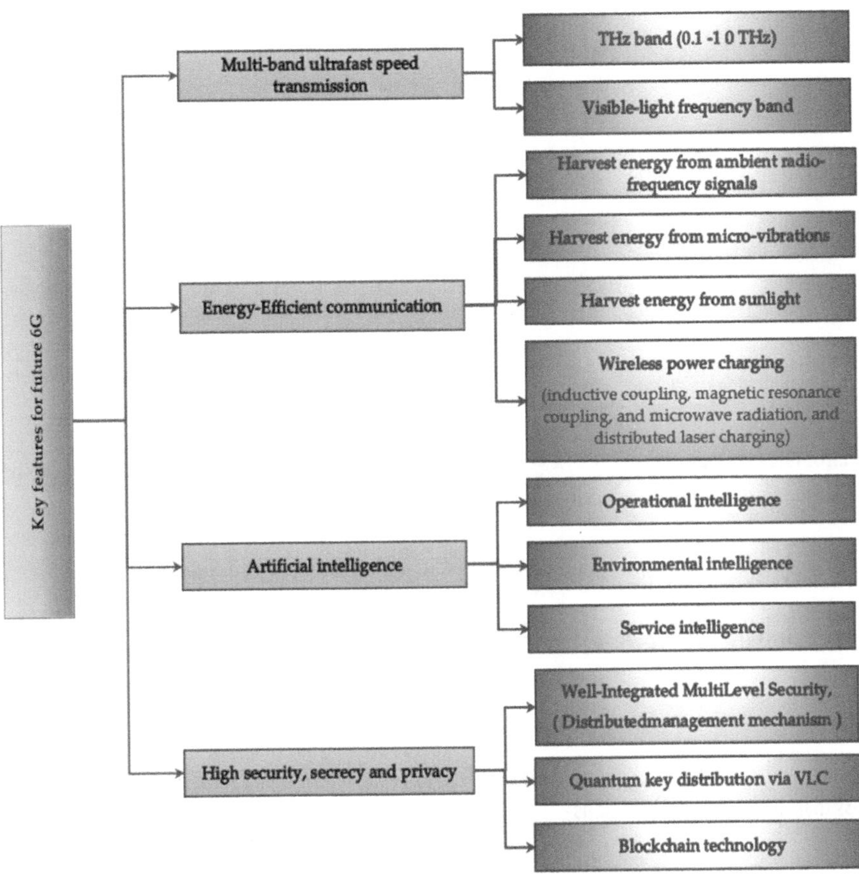

Figure 12.2 Features of future 6G [I]. Open Access.

ten over 5G for 6G to be feasible [14]. For devices with limited resources, it is expected to provide communications with very low power consumption. The ability to travel quickly at 1000 km/h will be made possible by active and advanced mobility management systems. The work in [14] has assessed the impact of security procedures on latency in order to ensure ERLLC service quality. In a similar vein, robust security measures are necessary to guarantee the availability of services and resources when demands are high. Using the latest distributed intelligent AI and ML security methods might be challenging due to the IoE. One important part is finding ways to make devices with limited resources use new security features. The data speeds, dependability, latency, and localisation accuracy comparison between 5G and 6G is summarised in Figure 12.3.

There are lot of issues with the 5G. Some of them are as follows:

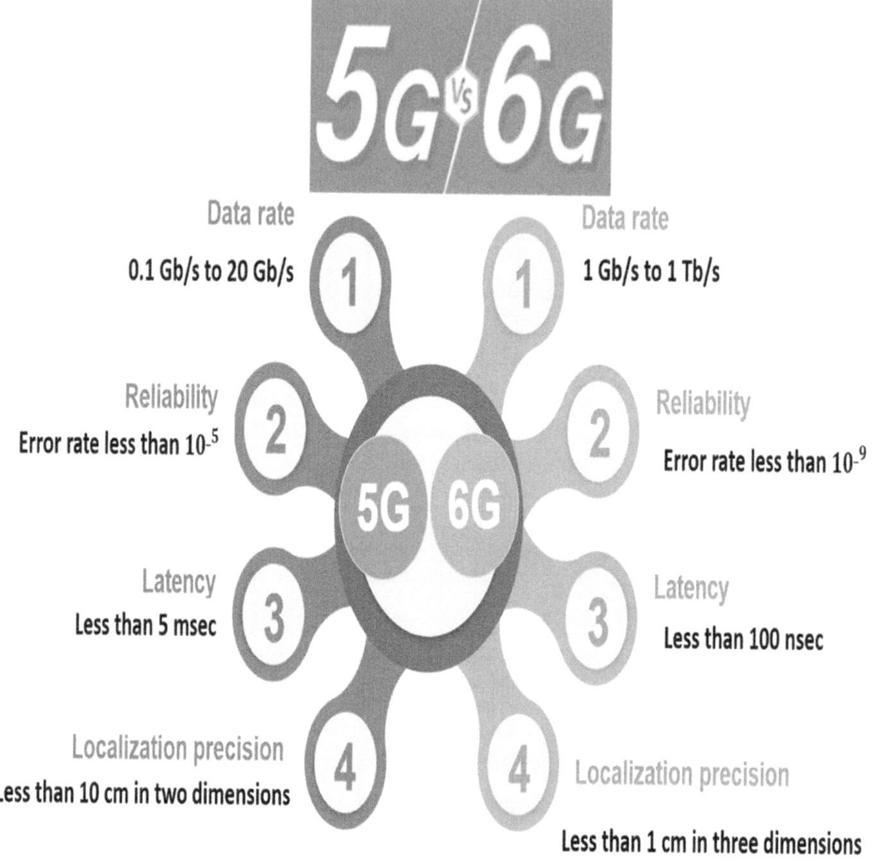

Figure 12.3 5G vs. 6G [14]. Open Access.

a. Restrictions on Coverage: 5G uses higher frequency spectrum bands to provide faster speeds, but these bands have a limited range and are more easily blocked by obstacles like buildings and trees. Deploying more tiny cells is a costly need for expanding coverage [9].

b. It is the disjointed allocation of spectrum bands, which makes equipment more complicated to accommodate various frequencies in different regions since there has been no worldwide harmonisation of 5G spectrum bands. Scale economies are also hindered by fragmentation.

c. Updating core networks and densifying networks with tiny cells are necessary for 5G, which necessitates new network infrastructure. Before monetisation can start, huge expenditures in infrastructure are needed [10].

d. There is a lack of uniformity in the standards; many nations have approached 5G in different ways. Delays in rollouts and reduced interoperability in network equipment may be caused by variations in standards [11].

e. Data Protection: 5G poses new security threats due to the proliferation of linked devices and advancements in virtualisation. Improving 5G security is a challenging task that requires network-wide software and operational upgrades.

f. Concerns about health have been voiced in relation to 5G equipment's higher frequency electromagnetic emissions, however no definitive evidence has been shown as of yet. Delays in 5G implementation may occur due to perception difficulties [12].

g. Power Consumption:The first 5G gear was rather power hungry, which made widespread deployment seem unrealistic. There is always room for improvement when it comes to energy efficiency.

6G may be great for connecting more people and providing more experiences, but it also opens the door to more cyberattacks, making people more susceptible to malware, smart inference, and privacy abuses. The new 6G features, such as ubiquitous sensing, integrated AI coding, and dynamic network topologies, increase the attack vectors. The integration of sensors and data networks allows for intrusive surveillance of behaviours, biometrics, and background, which inherently endangers people's rights, privileges, and sensitivities. Furthermore, complicated model inversion, data poisoning, and trojan assaults are threats associated with AI/ML's ubiquity and are notoriously difficult to protect against. Adaptive network designs, on the other hand, make linked systems more susceptible to exploit chains and bogus node insertion. The proliferation of ambient gadgets and apps that can monitor users' surroundings, states, and actions opens up new entry points for malicious actors to collect information and build intrusive user profiles. The lack of transparency in AI systems that analyse diverse cross-domain data is adding fuel to the fire. Plausibility and denial of real occurrences are also made possible by deep fakes. The hyper-connected fabric of 6G greatly amplifies cyber-physical threats by enabling embedded sensing/actuation, AI-based decision-making, and control point manipulation. By impersonating autonomous systems, attackers might use these capabilities to do harm, distort perceptions, or violate sensibilities. Consequently, 6G complexes, unless they include principled security into their design, disturb fundamental assumptions around consent, agency, and choice. The various challenges in 6G are shown in Figure 12.4. The idea is inspired from [13]. Attacks on security systems are a common driving force behind recommendations for next-generation security system changes meant to prevent the exploitation of known flaws. In many cases, these kinds of attacks expose vulnerabilities in systems or protocols that the designers could have overlooked during the planning phase. A simple way to learn about the differences in mobile network security advancements (often for resolving established weaknesses exposed by attacks in previous

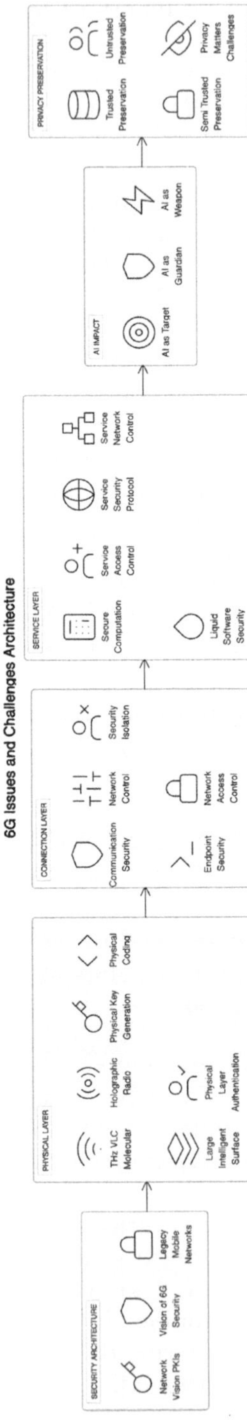

Figure 12.4 6G issues and challenges.

generations) is to study these types of attacks. Security assaults typically expose vulnerabilities in authentication protocols, which need modifications to essential network functionality. The steep price tag, however, has most merchants fighting against the idea of such a substitute. These unresolved concerns may be used as a foundation for future studies or as objectives for 6G enhancements.

Fundamentals learnt for 6G security from 1G to 5G safety issues and improvements are as follows:

i. Security has to be improved since new apps are a common source for privacy vulnerabilities. Brand new capabilities offered by apps showcase the improvements made to network standards from previous versions. Having said that, they may lead to further security holes. A security issue with keystream reuse in two consecutive conversations is present in the VoLTE protocol, for instance. An attacker may eavesdrop on phone conversations and decode VoLTE contents by exploiting this vulnerability.

ii. A slew of new apps, including mixed reality and self-driving cars, are expected to launch on 6G, according to many studies. Unfortunately, these applications are likely to be vulnerable to denial-of-service and identification assaults. So, when technological solutions are put into action, it is crucial to improve their safeguarding credentials.

iii. Bringing back support for an old protocol during the rollout of a new protocol could open up existing security holes. Problems stem from incompatibilities between two network standards' fundamental security features and the difficulty of safeguarding devices connected to outdated network components with inadequate security abilities. Because of these incompatibilities, it is common for the latest norm to request that older devices be authenticated using the older design. Old security holes in a previous standard may be rediscovered by using this access control paradigm. Forcing 4G-LTE devices to connect to 2G or 3G networks is one form of a downgrade attack. Afterwards, the attacker may easily gather the UE's IMSI and track its position by taking advantage of the 2G/3G standards' flaws, which include a lack of mutual verification between the UE and authentication servers. Concerns about the interoperability of 6G with older 4G/5G devices for identification and control of identities need to be carefully explored.

iv. To correct more current flaws quicker and create fewer new ones, it is better to make more modifications to protocol executions and less modification to protocol designs. To begin, it is very uncommon for end-user devices to need extensive core equipment changes in order to solve security architectural and protocol weaknesses, such as those in AKA and subscription login management.

v. Several businesses and consumers may be unprepared to bear the financial hardship that this transition might bring. To prevent the introduction of new vulnerabilities into a real-world setting, it takes a long time to test the security capabilities of a newly designed architecture and protocol. Implementing security patches for protocols or updating security appliances at endpoints is a good way to prioritise achievable schedules that may reduce the impact of current vulnerabilities. But in the grand scheme of things, it is still essential to improve the design so that the previous architecture's shortcomings are completely removed.

vi. Addressing the ongoing challenges of mutual authentication and end-to-end encryption is an area that requires significant progress. Many infamous attacks, with the value eavesdropping, tracing attacks, and fake operators, stem from a lack of these two qualities. Because of the difficulties in implementing these two characteristics due to communication congestion and expensive computation, not even 5G will be able to achieve these safety aims. A need of robust end-to-end encryption and mutual identification in 6G might affect several latency-sensitive applications unless there is an improvement in computational capability or administration approaches. If these functionalities aren't included in 6G at some point, the possibility of fully resolving the current security vulnerabilities would be severely diminished.

Putting into practice flawless privacy protection mechanisms—at least to meet the requirements of laws and regulations—is much more difficult than mere rhetoric. These are some things that may need to be thought about. First, in order to keep data private, deployment might put a strain on a company's budget. This amount is in addition to the expense of tailoring software to meet all legal demands, as well as the cost of investing in comprehensive safety machinery to encrypt and anonymise data. Unfortunately, few small businesses possess the financial means necessary to make such investments. In this age of data explosion, with technologies like 6G, businesses may find themselves unable to afford the commodity equipment needed to analyse vast amounts of data.

A further obstacle is the lack of resources for protecting privacy in large-scale IoT devices. There is a dearth of solid authorisation and safety features in wearable devices and inexpensive Internet of Things (IoT) sensors that monitor users' whereabouts and health data. If there isn't enough top-tier security, the vast data-collecting shield may be weakened. The problems of network security will grow dramatically in 6G when both people and smart objects are expected to be linked. Unfortunately, mobile consumers may be oblivious to such dangers until it's too late, and they may willingly embrace risks in order to enjoy services.

Many consumers also fall into the trap of believing that data collection is pointless because they believe that their private information is either not in danger or of no significance. It gets more serious: a lot of data-driven companies either don't warn consumers or deliberately deceive them about the possible implications of data gathering in all its forms and intensities. A corporation may request personal information about its workers on the grounds that it might have an effect on their productivity on the job. In order to provide users with a better experience and enable them to view more relevant advertisements, social media sites collect data.

The creation of comprehensive plans for the 6G wireless network's future is now the focus of all efforts. Between 2017 and 2025, studies on B5G and 6G will be given around EUR 95 million. Horizon2020, the European Union's programme for research and innovation, is funding these endeavours. Plus, the majority of them have only just begun to develop. Here, we'll take a look at one or two of these clinical trials and the lessons they've taught us with regard to 6G security [16–20].

12.2.1.1 Hexa-x

Ericsson began working on the Hexa-x project in 2021 [14]. In this collaboration, researchers from many universities are working together to bring cutting-edge technology to market. The 6G networks are intended to be established by the Hexa-x project. Its secondary objective is to spearhead R&I efforts into the future on a global scale. The primary goal of this project is to enhance the necessary technologies for bringing 6G networks to Europe. Connected intelligence, a network of networks, sustainability, worldwide services protection, reliability, and extraordinary experiencing are the six difficulties that will be addressed in innovative ways. In order to tackle these problems, Hexa-x will build many axes.

Improving the quality of connections between gadgets and people requires the use of new technologies like AI and ML. A unified network of connections is essential for the world's digital ecology. The ideal characteristics of this network include diversity, intelligence, and adaptability. For a network to endure, its resources need to be used effectively. We need to find realistic and affordable ways to ensure that the 6G network can reach the whole world. Data privacy, transmission honesty, secrecy, and operational resiliency ought to be guaranteed by the future generation to provide high security. To further improve 6G's performance, several technologies have to be created, including networking virtualisation, THz radio access, an AI-driven air interface, and network design. In order to bring the digital, physical, and human realms closer together, the project will focus on these innovative tools for communication.

12.2.1.2 RISE 6G

A notable project that was initiated in 2021 is RISE 6G, which stands for Reconfigurable Intelligent Sustainable Environments for 6G wireless networks [14]. One of the most influential emerging technologies, RIS will soon be a reality. RIS is concerned with the ever-changing regulation of radio radiation propagation. It opens the door for wireless environments to be seen as services. By using RIS, RISE 6G aims to enhance 6G capabilities for an intelligent, adaptable, and environmentally friendly wireless ecosystem.

Four RIS-related obstacles will be encountered by the project. The real transmission of the signal helped by RIS will first be modelled. The second step is to include various RISs in the redesigned network design. Third, to enhance the quality of service, several use cases will be developed. These use cases will address issues such as precise location, environmentally friendly communication, power consumption, and large capacity within an expanding wireless programming network. The fourth step is to propose an innovation prototype benchmark after two supplementary processes have been considered. As part of its efforts to standardise, the project is bringing its technological vision to life in an industrial setting.

12.2.1.3 New 6G

The nano-world will be the focal point of the NEW-6G project. "Microelectronic with telecommunications, networking with machinery, and computer with equipment" are the three connections that the project fosters. The primary goal of the project is to improve the network's efficiency via the creation of new techniques and technologies, such as:

- Optimisation and design of networks.
- Protocols and data flow.
- Protection of data and systems.
- Low power use, digital parts, circuits with integrated circuits, and powerful radio frequencies.
- Technology for semiconductors that is both environmentally friendly and very efficient.
- To take use of nano-electronics the internet, NEW-6G will provide new mechanisms. The investigation of nano-electronics technology will give rise to novel questions in both academic and business circles.

12.2.1.4 Next G alliance

As 2020 came to a close, the Next G Alliance was officially announced in the US by ATIS (Alliance for Telecommunications Industry Solutions). By implementing 6G's foundational components across North America, ATIS hopes to facilitate 6G leadership. Development of technology, manufacturing, standardisation, and being ready to sell are all parts of technical commercialisation. When it comes to future standards, the influence of member organisations on big companies in mobile communications may be enormous. Industry standards and new developments will be the focus of the Next G Alliance's tactical evaluation. Their goal is to spark a global dialogue on standards and the need of government-business partnerships. The expansion of several large companies is directly related to the rise of mobile technologies. Due to the proliferation of mobile devices, the United States is becoming more reliant on several industries, including aerospace, agriculture, defence, healthcare, schooling, production, the press, power, and shipping. In these vital sectors, North America must continue to lead the world in mobile technology.

12.3 TYPES OF PRIVACY THREATS

All 6G network technologies now include AI and ML as vital parts of their infrastructure design. Therefore, 6G networking greatly prioritised AI. Areas with abundant train information and efficient processing cores are where 5G networks deploy AI/ML. But AI and ML are now major players in 6G networks. Multiple 6G safety frameworks use AI and ML for defence and security. When applied to security, AI and ML provide greater autonomy and comprehensive strategies, as well as predictive analytics for security.

Liability Insurance: In the context of artificial intelligence (AI) managing network security, the dependability of machine learning (ML) algorithms and elements becomes critical. In order to maintain control and credibility, it is important to monitor security functions that rely on AI and ML in real time.

Concerns from a Moral and Legal Standpoint: AI-powered optimisation methods may exclude some users or uses.

Practicality and Scalability: Protecting the personal information of federated learners requires secure data transfers. A hurdle for AI/ML is making sure the necessary computation, networking, and storage capabilities can scale.

Managing Security Duties: When big data operations are linked to AI/ML security solutions, a lot of extra work can be in store. The study and interpretation phases of a model's lifecycle should be safe and accommodating.

In order to meet the demands of new use cases, advanced AI approaches [20,21] and methods are anticipated to be a part of the adaptive 6G system. The various threat types are shown in Figure 12.5 [14].

12.3.1 Radio layer

This layer is responsible for gathering data from the real world [22]. To keep tabs on physical occurrences, a network of interconnected devices shares data wirelessly. A higher layer is then notified of the acquired data and instructed to process it further. The tiny linked IoE devices are vulnerable to assaults introduced by this layer. Physical assaults, data theft from

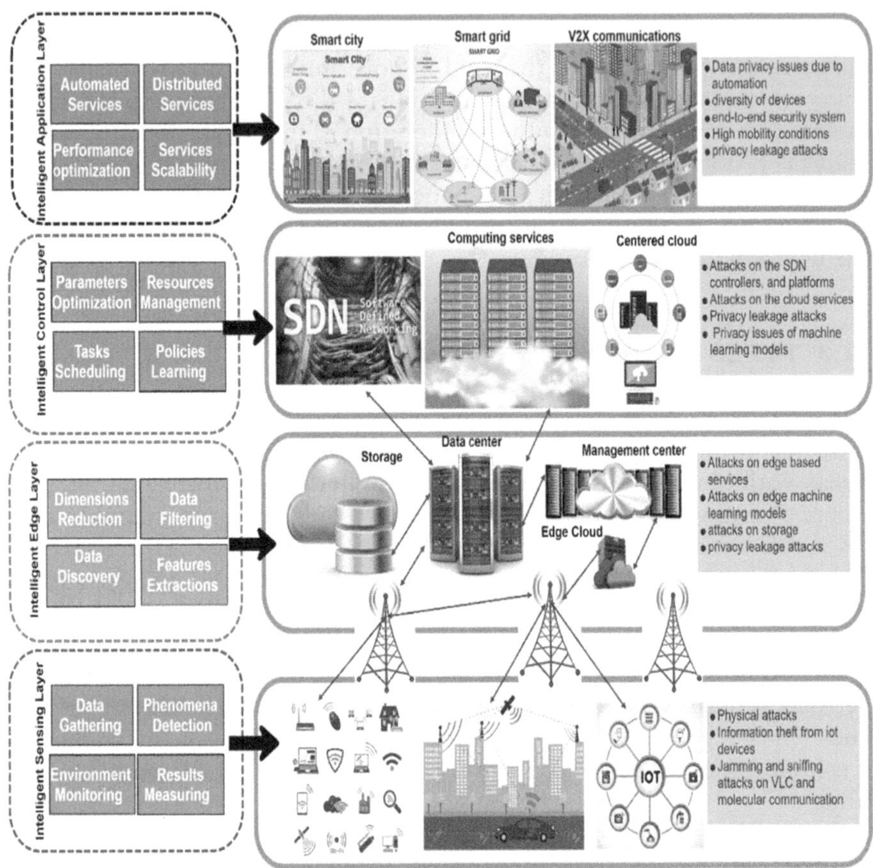

Figure 12.5 An AI/ML security architecture for 6G networks, with distinct layers for various types of threats [14]. Open Access.

gadgets, intrusions on telecommunications using visible light, and sniffer attacks all fall under this category. The fast evolution of artificial intelligence processors, together with developments in circuits and antennas, and designs based on meta-materials, has shown an unprecedented change in the structure of 6G transceivers, enabling hardware to be separated from algorithms used by transceivers. Consequently, the transceiver methods may adapt to their surroundings and hardware by updating and configuring oneself constantly. To address challenges in the wireless realm, intelligent radio will employ state-of-the-art AI/ML methods. These methods will include autonomous coordination, dynamic spectrum access, flexible infrastructure layout, optimisation, and installation of complex networks. Therefore, it is essential to anticipate unusual conduct from hostile sites while implementing secure radio protocols for interaction.

12.3.2 A smart edge layer

The edge layer sorts and analyses the gathered data after extracting a characteristic from it. ML models and edge services are the targets of assaults that occur at the edge layer. Concerns about personal information and data preservation also arise in the context of cybersecurity. Gathering, storing, or analysing data at the network's periphery is the domain of edge intelligence (EI), which employs AI/ML algorithms. Using the edge server in EI has several benefits, including reduced latency, faster feedback, and lower expenses. This is because it collects data from numerous connected equipment and distributes it with extra edge servers so that models may be trained. Due to the data dependence of AI/ML algorithm output, EI is especially susceptible to a number of security flaws. The outputs of the AI/ML applications may be compromised and the benefits of EI undermined if various attacks, including data poisoning, privacy violations, or fraud, took exploit of this dependence.

12.3.3 A layer of intelligent control

Task planning, handling resources, variable optimisation, and policy learning—all of which lead to different types of attacks—are controlled by this layer. Service in the cloud, software-defined networking (SDN), and centralised cloud computing are the targets of control layer assaults. Critical security challenges in the forthcoming 6G networks include SA cyberattacks on intelligent learning techniques and ML attacks on SDN controls and connections.

12.3.4 A smart layer for applications

There are a number of threats introduced by application layer services that are mechanised and dispersed. Smart cities and connectivity between vehicles raise concerns about personal data because of automation. A high degree of security is necessary due to the variety of devices that are integrated into the application layer. A major change in 6G architecture is necessary for the orchestration and administration of network services due to the fact that 6G is planned to completely automate end-to-end (E2E) network and managing services with AI and has a diverse set of requirements. There are a number of security concerns with those intelligence control network setups. Automated closed-loop networks are vulnerable to Man-in-the-Middle (MITM) and Denial-of-Service (DoS) attacks. One way to launch a denial-of-service attack is to artificially inflate demands for virtual network functions (VNFs) (VMs), which gradually increases the ability of virtual machines. An attack known as a man-in-the-middle (MITM) may redirect bandwidth via rogue devices by creating false events related to faults and stealing domain control messages. To conduct deceit attacks, it is possible to manipulate the sent data. Another scenario where attacks might arise is when 6G networks use intent-based interfaces like ZSM. These interfaces could be vulnerable to information leakage, undesired configuration, and unusual behaviour. There may be more attacks and compromises if unauthorised people have entry to system security objectives like privacy and confidentiality. It is possible to compromise the whole management system by altering the relationship among intention and act or by reducing the security level of intent-based interfaces. Likewise, the same might happen if the objective is incorrect [15].

Still, new, more sophisticated forms of attack have emerged, such as those targeting federated learning. 6G networks rely heavily on AI and ML technologies. Nonetheless, AI and ML will start AI/ML-related dangers. The training and testing stages are equally targeted by these assaults. If you look at, you can see a summary of the problems with 6G networks' security as well as traditional breaches and dangers using AI and ML. Poisoning attacks on AI training stages include manipulating trained data by injecting intentionally flawed samples, which impacts the learning method's outcome. The intelligence community runs the risk of overspending on resources and misclassifying services due to the introduction of prepared samples. Through the use of injected testing information, evasion attackers are able to sidestep the learnt model during testing. Model inversion attacks, last but not least, aim to steal training information from specific ML models.

12.4 EXAMINATION OF SECURITY MEASURES

12.4.1 Security measures designed for 6G systems

An important part of 6G networks is expected to include ML technology because of its potential to improve network efficiency, optimise the use of resources, and allow new services and applications. In 6G networks, ML technology might be used for the following purposes [23–28] as depicted in Figure 12.6.

- Optimisation of networks: Applying ML algorithms to analyse network traffic and optimise power and capacity may improve network efficiency and reduce latency. Network slowdowns may be avoided with the help of ML's congestion forecasting and real-time resource modification capabilities.
- Allocating network bandwidth (such as spectrum and frequency bands) effectively is one way ML algorithms may boost network efficacy and bandwidth. Due to the wide range of latency as well as bandwidth needs that 6G networks are expected to provide, this may play a pivotal role in these networks.

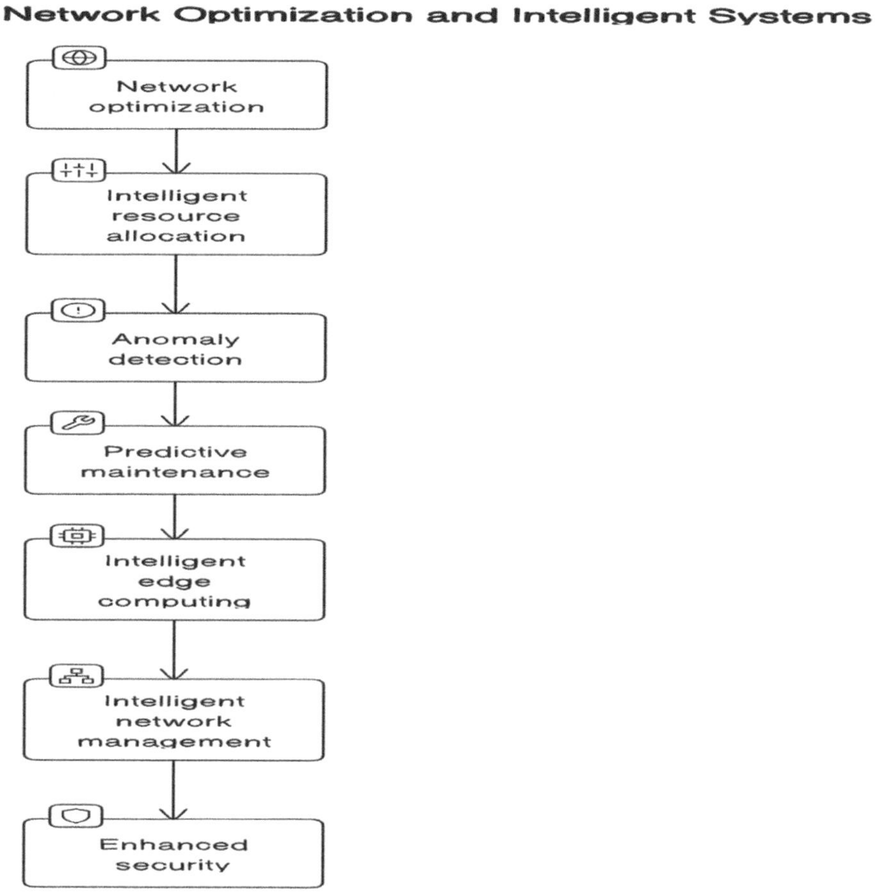

Network Optimization and Intelligent Systems

Figure 12.6 Few potentials use for ML technology in 6G networks.

- Anomaly Detection: Machine learning techniques may be used to spot peculiar patterns of data transfer in network traffic, which might be signs of security breaches or other issues. By identifying abnormalities in real-time, ML may facilitate faster responses to security risks and other networked challenges [29–31].
- One usage of ML algorithms is in predictive maintenance, which involves determining when components like base stations and antennas will require repair. Machine learning algorithms may detect trends in data collected from sensors and other sources, which might indicate impending issues or the need for repair [32,33]. Minimising downtime and maximising network efficiency may be achieved via preventive maintenance and repair.

- Smart Computing in the Network's Periphery: Machine learning algorithms may be pre-installed on gateways and routers, allowing for real-time data processing and examination. Consequently, network efficiency will improve and latency will decrease due to less data needing to be transferred over vast distances. Machine learning may also improve the efficiency and effectiveness of data processing by making the most of edge computing resources like memory and processing speed [34,35]
- Intelligent Network Management: Machine learning algorithms can analyse network performance data to identify improvement opportunities and simplify management processes. Examples of ML's practical applications include optimising network routing to reduce congestion and increase performance, or predicting network faults and suggesting proactive maintenance [36].
- Improved Safety: ML algorithms can make networks safer by finding and fixing security flaws faster. For instance, machine learning has the potential to detect suspicious network devices or traffic patterns and trigger automated security actions [37–39].

12.5 ADDITIONAL RESEARCH NEEDED TO ADDRESS EXPANDING 6G ATTACK CAPABILITIES

Many see the advent of the 6G era as the beginning of a sea change. The key to the future is learning from the past. Consequently, the work in [38] explores the historical roots of mobile wireless connections. We looked into the "beyond 5G era" because it may be a time of great promise at the crossroads of human history and technological advancement, bringing more people together and helping the world's most marginalised and neglected communities. They focused on the key technological challenges to the transition and spoke about the characteristics of 6G that will make it easier. As we pointed out in study, there have been some improvements, but there are still some obstacles. There are a lot of research "directions" leading up to the still-ticklish aim of universal, free, ultra-high-availability internet. Despite 5G's advancements, the dream of a cheap, fast, and latency-free internet with cyber-physical fusion for terrestrial communications (air, space) has not yet come to fruition. So, we looked at the big problems with 5G, which are going to be the biggest problems with 6G in terms of research. The work also draws the conclusion that 6G will solve 5G's shortcomings, paving the way for a seamless upgrade to a brand-new wireless communication ecosystem that incorporates cutting-edge equipment. We see a highly networked future where intelligent, autonomous gadgets and energy-efficient nodes charge themselves. It also predicts in this piece that 6G will usher in a new age of hyper-digitisation in a hyper-connected world, bringing with it technical advantages while also bolstering social megatrends like the United Nations Sustainable Development Goals, renewable energy, and environmental protection. In addition, they believe that 6G will be on more than just wireless data transmission; it will also be about helping people and companies thrive in an intelligent, automated world where everything is interconnected. In the next years, it can be expected as a new era to begin, one that will need substantial study. Extensive and convincing research is needed to determine the roles that MC, the Thz regime, quantum computing, deep federated learning, blockchain, and molecular computing will play in 6G [38–40].

Here we lay out the plans for where 6G intelligent healthcare research is headed in the future.

12.5.1 AI in smart healthcare powered by 6G

To choose the most appropriate ML strategy for 6G-based smart healthcare, it is necessary to examine the application's characteristics. ML is a vital component of this framework. The creation of massive volumes of info in cognitive healthcare, for instance, necessitates real-time involvement due to the IoE. With only one training session, centralised ML may be implemented. The constant influx of fresh data, however, means that the model trained via centralised ML could not provide satisfactory outcomes. Federated learning outperforms centralised ML in this case. This method sends the model's parameters to the end devices after training it on a distributed dataset. Lastly, the end devices refresh the global learning models with data from their local datasets. One advantage with distributed ML is the fact that it eliminates resources unfairness concerns by sharing learning model parameters in real-time between the final devices and the centralised server. There is some loss of privacy with distributed computation, but it's less severe than with centralised ML since it relies on a dataset at a centralised location with enough training data, which wouldn't be able to include every gadget details [41–43].

12.5.2 Intelligent healthcare on 6G is made possible by blockchain technology

An intriguing new technology, blockchain allows for the secure storage of events in a distributed ledger that cannot be altered. Logistics, smart healthcare, and smart transportation are just a few examples of the smart services that have been made possible by blockchain technology. With a 6G system, you may expect much lower latency and energy consumption, as well as greater reliability and flexibility. The present consensus mechanisms for blockchains, however, have issues with the usage of energy, delay, dependability, durability, and delay. Core goals of 6G-based advanced healthcare systems, such as security, fault-tolerance, lowered latency, and decentralisation, are repeatedly obstructed by blockchain implementation, which poses substantial challenges to scalability and dependability. A novel consensus method that enhances scalable as well as dependability while combining latency, security, and latency must be proposed in order to reap the benefits of blockchain deployment [43–46].

12.5.3 Stronger protections for sophisticated healthcare systems powered by 6G

Health 6G Security Enhanced Approaches should focus on creating advanced, situation-specific security measures. Making encryption methods that are more resistant to 6G-specific assaults is part of this process. The use of sophisticated ML models for real-time risk prediction and neutralisation ought to be prioritised in order to improve intrusion detection systems. Furthermore, it is critical to study effective privacy-preserving techniques that support 6G networks' speed and efficiency without sacrificing data integrity or confidentiality. For private medical records to be sent and handled securely over upcoming 6G networks, these technological initiatives are crucial.

12.5.4 6G-based intelligent healthcare is made possible by terahertz communication

Eliminating inherent obstacles to efficient adoption is the primary emphasis of the highly technical field of terahertz connectivity as it pertains to 6G healthcare. To do this, advanced methods of signal creation and detection that are adapted to healthcare settings must be developed. To further address the concerns of signal attenuation and interference, further research on the propagation of terahertz waves in different healthcare environments is essential. To guarantee optimal transmission and reception, it is crucial to develop new antenna designs tailored for terahertz frequencies. These areas of technological study are important for real-time medical applications in the 6G era because they seek to tap into the ultra-high-speed capabilities of terahertz transmission.

12.5.5 An intelligent healthcare system that can sustain itself

The administration, optimised, arrangement, and planning processes may be expedited with the help of a self-organising network. Release 8 of the 3GPP standard included extensive documentation on self-organising networks. But 6G systems operate in a complex and ever-changing environment, thus the conventional self-organising networking design may not work. Consequently, it is critical to create a ground-breaking, autonomous 6G network architecture. The ability to adjust to new circumstances is crucial for autonomous 6G systems. Furthermore, in order for the 6G-based smart healthcare system to be self-sustaining, it is necessary to use evolving machine learning algorithms [47,48].

12.5.6 A healthcare system that incorporates 6G technology and requires zero energy

Hybrid energy sources that harness both solar and wind power to support 6G wireless data transmission is essential for smart healthcare. If the amount of energy produced by the radio wave harvesting drops below what is needed to keep the system running, power must be drawn from the grid station. During periods of surplus radio wave harvesting, the zero-energy system should feed back into the grid a comparable amount of energy that it draws from it [42,49,50].

12.6 DISCUSSION AND FUTURE SCOPE

6G networks are a huge deal for hyperconnectivity and immersive tech. Having said that, we must not ignore the serious privacy and security concerns that these capabilities bring. While explaining the concerns for data privacy, this analysis of 6G networks uncovered the new characteristics that would characterise 6G. The threats of cyberattacks that reveal private information about users in novel ways were discussed in detail, including malware, intelligent inference, and others. There is still a lot of work to be done in order to tailor security mechanisms like physical-layer encryption, AI-enabled authentication, and blockchain protocols to the ever-increasing 6G attack surfaces, but these approaches show promise. In the end, there is still an unresolved multidisciplinary issue of how to protect user rights while yet achieving the tremendous potential of 6G connection. With the advent of ambitious 6G communication technologies on the horizon, it is more important than ever to put privacy and security precautions at the forefront. If this doesn't happen, user rights, safety, and trust will be compromised in the pursuit of a hyperconnected future. The advent of 6G, however, may bring about a new age of inclusion, equality, and opportunity supported by next-generation digital experiences—if we work together, think forward, and collaborate across disciplines.

There is much hope for expanded opportunities, empowerment, and inclusivity brought about by technology as the 6G age begins. But to tap into this potential to its fullest, we must solve complicated multi-faceted problems in fields as diverse as technology, politics, the field of economics ethics, and engineering. A lot of multidisciplinary effort has to be done to tailor advanced security for developing 6G systems, even if looking at current security frameworks sets the basis. Technologies like as quantum communication, blockchain systems, artificial intelligence and machine learning, and advanced cryptography all contribute to the picture. It is still not clear how to include these diverse technologies into comprehensive authenticated, robust, and self-healing network designs. The growing demand for data regulation in immersive settings necessitates that policy discussions progress in tandem. Earning user trust requires governance structures that prioritise privacy, consent, and safeguarding vulnerable groups. To ensure that security is not an afterthought but rather an integral part of 6G, further techno-economic research is needed to determine adoption roadmaps and incentives. On the path to mature 6G deployment, the primary contradiction is resolving the ethical imperatives around safety while simultaneously pursuing ambitious hyperconnectivity goals. For the benefit of everybody, we must unleash a digital society of the future by constructively tackling the interrelated components of technology, legislation, economics, and society. Although 6G has tremendous potential, achieving its goals will need an even higher level of accountability.

REFERENCES

1. Alsharif, M.H.; Kelechi, A.H.; Albreem, M.A.; Chaudhry, S.A.; Zia, M.S.; Kim, S. Sixth Generation (6G) Wireless Networks: Vision, Research Activities, Challenges and Potential Solutions. *Symmetry* 2020, 12, 676. https://doi.org/10.3390/sym12040676.
2. Holland, O.; Steinbach, E.; Prasad, R.; Liu, Q.; Dawy, Z.; Aijaz, A. The IEEE 1918.1 "Tactile Internet" Standards Working Group and Its Standards. *Proc. IEEE* 2019, 107, 256–279.
3. Khalid, M.; Amin, O.; Ahmed, S.; Shihada, B.; Alouini, M.-S. Communication through Breath: Aerosol Transmission. *IEEE Commun. Mag.* 2019, 57, 33–39.
4. Yastrebova, A.; Kirichek, R.; Koucheryavy, Y.; Borodin, A.; Koucheryavy, A. Future Networks 2030: Architecture & Requirements. In *Proceedings of the 2018 10th International Congress on Ultra-Modern Telecommunications and Control Systems and Workshops (ICUMT)*, Moscow, Russia, 5–9 November 2018, pp. 1–8.
5. Rappaport, T.S.; Xing, Y.; Kanhere, O.; Ju, S.; Madanayake, A.; Mandal, S.; Alkhateeb, A.; Trichopoulos, G.C. Wireless Communications and Applications above 100 GHz: Opportunities and Challenges for 6G and Beyond. *IEEE Access* 2019, 7, 78729–78757.
6. Elayan, H.; Amin, O.; Shihada, B.; Shubair, R.M.; Alouini, M.-S. Terahertz Band: The Last Piece of RF Spectrum Puzzle for Communication Systems. *IEEE Open J. Commun. Soc.* 2019, 1, 1–32.
7. Elmeadawy, S.; Shubair, R.M. Enabling Technologies for 6G Future Wireless Communications: Opportunities and Challenges. arXiv 2020, arXiv:2002.06068.
8. Yang, P.; Xiao, Y.; Xiao, M.; Li, S. 6G Wireless Communications: Vision and Potential Techniques. *IEEE Netw.* 2019, 33, 70–75.

9. Taheribakhsh, Morteza, AmirHossein Jafari, Mahdi Moazzami Peiro, and Nasrin Kazemifard, 5g Implementation: Major Issues and Challenges. In *2020 25th International Computer Conference, Computer Society of Iran (CSICC)*, 2020.

10. Cengiz, K.; Aydemir, M. Next-Generation Infrastructure and Technology Issues in 5G Systems. *J. Commun. Softw. Syst.* 2018, 14(1), 33–39.

11. Qamar, F., Hindia, M.H.D.N., Dimyati, K. et al. Interference Management Issues for the Future 5G Network: A Review. *Telecommun. Syst.* 2019, 71, 627–643.

12. Humayun, M.; Hamid, B.; Jhanjhi, N.Z.; Suseendran, G.; Talib, M.N. 5G Network Security Issues, Challenges, Opportunities and Future Directions: A Survey. *J. Phys. Conf. Ser.* 2021, 1979(1), 012037.

13. V. -L. Nguyen, P. -C. Lin, B. -C. Cheng, R. -H. Hwang and Y. -D. Lin, "Security and Privacy for 6G: A Survey on Prospective Technologies and Challenges. *IEEE Commun. Surv. Tutor.* 2021, 23(4), 2384–2428.

14. Abdel Hakeem, S.A.; Hussein, H.H.; Kim, H. Security Requirements and Challenges of 6G Technologies and Applications. *Sensors* 2022, 22, 1969. https://doi.org/10.3390/s22051969.

15. Uusitalo, M.A.; Rugeland, P.; Boldi, M.R.; Strinati, E.C.; Demestichas, P.; Ericson, M.; Fettweis, G.P.; Filippou, M.C.; Gati, A.; Hamon, M.H.; et al. 6G Vision, Value, Use Cases and Technologies from European 6G Flagship Project Hexa-X. *IEEE Access* 2021, 9, 160004–160020.

16. Strinati, E.C.; Barbarossa, S. 6G Networks: Beyond Shannon towards Semantic and Goal-Oriented Communications. *Comput. Netw.* 2021, 190, 107930.

17. Strinati, Emilio Calvanese and Alexandropoulos, George C. and Sciancalepore, Vincenzo and Di Renzo, Marco and Wymeersch, Henk and Phan-Huy, Dinh-Thuy and Crozzoli, Maurizio and D'Errico, Raffaele and De Carvalho, Elisabeth and Popovski, Petar and Di Lorenzo, Paolo and Bastianelli, Luca and Belouar, Mathieu and Mascolo, Julien Etienne and Gradoni, Gabriele and Phang, Sendy and Lerosey, Geoffroy and Denis, Benoît, Wireless Environment as a Service Enabled by Reconfigurable Intelligent Surfaces: The RISE-6G Perspective. In *2021 Joint European Conference on Networks and Communications & 6G Summit (EuCNC/6G Summit)*, Porto, Portugal, 2021, pp. 562–567. https://doi.org/10.1109/EuCNC/6GSummit51104.2021.9482474.

18. Strinati, E.; Alexandropoulos, G.C.; Wymeersch, H.; Denis, B.; Sciancalepore, V.; D'Errico, R.; Clemente, A.; Phan-Huy, D.-T.; De Carvalho, E.; Popovski, P. Reconfigurable, Intelligent, and Sustainable Wireless Environments for 6G Smart Connectivity. *IEEE Commun. Mag.* 2021, 59, 99–105.

19. Di Renzo, M.; Debbah, M.; Phan-Huy, D.T.; Zappone, A.; Alouini, M.S.; Yuen, C.; Fink, M. Smart Radio Environments Empowered by Reconfigurable AI Meta-Surfaces: An Idea Whose Time Has Come. *EURASIP J. Wirel. Commun. Netw.* 2019, 2019, 1–20.

20. Castro, C. 6G Gains Momentum with Initiatives Launched across the World. 6GWorld, 18-Mar-2021. [Online]. Available: https://www.6gworld.com/exclusives/6g-gains-momentum-with-initiatives-launched-across-the-world/. [Accessed: 29-Jan-2024].

21. Giordani, M.; Zorzi, M. Non-Terrestrial Networks in the 6G Era: Challenges and Opportunities. *IEEE Netw.* 2021, 35, 244–251.

22. Saeed, M.M.; Saeed, R.A.; Abdelhaq, M.; Alsaqour, R.; Hasan, M.K.; Mokhtar, R.A. Anomaly Detection in 6G Networks Using Machine Learning Methods. *Electronics* 2023, 12, 3300. https://doi.org/10.3390/electronics12153300.

23. Plastiras, G.; Terzi, M.; Kyrkou, C.; Theocharidcs, T. Edge Intelligence: Challenges and Opportunities of Near-Sensor Machine Learning Applications. In *Proceedings of the 2018 IEEE 29th International Conference on Application-Specific Systems, Architectures and Processors (ASAP)*, Milan, Italy, 10–12 July 2018, pp. 1–7.

24. Peng, H.; Wang, Z.; Han, S.; Jiang, Y. Physical Layer Security for MISO NOMA VLC System under Eaves-Dropper Collusion. *IEEE Trans. Veh. Technol.* 2021, 70, 6249–6254.

25. Chih-Lin, I. AI as an Essential Element of a Green 6G. *IEEE Trans. Green Commun. Netw.* 2021, 5, 1–3.

26. Ahmed, E.S.A.; Mohammed, Z.T.; Hassan, M.B.; Saeed, R.A. Algorithms Optimization for Intelligent IoV Applications. In *Handbook of Research on Innovations and Applications of AI, IoT, and Cognitive Technologies*, Zhao, J.; Kumar, V.V., Eds. Hershey, PA: IGI Global, 2021, pp. 1–25.

27. Zhang, S.; Zhu, D. Towards Artificial Intelligence Enabled 6G: State of the Art, Challenges, and Opportunities. *Comput. Netw.* 2020, 183, 107556.

28. Khalifa, O.O.; Roubleh, A.; Esgiar, A.; Abdelhaq, M.; Alsaqour, R.; Abdalla, A.; Ali, E.S.; Saeed, R. An IoT-Platform-Based Deep Learning System for Human Behavior Recognition in Smart City Monitoring Using the Berkeley MHAD Datasets. *Systems* 2022, 10, 177.

29. Mika Ylianttila, Raimo Kantola, Andrei Gurtov, Lozenzo Mucchi, Ian Oppermann, Zheng Yan, Tri Hong Nguyen, Fei Liu, Tharaka Hewa, Madhusanka Liyanage, Ahmad Ijaz, Juha Partala, Robert Abbas, Artur Hecker, Sara Jayousi, Alessio Martinelli, Stefano Caputo, Jonathan Bechtold, Ivan Morales, Andrei Stoica, Giuseppe Abreu, Shahriar Shahabuddin, Erdal Panayirci, Harald Haas, Tanesh Kumar, Basak Ozan Ozparlak, Juha Röning. 6G White Paper: Research Challenges for Trust, Security and Privacy. arXiv 2020, arXiv:2004.11665.

30. Gür, G. Expansive Networks: Exploiting Spectrum Sharing for Capacity Boost and 6G Vision. *J. Commun. Netw.* 2020, 22(6), 444–454.

31. Ramezanpour, K.; Jagannath, J.; Jagannath, A. Security and Privacy Vulnerabilities of 5G/6G and WiFi 6: Survey and Research Directions from a Coexistence Perspective. *Comput. Netw.* 2023, 221, 109515.

32. Y. Siriwardhana, P. Porambage, M. Liyanage and M. Ylianttila. AI and 6G Security: Opportunities and Challenges. In *2021 Joint European Conference on Networks and Communications & 6G Summit (EuCNC/6G Summit)*, 2021.

33. P. Porambage, G. Gür, D. P. M. Osorio, M. Liyanage, A. Gurtov and M. Ylianttila. The Roadmap to 6G Security and Privacy. *IEEE Open J. Commun. Soc.* 2021, 2, 1094–1122.

34. B. Mao, J. Liu, Y. Wu and N. Kato. Security and Privacy on 6G Network Edge: A Survey. *IEEE Commun. Surv. Tutor.* 2023.

35. Shen, L.-H.; Feng, K.-T.; Hanzo, L. Five Facets of 6G: Research Challenges and Opportunities. *ACM Comput. Surv.* 2023, 55(11), 1–39.

36. P. Porambage, G. Gür, D. P. Moya Osorio, M. Livanage and M. Ylianttila. 6G Security Challenges and Potential Solutions. In *2021 Joint European Conference on Networks and Communications & 6G Summit (EuCNC/6G Summit)*, 2021.

37. M. Mitev, A. Chorti, H. V. Poor and G. P. Fettweis. What Physical Layer Security Can Do for 6G Security. *IEEE Open J. Veh. Technol.* 2023.

38. Asghar, M.Z.; Memon, S.A.; Hämäläinen, J. Evolution of Wireless Communication to 6G: Potential Applications and Research Directions. *Sustainability* 2022, 14(10), 6356.

39. W. Jiang, B. Han, M. A. Habibi and H. D. Schotten. The Road towards 6G: A Comprehensive Survey. *IEEE Open J. Commun. Soc.* 2021, 2, 334–366.

40. Bhat, J.R.; Alqahtani, S.A. 6G Ecosystem: Current Status and Future Perspective. *IEEE Access* 2021, 9, 43134–43167.

41. Ahad, A.; Jiangbina, Z.; Tahir, M.; Shayea, I.; Sheikh, M.A.; Rasheed, F. 6G and Intelligent Healthcare: Taxonomy, Technologies, Open Issues and Future Research Directions. *Internet of Things* 2024, 25(101068), 101068.

42. Bale, A.S.; Soni, S.U.; Rajput, D.S.R.H.; Saranya, S.; Joy, S.; Chithra, R.B.; Savadatti, M.B. 5G Wireless Communication and Its Adverse Effects on the Human Body: Distinguishing Falsehoods from Reality. *Int. J. Intell. Syst. Appl. Eng.* 2023, 12(5s), 101–112. Retrieved from https://ijisae.org/index.php/IJISAE/article/view/3870.

43. Savadatti, M.B.; Kulkarni, N.J.; Bansod, G.D.; Sarkar, P.; Kumar, P.; Sargam, K.; Gulati, P. Keyboard Acoustic Side Channel Attacks on Android Phones in the AI Era of Digital Banking: An Elementary Review. *Int. J. Intell. Syst. Appl. Eng.* 2023, 12(4s), 292–300. Retrieved from https://www.ijisae.org/index.php/IJISAE/article/view/3792.

44. Bale, A.S.; Biswas, A.; Malik, S.; Uchoi, E.; Soni, S.U.; Soni, A. IoT Applications in Blockchain Technology. In *2023 International Conference on Computer Science and Emerging Technologies (CSET)*, Bangalore, India, 2023, pp. 1–6. https://doi.org/10.1109/CSET58993.2023.10346695.

45. Bale, A.S.; Prakash, H.R.B.; Babu, P.S.; Gupta, R.; Malik, S. Recent Scientific Achievements and Developments in Software Defined Networking: A Survey. In *2023 1st International Conference on Circuits, Power and Intelligent Systems (CCPIS)*, Bhubaneswar, India, 2023, pp. 1–6. https://doi.org/10.1109/CCPIS59145.2023.10291262.

46. Bale, A.S.; Purohit, T.P.; Hashim, M.F.; Navale, S. Blockchain and Its Applications in Industry 4.0. In *A Roadmap for Enabling Industry 4.0 by Artificial Intelligence*. Wiley, 16-Dec-2022, pp. 295–313.

47. Sumit Singh Dhanda, Brahmjit Singh, Poonam Jindal, Tarun Kumar Sharma, Deepak Panwar. 6G-Enabled Internet of Medical Things. *Expert Systems* 2024, 41(1), e13472.

48. Tanabe, Y. Wireless Power Transfer Application for Healthcare and Treatments. Institute of Biomaterials and Bioengineering, Tokyo Medical and Dental University, Tokyo, Japan, Kohji Mitsubayashi. In *Wearable Biosensing in Medicine and Healthcare.* Singapore: Springer Nature Singapore, 2024, pp. 421–437.
49. Vu Khanh Quy, Dinh C. Nguyen, Dang Van Anh, Nguyen Minh Quy. Federated Learning for Green and Sustainable 6G IIoT Applications. *Internet of Things*, Volume 25, 2024, 101061.
50. Nahyun Kim, Gayeong Kim, Sunghoon Shim, Sukbin Jang, Jiho Song and Byungju Lee. Key Technologies for 6G-Enabled Smart Sustainable City. Electronics 2024, 13(2), 268.

Chapter 13

Hybrid optimization-based deep learning classifier for cluster head selection in 6G IoT network

Nitesh Chouhan, Awanit Kumar, and Naresh Kumar

13.1 INTRODUCTION

6G IoT networks is the future of connectivity. The cutting-edge sixth generation of wireless technology poised to redefine our digital landscape. Envisioned [1] for commercial availability in the 2030s, 6G promises terabit-per-second data rates, ultra-low latency, and unparalleled capacity, setting the stage for transformative advancements. This next evolution in wireless communication not only aims to deliver lightning-fast speeds but also prioritizes seamless integration with the Internet of Things (IoT). Anticipated features include the ability to support a massive number of interconnected devices simultaneously, enhanced reliability, and the incorporation of artificial intelligence for optimized network management. As global [2] collaboration continues to drive research and development, the 6G era holds the potential to revolutionize connectivity and unlock new possibilities for innovation across various industries.

Cluster Head selection in 6G IoT networks [3] represents a sophisticated approach to enhance the efficiency and reliability of communication by selecting the proper cluster head selection. In the context of 6G, which is expected to deliver ultra-high data rates and support a massive number of interconnected devices, multipath routing becomes crucial for optimizing network performance. This technique [4] involves the selection of cluster heads that can be randomly or based on one or more criteria. The lifetime of 6G WSNs is largely determined by the cluster head choice. The cluster head that offers advantages like load balancing, enhanced fault tolerance, and decreased latency is the ideal one; it has the highest residual energy, the greatest number of neighbour nodes, and the smallest distance from the base station. In the 6G IoT landscape [5], where a diverse range of devices with varying communication requirements coexist, a cluster head selection algorithm can be particularly advantageous.

Combining cutting-edge technologies like machine learning, deep learning, and artificial intelligence, in 6G networks [6] further enhances the capabilities of cluster head selection. AI algorithms and Deep Learning

DOI: 10.1201/9781003522003-15

can intelligently analyse network conditions, device characteristics, and real-time data to make dynamic cluster head selection decisions, optimizing the utilization of available conditions based on the specific requirements of diverse IoT applications [7–8].

13.2 PROPOSED METHOD

This section explains the developed Taylor-PO method for the choosing the routing mechanism and cluster head for the 6G IoT-based WSN.

13.2.1 System model

The IoT network is made up of various items that are smart devices [9]. These objects are connected to other nodes in the network to exchange data packets. A variety of resource-constrained devices [10] are present in the network, providing both communication and processing capabilities for information exchange. The system model for the 6G IoT network is shown in Figure 13.1.

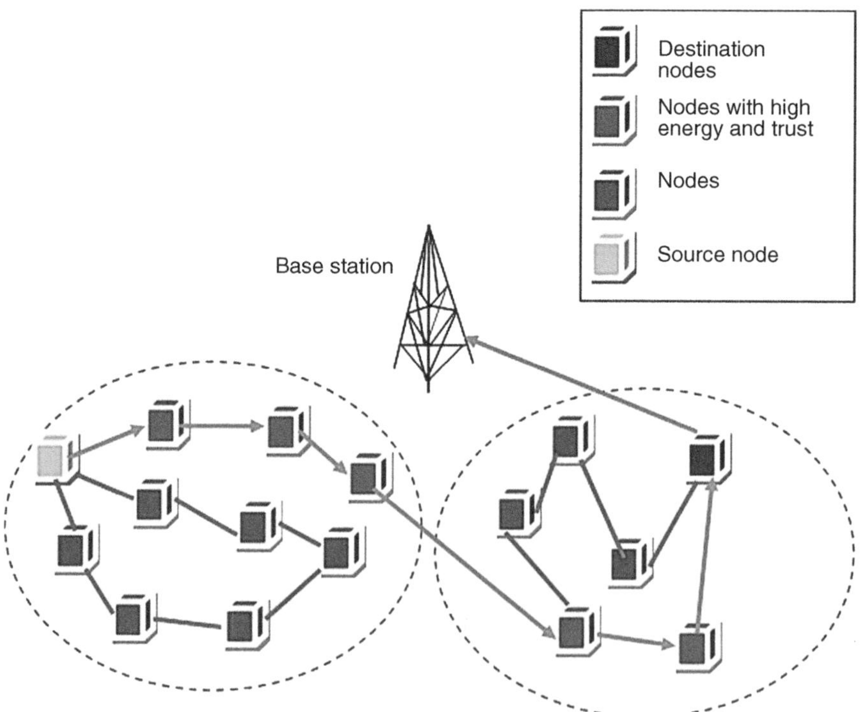

Figure 13.1 6G IoT network model.

The 6G IoT-based WSN model comprises [11] different devices in such a way that the devices are connected through internet. Accordingly, the IoT-based network contains a Base Station (BS) B, CHs (CHs) C, and the k count of cluster nodes. In this case, wireless connectivity between IoT nodes demonstrates that communication is dispersed uniformly in a radio transmission range. In this case, each network node has a unique ID, and a cluster is created in the network area by a collection of nodes. Data is received between IoT nodes via the BS. Additionally, nodes use CHs to send data to the BS. To make network nodes equal to each other, the network is divided into multiple clusters [17]. Data packets are routed from the node to CH once cluster groups are formed in the network. As a result, CH has all the data sent to the BS. As a result, complete nodes are positioned between the BS and CH in a special spot [18].

13.2.1.1 Energy model

6G IoT network comprises a set of nodes [12] such that each node has the initial energy as (Q_0), whereas the energy [21] of the nodes is rejuvenated. Here, packets are transmitted from normal nodes to oth CH. During the communication of data packets between nodes, some energy is lost in the network and it follows the free space mechanism and the multi-path fading scheme [19]. When transmission of data bytes is carried out from the normal nodes, the energy loss that occurs from the normal node is specified as,

$$Q^{\text{emt}}\left(H_{\text{M}}^{\text{y}}\right) = Q^{\text{ele}} \times R_{\text{L}} + S^{\text{pa}} \times R_{\text{L}} \times \left\|H_{\text{M}}^{\text{y}} - H_{\text{A}}^{\text{m}}\right\|^4 ; \text{ if } \left\|H_{\text{M}}^{\text{y}} - H_{\text{A}}^{\text{m}}\right\|^4 \geq t_0 \quad (13.1)$$

$$Q^{\text{emt}}\left(H_{\text{M}}^{\text{y}}\right) = Q^{\text{ele}} \times R_{\text{L}} + S^{\text{fs}} \times R_{\text{L}} \times \left\|Q_{\text{M}}^{\text{y}} - Q_{\text{A}}^{\text{m}}\right\|^4 ; \text{ if } \left\|Q_{\text{M}}^{\text{y}} - Q_{\text{A}}^{\text{m}}\right\|^4 < t_0 \quad (13.2)$$

Here, $t_0 = \sqrt{\dfrac{S^{\text{fs}}}{S^{\text{pa}}}}$, S^{fs} indicates free space energy, S^{pa} denotes multipath fading amplification, electronic energy created by considering digital coding, modulation, filtering, and spreading is indicated as Q^{ele}, and $\left\|H_{\text{M}}^{\text{y}} - H_{\text{A}}^{\text{m}}\right\|$ indicates distance between the CH and the normal node. Accordingly, the electronic energy of node [13] is given as,

$$Q^{\text{ele}} = Q^{\text{txr}} + Q^{\text{DA}} \quad (13.3)$$

Here, Q^{txr} indicates transmitter energy, Q^{DA} denotes data aggregation energy. During the reception of R_{L} data bytes by CH, the energy dissipated is [20],

$$Q^{\text{emt}}\left(H_{\text{A}}^{\text{y}}\right) = Q^{\text{ele}} \times R_{\text{L}} \quad (13.4)$$

After transmitting the data packets, energy values can be updated. Suppose if the node transmits B_L data packets to CH, the energy updated in normal node is $Q_{a+1}\left(H_M^y\right)$, and so the energy update in the CH is given as $Q_{a+1}\left(H_A^y\right)$.

$$Q_{a+1}\left(H_M^y\right) = Q_a\left(H_M^y\right) - Q^{emt}\left(H_M^y\right) \tag{13.5}$$

$$Q_{a+1}\left(H_A^y\right) = Q_a\left(H_A^y\right) - Q^{emt}\left(H_A^y\right) \tag{13.6}$$

Until the node dies, the aforementioned process can be repeated.

13.2.1.2 Mobility model

The mobility model [23] is used to display node movement within a network. It calculates node position changes by taking acceleration, velocity, and unique movement location into account. In this case, node movement is determined using the mobility model. Accordingly, the node b_f changes the distance $D_{b,f}$ and the node b_l sends the distance $D_{b,l}$ with the time $z = 0$. Let us assume (u_f, v_f) and (u_l, v_l) shows the updated location placed by the nodes b_f and b_l with the time $z = 1$.

However, Euclidean distance among nodes b_f and b_l with the time $z = 0$ is specified as [22],

$$D(b_f, b_l, 0) = \sqrt{\left|u - u^*\right|^2 + \left|v - v^*\right|^2} \tag{13.7}$$

Here, (u, v) and (u^*, v^*) indicates the new location of the nodes b_f with $z = 0$. Let us consider the nodes b_f and b_l sends in the velocity $x_{b,f}$ and $x_{b,l}$ through the angle $\varphi_{b,f}$ and $\varphi_{b,l}$ at x-axis. Therefore, the distance placed by the node at certain time z is represented as [24],

$$D_{b,f} = x_{b,f} \times z \tag{13.8}$$

$$D_{b,l} = x_{b,l} \times z \tag{13.9}$$

When $z = 1$, the node b_f sends at the distance $D_{b,f}$ and the angle $\varphi_{b,f}$, and so new location of the node b_f with the time z is specified as [26],

$$u_f = u + x_{b,f} \times z \times \cos(\varphi_{b,f}) \tag{13.10}$$

$$v_f = v + x_{b,f} \times z \times \cos(\varphi_{b,f}) \tag{13.11}$$

Here, $\varphi_{b,f}$ shows angle such that node b_f sends to a new location. Similarly, whenever the node b_l sends at the distance $D_{b,l}$ with the angle $\varphi_{b,l}$, new location of the node b_l with the time z is indicated as [28],

$$u_l = u^* + x_{b,l} \times z \times \cos(\varphi_{b,l}) \tag{13.12}$$

$$v_l = v^* + x_{b,l} \times z \times \cos(\varphi_{b,l}) \tag{13.13}$$

At time z, distance amongst the nodes at b_f placed at location (u_f, v_f) and (u_l, v_l) is signified as [30],

$$D(b_f, b_l, z) = \sqrt{\left|u_f - u_l\right|^2 + \left|v_f - v_l\right|^2} \tag{13.14}$$

where, (u_f, v_f) and (u_l, v_l) indicates new location of the nodes b_f and b_l.

13.2.1.3 Link lifetime model

LLT [25] shows the network lifetime that connects two sensor nodes in the network. Accordingly, network link breakage shows routing failure in network. Here, the lifespan of network is calculated based on constraints, like coordinate between the nodes, their direction of mobility, and their own mobility. Let us assume D_1 and D_2 indicates two mobile nodes placed at E_{D_1}, F_{D_1} and E_{D_2}, F_{D_2} such that the LLT is measured as [32]:

$$\text{LLT} = \frac{-(mn + pq) + \sqrt{\left(m^2 + q^2\right)l^2 - \left(mq - pn\right)^2}}{\left(m^2 + q^2\right)} \tag{13.15}$$

Here,

$$\left. \begin{aligned} m &= C_{D_1} \cos\theta_{D_1} - C_{D_2} \cos\theta_{D_2} \\ n &= E_{D_1} - E_{D_2} \\ p &= C_{D_1} \sin\theta_{D_1} - C_{D_2} \sin\theta_{D_2} \\ q &= F_{D_1} - F_{D_2} \end{aligned} \right\},$$

where l indicates the range of data communication, C_{D_1} indicates the mobility speed at the sensor node D_1, C_{D_2} indicates the mobility speed at the sensor node D_2, E_{D_2}, F_{D_2} indicates coordinate of node D_2, E_{D_1}, F_{D_1} signifies the coordinate of node D_1, θ_{D_1} represents the motion direction, D_1, and θ_{D_2} represents the motion direction at the node D_2.

13.2.2 Taylor-PO and deep neuro fuzzy for selection of cluster head and multipath routing prediction in internet of things-based WSN

In recent decades, CH is selected in the network for improving the performance of network and to improve quality of service in the transmission range. Various methods are modelled to achieve the multi-path routing mechanism, but a major problem in the generation of low interference path and the inter path interference is more expensive. Hence, energy-based mechanism is modelled to develop the protocol to improve the network's performance. This chapter is designed to model a technique for choosing the best CH for an IoT-based WSN network while utilizing a multipath routing strategy. At first, nodes are distributed in network and then CH is selected utilizing the optimization algorithm known as Taylor-PO that is created by combining PO [27] and the Taylor series [29], respectively. However, by taking into account metrics like LLT, energy, intra- and inter-cluster distance, predicted delay, and energy, the fitness measure is used to accomplish the selection strategy. Accordingly, the energy is predicted with the Deep neuro fuzzy network. As a result, nodes with the highest energy are designated as CH, and the communication process uses the CH that has been chosen the multipath routing mechanism is carried out with the TSGWO algorithm. The schematic view of the developed Taylor-PO in the 6G IoT network is shown in Figure 13.2.

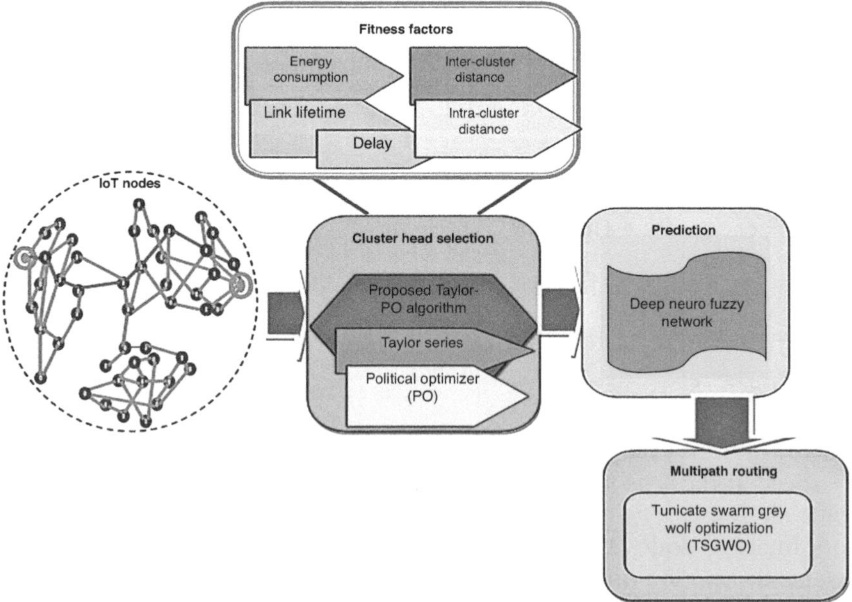

Figure 13.2 Schematic view of proposed Taylor-PO 6G IoT network.

13.2.2.1 Selection of CH with proposed Taylor-PO

The choice of CH in the IoT-based WSN with the developed Taylor-PO is explained in this section in order to ensure that data transmission between nodes in the 6G IoT network is efficient and effective. Let's assume that the number of nodes in the network is such that the best CH is chosen within it. The nodes' initial energy is used to specify the network. Here, the selection of CH is made with the developed method named Taylor-PO which is employed to lengthen the network's lifespan and to achieve higher throughput by consuming lesser energy. However, the Taylor-PO is designed by the incorporation of PO with the Taylor series. However, the selection of CH is made using attributes, like delay, LLT, distance, and energy.

13.2.2.1.1 Solution encoding

To find the optimal solution representation using the developed optimization approach, one uses the solution encoding [14]. Here, the solution of the proposed method shows a number of nodes using the selection of optimal CH with the fitness measure. Figure 13.3 depicts solution encoding of proposed Taylor-PO. Here, t denotes optimal CHs, and h indicates the total nodes. Accordingly, total CH should be less than nodes.

13.2.2.1.2 Fitness function

The factors distance, delay, LLT, and node energy are analyzed along with the total number of nodes and CH in the network to determine the fitness factor [15]. On the other hand, the developed Taylor-PO is used to compute optimal CH using fitness. Thus, the fitness metric is calculated as,

$$\text{Fitness} = \frac{1}{N}\left[E + (1 - D^{\text{ec}}) + D^{\text{ac}} + X + L + E^{\text{P}} + X^{\text{P}} \right] \tag{13.16}$$

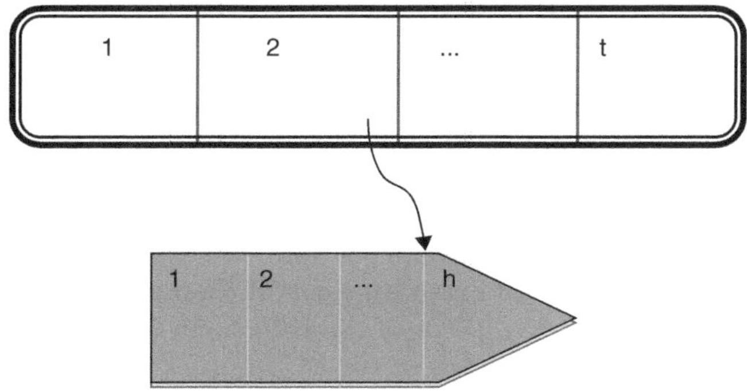

Figure 13.3 Solution encoding.

where N implies normalization factor, E indicates energy consumption, D^{ec} represents the inter-cluster distance, D^{ac} signifies the intra-cluster distance, X denotes delay, L indicates LLT, E^p implies predicted energy, and X^p denotes predicted delay.

Inter cluster distance: It is calculated using the following equation and is known as distance [16] between particular clusters.

$$D^{ec} = \frac{1}{Y \times n \times m} \sum_{i=1}^{m} \sum_{\substack{j=1 \\ l \in j}}^{n} D_{l,j} \qquad (13.17)$$

where Y denotes normalization factor, n indicates the count of CH and m implies number of nodes and $D_{l,j}$ shows distance among lth node and jth CH.

Accordingly, the measurement of intra-cluster distance is,

$$D^{ac} = \frac{1}{Z \times n} \sum_{j=1}^{n} \sum_{k=1}^{n} D_{j,k} \qquad (13.18)$$

where, Z denotes the normalization factor, $D_{j,k}$ indicates the distance among jth CH and the kth CH.

Accordingly, the node present in the cluster group is defined using the below equation as

$$X = \frac{\sum_{i=1}^{m} X_i}{m} \qquad (13.19)$$

where X_i indicates the ith node in the cluster.

13.2.2.2 Proposed Taylor-PO

This section presents the algorithmic steps present in the Taylor-PO algorithm. However, the proposed Taylor-PO is designed with the incorporation of PO and Taylor series, respectively. The multiphase process of politics, which is structured with different phases—party switching, inter-party elections, constituency allocation, parliamentary affairs, and the election campaign—encourages the PO [27]. The algorithm's two main purposes are to boost the convergence rate and regulate solution diversity. The Taylor series is an extended form of the function in the infinite sum of the variables that illustrates the behaviour of complex variables. The Taylor series is integrated with the PO in order to improve both performance and efficiency. The suggested Taylor-PO's pseudocode is represented by Algorithm 13.1.

Algorithm 13.1 Pseudo code of developed Taylor-PO

Input: Solutions list F, overall iterations $d_{\max}$

Output: optimal resolution F^*

Start

Start with a random solution

   Use Equation (13.16) to calculate fitness

Keep track of each member's fitness

Determine the winning constituencies and the group of party leaders

   While ($k <$ MaxDays)

$F_{\text{temp}} = F$

$f(F_{\text{temp}}) = f(F)^\#$

For each $F_l \in F$ do

For each $F_l^m \in F_l$ do

$F_l^m = \text{ElectionCampaign}(F_l^m, F_l^m(d-1), F_l^*, m^*)$

End

      End

Party Switching (F, ϑ)

Using Equation (13.20), determine the group of party leaders and the constituency winners.

Reassess and maintain each member's level of fitness

$d = d + 1$

Return F^*

End

13.2.2.3 Prediction with deep neuro fuzzy network

The input passed to the DNFN [31] classifier is the energy E_i and the delay X_i. However, the DNFN is the integration of fuzzy logic theory and deep neural network. This classifier uses two distinct processes: one that analyzes the system objective using fuzzy logic and the other that uses a deep network for execution.

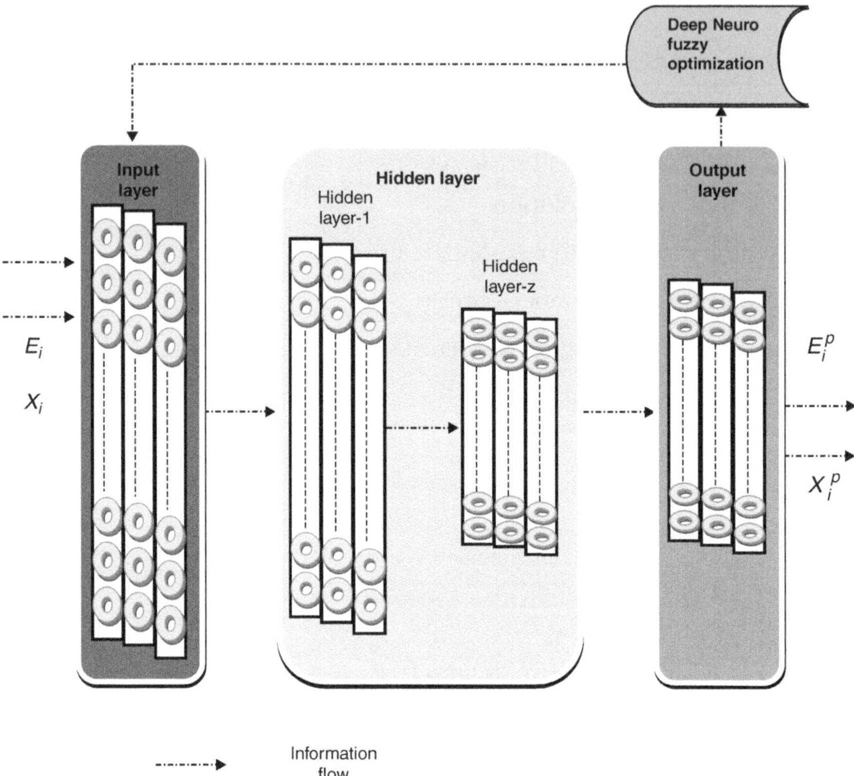

Figure 13.4 Structure of deep neuro fuzzy network classifier.

Three layers make up this structure: input, hidden, and output. Figure 13.4 shows the DNFN structure. The data processing using the fuzzy network is described by mapping each output to its corresponding input. In this case, the fuzzy model displays a value between 0 and 1 for each degree of the input unit. Let us assume the two premises as z and a, and one consequent as y, that is specified as,

$$F_{f,g}^e(d+1) = \frac{1.3591}{2h+0.3591}$$

$$\left[2hc^* + \frac{(2h-1)}{1.3591} \begin{bmatrix} 0.5F_{g,h}^e(d) - 1.3591F_{g,h}^e(d-2) + 0.6795F_{g,h}^e(d-3) \\[2mm] -0.2259F_{g,h}^e(d-4) + 0.0555F_{g,h}^e(d-5) \\[2mm] -0.0104F_{g,h}^e(d-6) + 1.38e^{-3}F_{g,h}^e(d-7) \\[2mm] -9.92e^{-5}F_{g,h}^e(d-8) \end{bmatrix} \right] \tag{13.20}$$

$$X_{1,\omega}(z) = fB_\omega(z) \text{ or } X_{1,\omega} = fP_{\omega-2}(a), \; \forall \omega = 1, 2, 3 \tag{13.21}$$

where a and z denotes the input to all ωth entity, fB_ω and $fP_{\omega-2}$ shows the antecedent membership function, and $X_{1,\omega}$ indicates the function of membership degree. Consequently, the membership function is listed as [34]

$$fB_\omega(z) = \cfrac{1}{1 + \left|\dfrac{z = D_\omega}{N_\omega}\right|^{2R_\omega}} \tag{13.22}$$

where R_ω, D_ω and N_ω denotes the membership function of the attribute of the premises. The multiplication of the membership demonstrates how strong the rule is.

$$X_{2,\omega} = \mu_\omega = fB_\omega(z)fP_{\omega-2}(a), \; \forall \omega = 1, 2 \tag{13.23}$$

The output of each rule is normalized with the firing strength of rule and is given as,

$$X_{3,\omega} = \bar{\mu}_\omega = \frac{\mu_\omega}{\mu_1 + \mu_2}, \; \forall \omega = 1, 2 \tag{13.24}$$

The defuzzification layer is given as,

$$X_{4,\omega} = \bar{\mu}_\omega I_\omega = \bar{\mu}_\omega \left(E_\omega z + F_\omega a + C_\omega\right), \; \forall \omega = 1, 2 \tag{13.25}$$

where E, F and C denotes the consequent parameter set. However, the resultant output is provided as

$$X_{5,\omega} = \sum_\omega \bar{\mu}_\omega I_\omega = \frac{\sum_\omega \mu_\omega I_\omega}{\sum_\omega \mu_\omega} \tag{13.26}$$

The output acquired from DNFN is the predicted energy E_i^p and the predicted delay X_i^p.

13.2.2.4 Multipath routing using TSGWO

After the selection of CH, the predicted delay and the energy is computed with DNFN. In this case, TSGWO—which was created in collaboration with the TSA [33] and the GWO [35]—is used to carry out multipath routing.

In the TSGWO, the update solution is performed by adding the term A_4. Therefore, the update solution is represented as,

$$\vec{K}(o+1) = \frac{\vec{K}_1 + \vec{K}_2 + \vec{K}_3 + \vec{K}_4}{4} \tag{13.27}$$

Here, K_1, K_2, K_3, and K_4 indicate the first best, second best, third best, and the fourth best solution and is expressed as

$$\vec{K}_1 = \vec{K}_\alpha - B_1(\vec{W}_\alpha) \tag{13.28}$$

where K_α denotes first best search agent, B_1 represents the coefficient vector, and W_α indicates the distance travelled by the first best search agent.

$$\vec{K}_2 = \vec{K}_\beta - B_2(\vec{W}_\beta) \tag{13.29}$$

where K_β denotes the first best search agent, B_2 represents the coefficient vector, and W_β implies the distance travelled by the first best search agent.

$$\vec{K}_3 = \vec{K}_\gamma - B_3(\vec{W}_\gamma) \tag{13.30}$$

where K_γ denote the first best search agent, B_3 represents the coefficient vector, and W_γ shows distance travelled by the first best search agent. Moreover, K_4 is used by TSA. Hence, the update solution is given as,

$$\vec{K}_4 = L + \vec{B} \cdot M \tag{13.31}$$

where L denotes the food source, B indicates the conflict attribute and M indicates the distance between food source and the search agent.

13.3 RESULTS AND PERFORMANCE ANALYSIS

This section presents discussion of the results of developed Taylor-PO by varying number of nodes in network.

13.3.1 Experimental Setup

The implementation of the developed method is carried out in NS-2 with 50 nodes, 100 nodes, 150 nodes, and 200 nodes.

13.3.2 Evaluation Metrics

The developed scheme's performance is analyzed using metrics such as throughput, energy, and delay.

Energy: It is known as the node energy that is required to finish the data transmission process.

Delay: It refers the time taken to transmit data packets to the destination side.

Throughput: It is known as the ratio of the total number of packets received by the recipient end over a given period of time.

13.3.3 Performance Analysis

This section elaborates the performance analysis of developed scheme based on the number of nodes.

13.3.3.1 Analysis with 50 nodes

Figure 13.5 demonstrates the delay analysis. When the number of rounds=40, the delay achieved by Taylor-PO with cluster size 2 is 0.0722 s, cluster size 3 is 0.0626 s, cluster size 4 is 0.0540 s, and cluster size 5 is 0.0406 s. By considering the number of rounds as 50, the delay measured by Taylor-PO with cluster size 2 is 0.0704 s, cluster size 3 is 0.0595 s, cluster size 4 is 0.0489 s, and cluster size 5 is 0.0373 s. For 60 rounds, the delay measured by Taylor-PO with cluster size 2 is 0.0677 s, cluster size 3 is 0.0561 s, cluster size 4 is 0.0467 s, and cluster size 5 is 0.0345 s.

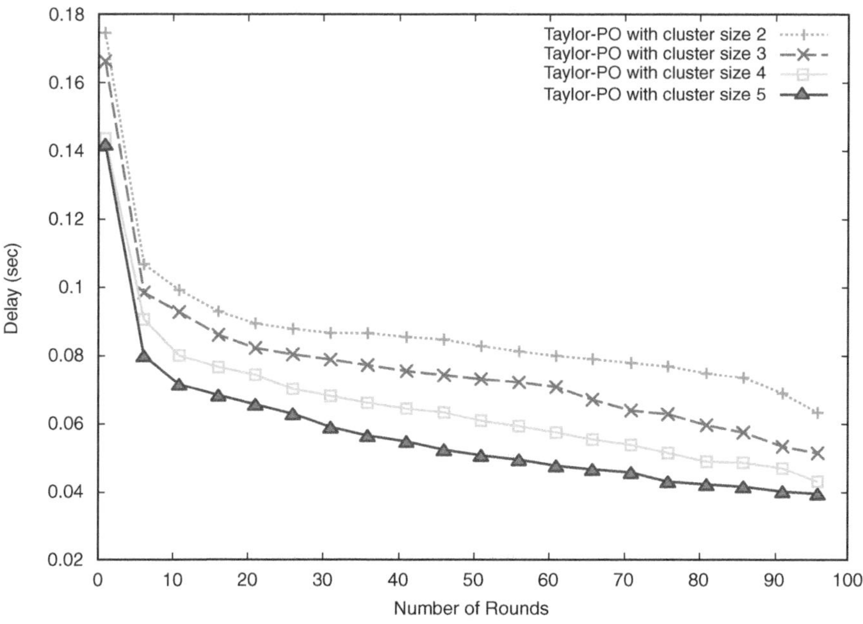

Figure 13.5 Analysis of Taylor-PO in terms of delay with 50 nodes.

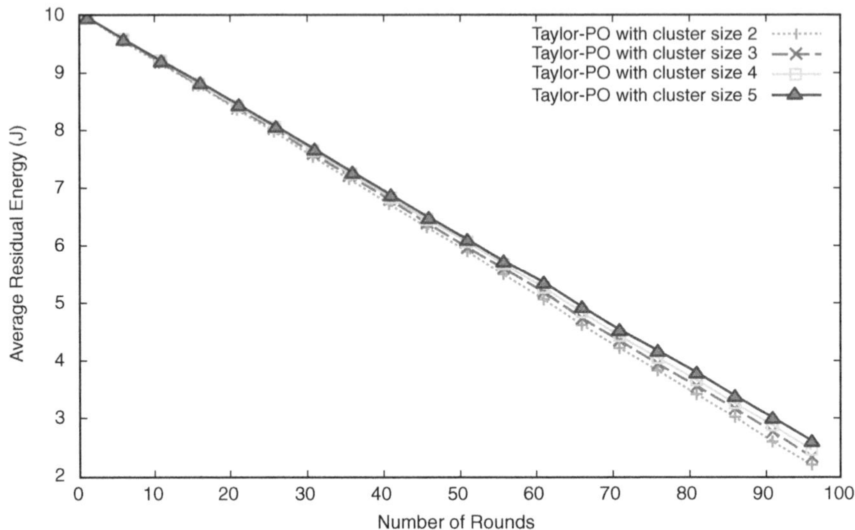

Figure 13.6 Analysis of Taylor-PO based on average residual energy with 50 nodes.

Figure 13.6 portrays analysis of average residual energy. When rounds=40, the average residual energy measured by Taylor-PO with cluster size 2 is 6.781J, cluster size 3 is 6.846J, cluster size 4 is 6.905J, and cluster size 5 is 6.953J. By considering number of rounds as 50, the average residual energy computed by Taylor-PO with cluster size 2 is 5.974J, cluster size 3 is 6.056J, cluster size 4 is 6.130J, and cluster size 5 is 6.190J. For 60 rounds, the average residual energy measured by Taylor-PO with cluster size 2 is 5.157J, cluster size 3 is 5.256J, cluster size 4 is 5.345J, and cluster size 5 is 5.418J.

Figure 13.7 portrays the analysis of throughput. When rounds=40, throughput measured by Taylor-PO with cluster size 2 is 417.91 kbps, cluster size 3 is 424.03 kbps, cluster size 4 is 431.82 kbps, and cluster size 5 is 439.18 kbps. By considering number of rounds as 50, throughput measured by Taylor-PO with cluster size 2 is 426.97 kbps, cluster size 3 is 435.76 kbps, cluster size 4 is 441.22 kbps, and cluster size 5 is 449.40 kbps. For 60 rounds, the throughput computed by Taylor-PO with cluster size 2 is 438.58 kbps, cluster size 3 is 445.07 kbps, cluster size 4 is 450.75 kbps, and cluster size 5 is 458.34 kbps.

13.3.3.2 Analysis with 100 nodes

Figure 13.8 illustrates the analysis of delay. By considering the number of rounds as 40, the delay achieved by Taylor-PO with cluster size 2 is 0.0856 s, cluster size 3 is 0.0759 s, cluster size 4 is 0.0645 s, and cluster size 5 is 0.0549 s. By considering the number of rounds as 50, the delay measured

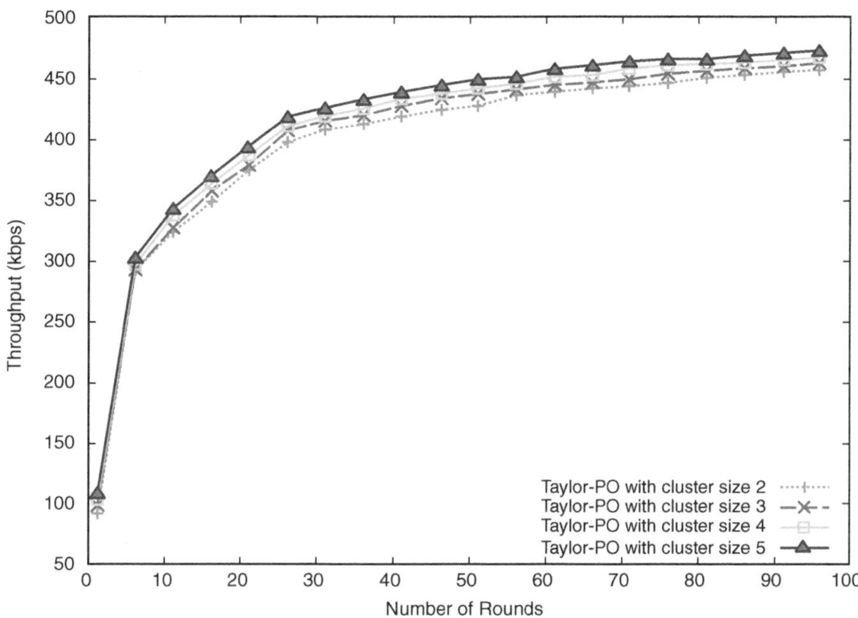

Figure 13.7 Analysis of Taylor-PO based on throughput with 50 nodes.

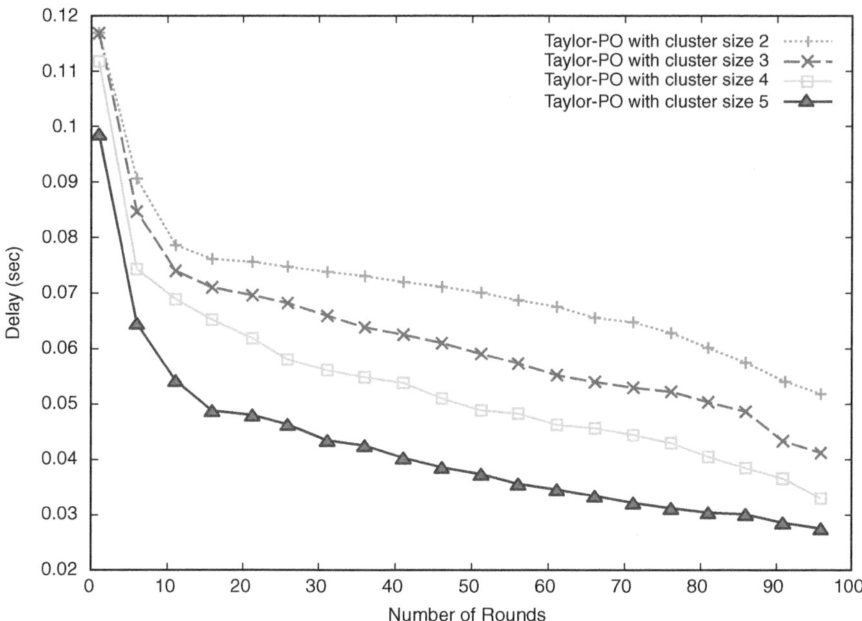

Figure 13.8 Analysis of Taylor-PO in terms of delay with 100 nodes.

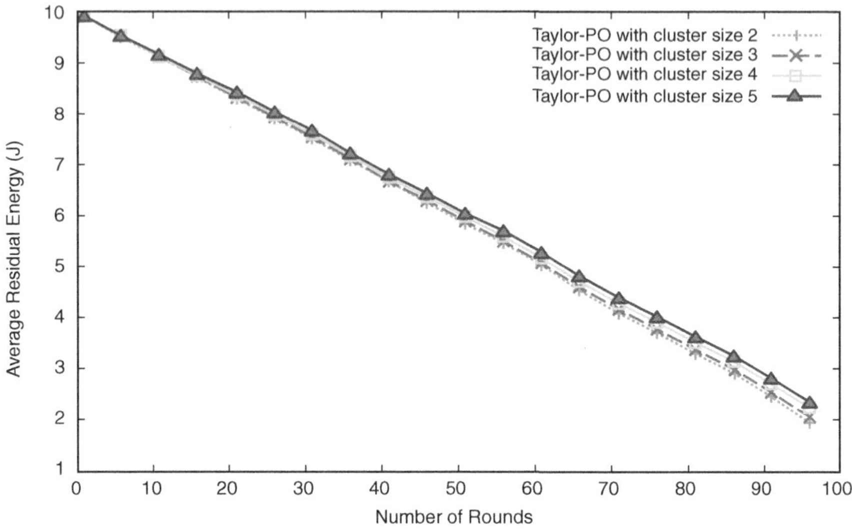

Figure 13.9 Analysis of Taylor-PO based on average residual energy with 100 nodes.

by Taylor-PO with cluster size 2 is 0.0836 s, cluster size 3 is 0.0733 s, cluster size 4 is 0.0609 s, and cluster size 5 is 0.0505 s. For 60 rounds, the delay measured by Taylor-PO with cluster size 2 is 0.0804 s, cluster size 3 is 0.0708 s, cluster size 4 is 0.0583 s, and cluster size 5 is 0.0486 s.

Figure 13.9 depicts the analysis of average residual energy. If number of rounds = 40, average residual energy measured by Taylor-PO with cluster size 2 is 6.7475 J, cluster size 3 is 6.7847 J, cluster size 4 is 6.8235 J, and cluster size 5 is 6.8623 J. By considering number of rounds as 50, the average residual energy obtained by Taylor-PO with cluster size 2 is 5.945 J, cluster size 3 is 5.992 J, cluster size 4 is 6.040 J, and cluster size 5 is 6.089 J. For 60 rounds, the average residual energy measured by Taylor-PO with cluster size 2 is 5.161 J, cluster size 3 is 5.216 J, cluster size 4 is 5.274 J, and cluster size 5 is 5.332 J.

Figure 13.10 portrays the analysis of throughput. When rounds = 40, throughput achieved by Taylor-PO with cluster size 2 is 435.11 kbps, cluster size 3 is 440.72 kbps, cluster size 4 is 448.91 kbps, and cluster size 5 is 456.15 kbps. By considering the number of rounds as 50, throughput measured by Taylor-PO with cluster size 2 is 438.84 kbps, cluster size 3 is 447.14 kbps, cluster size 4 is 453.25 kbps, and cluster size 5 is 460.03 kbps. For 60 rounds, the throughput computed by Taylor-PO with cluster size 2 is 441.3 kbps, cluster size 3 is 449.7 kbps, cluster size 4 is 455.9 kbps, and cluster size 5 is 462.7 kbps.

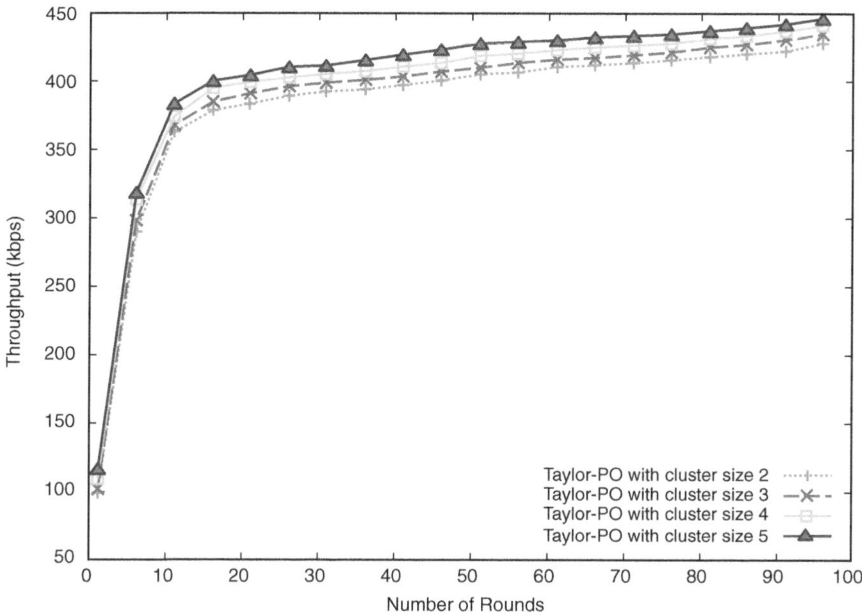

Figure 13.10 Analysis of Taylor-PO based on throughput with 100 nodes.

13.3.3.3 Analysis with 150 nodes

Figure 13.11 portrays the analysis of delay. By considering the number of rounds as 40, the delay obtained by Taylor-PO with cluster size 2 is 0.0838 second, cluster size 3 is 0.0741 second, cluster size 4 is 0.0612 second, and cluster size 5 is 0.0478 second. By considering the number of rounds as 50, the delay measured by Taylor-PO with cluster size 2 is 0.0818 second, cluster size 3 is 0.0700 second, cluster size 4 is 0.0579 second, and cluster size 5 is 0.0454 second. For 60 rounds, the delay measured by Taylor-PO with cluster size 2 is 0.0792 second, cluster size 3 is 0.0644 second, cluster size 4 is 0.0542 second, and cluster size 5 is 0.0410 second.

Figure 13.12 portrays the analysis of average residual energy. If number of rounds=40, the average residual energy measured by Taylor-PO with cluster size 2 is 6.6480J, cluster size 3 is 6.6498J, cluster size 4 is 6.7485J, and cluster size 5 is 6.7547J. By considering number of rounds as 50, the average residual energy achieved by Taylor-PO with cluster size 2 is 5.8170J, cluster size 3 is 5.8192J, cluster size 4 is 5.9428J, and cluster size 5 is 5.9506J. For 60 rounds, the average residual energy measured by Taylor-PO with cluster size 2 is 4.9839J, cluster size 3 is 4.9865J, cluster size 4 is 5.1351J, and cluster size 5 is 5.1445J.

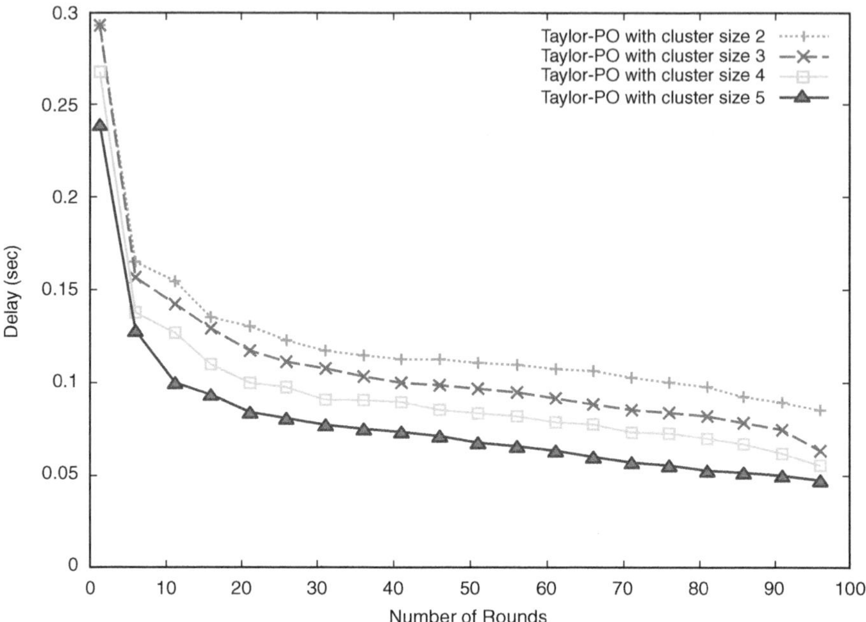

Figure 13.11 Analysis of Taylor-PO in terms of delay with 150 nodes.

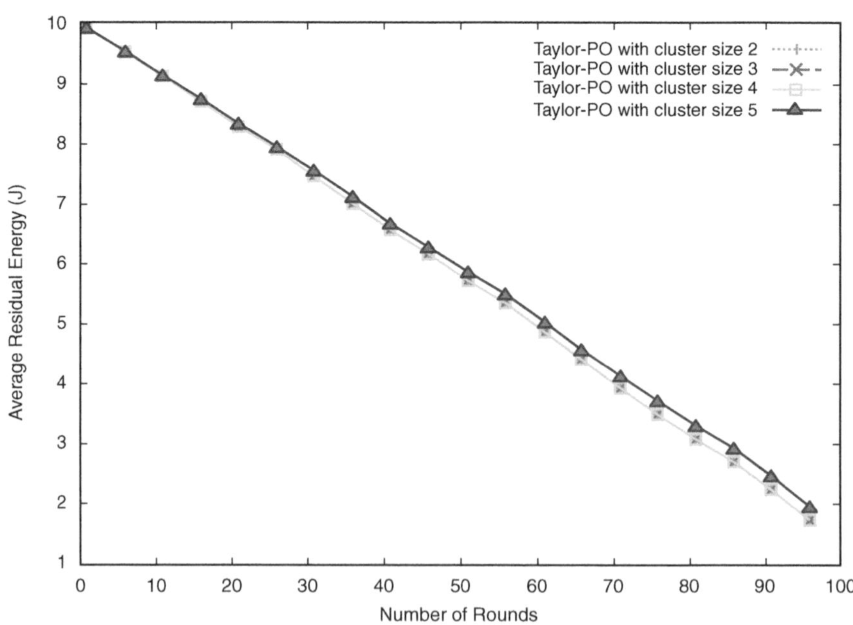

Figure 13.12 Analysis of Taylor-PO in terms of average residual energy with 150 nodes.

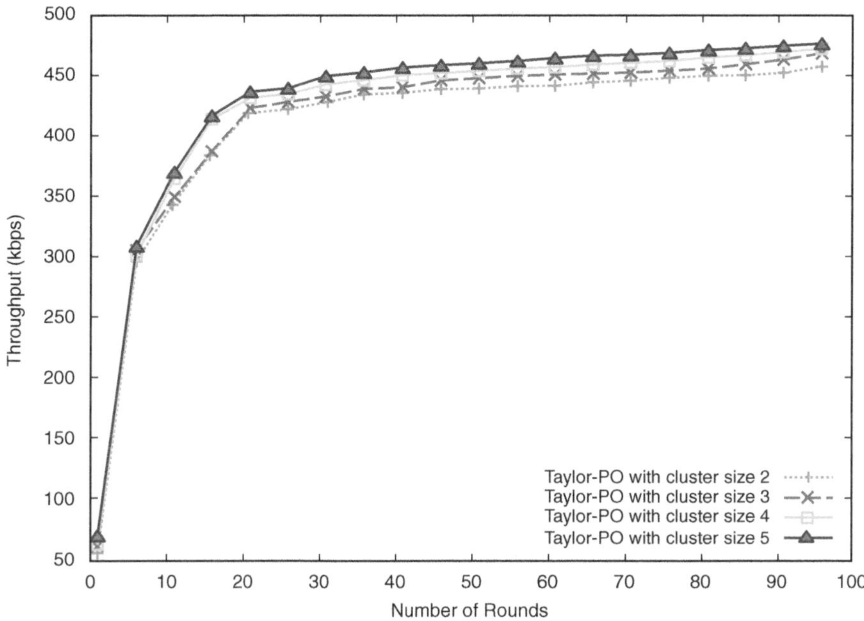

Figure 13.13 Analysis of Taylor-PO in terms of throughput with 150 nodes.

Figure 13.13 portrays the analysis of throughput. When number of rounds as 40, throughput achieved by Taylor-PO with cluster size 2 is 396.62 kbps, cluster size 3 is 403.87 kbps, cluster size 4 is 409.58 kbps, and cluster size 5 is 419.90 kbps. By considering the number of rounds as 50, throughput measured by Taylor-PO with cluster size 2 is 405.36 kbps, cluster size 3 is 409.97 kbps, cluster size 4 is 418.46 kbps, and cluster size 5 is 426.35 kbps. For 60 rounds, the throughput computed by Taylor-PO with cluster size 2 is 411.09 kbps, cluster size 3 is 414.93 kbps, cluster size 4 is 423.36 kbps, and cluster size 5 is 430.79 kbps.

13.3.3.4 Analysis with 200 Nodes

Figure 13.14 illustrates the analysis of delay. By considering the number of rounds as 40, the delay obtained by Taylor-PO with cluster size 2 is 0.1138 second, cluster size 3 is 0.1002 second, cluster size 4 is 0.0890 second, and cluster size 5 is 0.0742 second. By considering the number of rounds as 50, the delay measured by Taylor-PO with cluster size 2 is 0.1115 second, cluster size 3 is 0.0974 second, cluster size 4 is 0.0832 second, and cluster size 5 is 0.0681 second. For 60 rounds, the delay measured by Taylor-PO with cluster size 2 is 0.1078 second, cluster size 3 is 0.0928 second, cluster size 4 is 0.0791 second, and cluster size 5 is 0.0628 second.

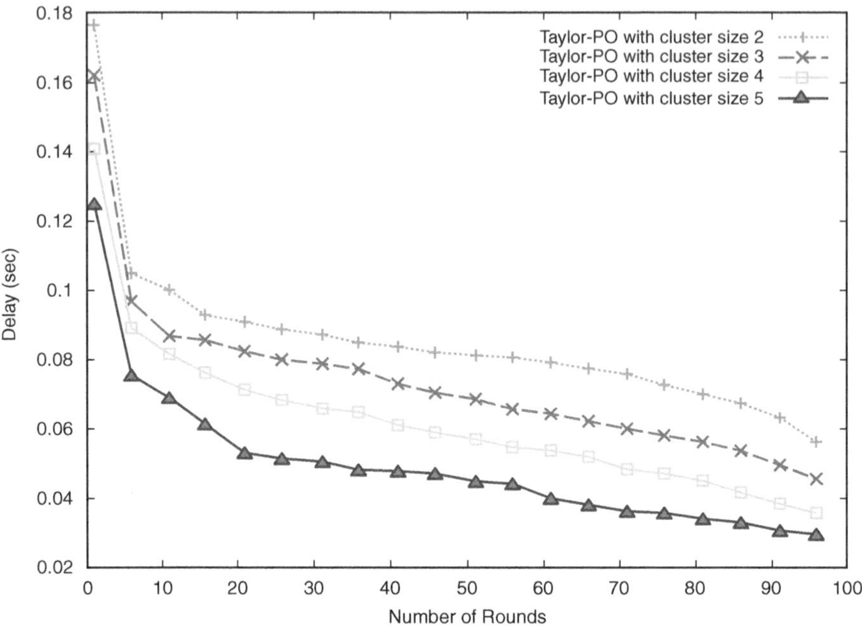

Figure 13.14 Analysis of Taylor-PO in terms of delay with 200 nodes.

Figure 13.15 portrays the analysis of average residual energy. If number of rounds=40, the average residual energy measured by Taylor-PO with cluster size 2 is 6.7491J, cluster size 3 is 6.7874J, cluster size 4 is 6.8686J, and cluster size 5 is 6.9088J. By considering number of rounds as 50, average residual energy computed by Taylor-PO with cluster size 2 is 5.944J, cluster size 3 is 5.992J, cluster size 4 is 6.093J, and cluster size 5 is 6.144J. For 60 rounds, the average residual energy measured by Taylor-PO with cluster size 2 is 5.136J, cluster size 3 is 5.194J, cluster size 4 is 5.316J, and cluster size 5 is 5.376J.

Figure 13.16 portrays the analysis of throughput. When rounds as 40, throughput achieved by Taylor-PO with cluster size 2 is 427.90 kbps, cluster size 3 is 434.54 kbps, cluster size 4 is 442.23 kbps, and cluster size 5 is 449.99 kbps. By considering the number of rounds as 50, throughput measured by Taylor-PO with cluster size 2 is 432.45 kbps, cluster size 3 is 440.36 kbps, cluster size 4 is 446.50 kbps, and cluster size 5 is 454.90 kbps. For 60 rounds, the throughput computed by Taylor-PO with cluster size 2 is 437.2 kbps, cluster size 3 is 445.0 kbps, cluster size 4 is 450.7 kbps, and cluster size 5 is 457.9 kbps.

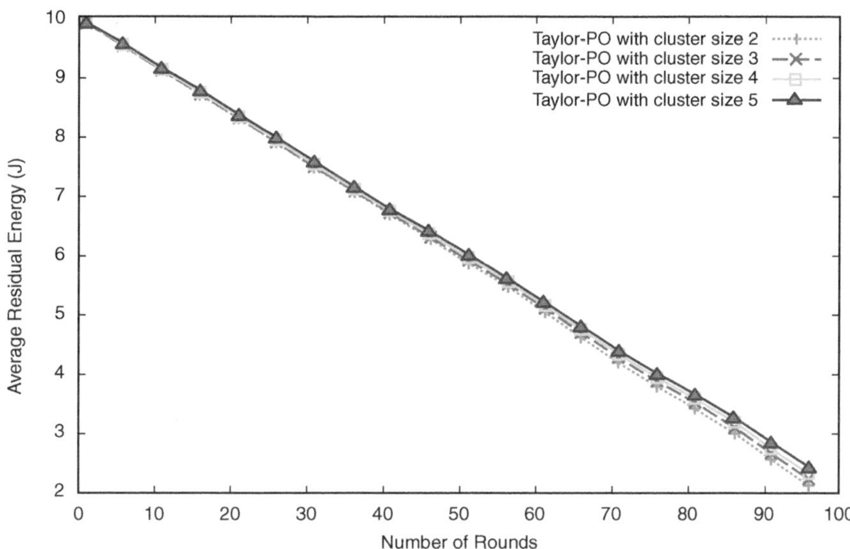

Figure 13.15 Analysis of Taylor-PO in terms of average residual energy with 200 nodes.

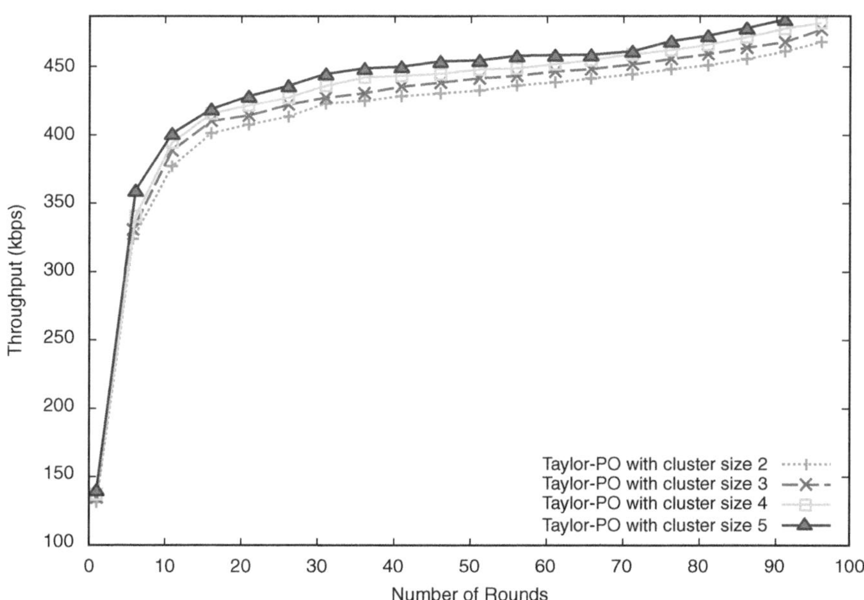

Figure 13.16 Analysis of Taylor-PO in terms of throughput with 200 nodes.

13.4 CONCLUSION

The approach for choosing a cluster head and completing the multipath routing process in a 6G IoT network is explained in this chapter. In this case, the Taylor-PO, which is obtained by combining the Taylor series and PO, is used to choose the CH. As a result, the DNFN is used to calculate both the energy prediction and the delay prediction. The constraints, which include delay, LLT, distance, inter- and intra-cluster distance, as well as the anticipated delay and energy, are used to calculate the fitness measure.

In conclusion, choosing the right cluster head is essential for overcoming the difficulties brought on by the size and complexity of IoT deployments in 6G networks. In the era of 6G connectivity, it provides a robust and effective mechanism for data transmission, improving the network's overall performance and dependability.

REFERENCES

[1] Dhumane, A.V. and Prasad, R.S, "Multi-objective fractional gravitational search algorithm for energy efficient routing in IoT," *Wireless Networks*, 25, 399–413 (2017).

[2] Chen, Z., He, M., Liang, W. and Chen, K., "Trust-aware and low energy consumption security topology protocol of wireless sensor network," *Journal of Sensors*, (1), p.716468, 2015.

[3] Zhu, J., "Wireless sensor network technology based on security trust evaluation model," *International Journal of Online and Biomedical Engineering (iJOE)*, vol. 14, no. 04, pp. 211–226, 2018.

[4] Solteiro Pires, E.J., Tenreiro Machado, J.A., de Moura Oliveira, P.B., Boaventura Cunha, J. and Mendes, L., "Particle swarm optimization with fractional-order velocity," *Nonlinear Dynamics*, vol. 61, no. 1–2, pp. 295–301, 2010.

[5] Rashedi, E., Nezamabadi-pour, H. and Saryazdi, S., "GSA: a gravitational search algorithm," *Information Sciences*, vol. 179, pp. 2232–2248, 2009.

[6] Ezhilarasi, M. and Krishnaveni, V, "An evolutionary multipath energy-efficient routing protocol (EMEER) for network lifetime enhancement in wireless sensor networks," *Soft Computing*, 23, 8367–8377 (2019).

[7] Dhumane, A.V. and Prasad, R.S., "Fractional gravitational grey wolf optimization to multi-path data transmission in IoT," *Wireless Personal Communications*, vol. 102, no. 1, pp. 411–436, 2018.

[8] Agarkhed, J., Dattatraya, P.Y. and Patil, S., "Multi-QoS constraint multipath routing in cluster-based wireless sensor network," *International Journal of Information Technology*, vol. 13, pp. 1–12, 2020.

[9] Augustine, S. and Ananth, J.P., "A modified rider optimization algorithm for multihop routing in WSN," *International Journal of Numerical Modelling: Electronic Networks, Devices and Fields*, vol. 33, no. 6, p. 2764, 2020.

[10] Rajashanthi, M. and Valarmathi, K., "Energy-efficient multipath routing in networking aid of clustering with OGFSO algorithm," *Soft Computing*, vol. 24, no. 17, pp. 12845–12854, 2020.

[11] Moridi, E., Haghparast, M., Hosseinzadeh, M. and Jassbi, S.J., "Novel fault-tolerant clustering-based multipath algorithm (FTCM) for wireless sensor networks," *Telecommunication Systems*, vol. 74, no. 4, pp. 411–424, 2020.

[12] Mahajan, S., Malhotra, J. and Sharma, S., "An energy balanced QoS based CH selection strategy for WSN," *Egyptian Informatics Journal*, vol. 15, no. 3, pp. 189–199, 2014.

[13] Jadhav, A.N. and Gomathi, N., "DIGWO: hybridization of dragonfly algorithm with improved grey wolf optimization algorithm for data clustering," *Multimedia Research*, vol. 2, no. 3, pp. 1–11, 2019.

[14] Deepa, O. and Suguna, J., "An optimized QoS-based clustering with multipath routing protocol for Wireless Sensor Networks," *Journal of King Saud University - Computer and Information Sciences*, vol. 32, 2017.

[15] Abazeed, M., Faisal, N. and Ali, A., "Cross-layer multipath routing scheme for wireless multimedia sensor network," *Wireless Networks*, pp. 4887–4901, 2018.

[16] Shen, J., Wang, A., Wang, C., Hung, P.C. and Lai, C.F., "An efficient centroid-based routing protocol for energy management in WSN-assisted IoT," *IEEE Access*, vol. 5, pp. 18469–18479, 2017.

[17] Robinson, Y.H., Julie, E.G. and Kumar, R., "Probability-based CH selection and fuzzy multipath routing for prolonging lifetime of wireless sensor networks," *Peer-to-Peer Networking and Applications*, vol. 12, no. 5, pp. 1061–1075, 2019.

[18] Sarkar, A. and Murugan, T.S., "CH selection for energy efficient and delay-less routing in wireless sensor network," *Wireless Networks*, vol. 25, no. 1, pp. 303–320, 2019.

[19] Bouaziz, M. and Rachedi, A., "A survey on mobility management protocols in Wireless Sensor Networks based on 6LoWPAN technology," *Computer Communications*, vol. 74, pp. 3–15, January 2016.

[20] Chouhan, N. and Jain, S.C., "Tunicate swarm grey wolf optimization for multi-path routing protocol in IoT assisted WSN networks," *Journal of Ambient Intelligence and Humanized Computing*, vol. 11, pp. 1–17, November 2020.

[21] Balachandra, M., Prema, K.V. and Makkithaya, K., "Multiconstrained and multipath QoS aware routing protocol for MANETs," *Wireless Networks*, vol. 20, no. 8, pp. 2395–2408, 2014.

[22] Chouhan, N., "Artificial intelligence-based energy-efficient clustering and routing in IoT-assisted wireless sensor network," In *Artificial Intelligence for Renewable Energy Systems*, John Wiley & Sons, Inc., pp. 79–91, February 2022.

[23] Ahmed, G., Zou, J., Fareed, M.M.S. and Zeeshan, M., "Sleep-awake energy efficient distributed clustering algorithm for wireless sensor networks," *Computers & Electrical Engineering*, vol. 56, pp. 385–398, 2016.

[24] Vjay Birchha, G Pradeepa, Madanachitran R, Nitesh Chouhan, G. Jiji, Y. Ramakrishna, P Kiran Kumar, "Machine Learning-Based Approach For Smart Agricultural Manlevelment", *European Chemical Bulletin*, 2023, 12 (S3), 661–668

[25] Dhumane, A.V. and Prasad, R.S., "Multi-objective fractional gravitational search algorithm for energy efficient routing in IoT," *Wireless Networks*, vol. 25, 2017.

[26] Chouhan, N., Saini, H.K. and Jain, S.C., "Internet of things: illuminating and study of protection and justifying potential countermeasures," *Soft Computing and Signal Processing: Proceedings of ICSCSP 2018*, vol. 2, pp. 21–27, April 2019.

[27] Askari, Q., Younas, I. and Saeed, M., "Political Optimizer: a novel socio-inspired meta-heuristic for globaloptimization," *Knowledge-Based Systems*, p. 105709, 2020.

[28] Jhajharia, A., Kumari, U., Chouhan, N. and Meena, Y., "Fault detection in power distribution," *Recent Advances in Electrical & Electronic Engineering*, vol. 14, no. 3, pp. 304–311, May 2021.

[29] Alamelu Mangai, S., Ravi Sankar, B. and Alagarsamy, K., "Taylor series prediction of time series data with error propagated by artificial neural network," *International Journal of Computer Applications*, vol. 89, no. 1, March 2014.

[30] Chouhan, N. and Jain, S.C., "An energy-efficient hybrid hierarchical clustering algorithm for wireless sensor devices in IoT," *Advances in Computing and Data Sciences*, pp. 1–14, April 2021.

[31] Javaid, S., Abdullah, M., Javaid, N., Sultana, T., Ahmed, J. and Sattar, N.A., "Towards buildings energy management: using seasonal schedules under time of use pricingtariff via deep neuro-fuzzy optimizer," In *Proceedings of 15th International Wireless Communications and Mobile Computing Conference (IWCMC)*, IEEE, pp. 1594–1599, 2019.

[32] Chouhan, N. et al., "A novel machine learning unmanned swarm intelligence based data security in defence system," *IEEE Xplore*, 2023, https://ieeexplore.ieee.org/document/10199709.

[33] Kaur, S., Awasthi, L.K., Sangal, A.L. and Dhiman, G., "Tunicate swarm algorithm: a new bio-inspired based metaheuristic paradigm for global optimization," *Engineering Applications of Artificial Intelligence*, vol. 90, p. 103541, 2020.

[34] Jarapunphol, P. and Chouhan, N., "Exploration of the impact of cybersecurity awareness on Small and Medium Enterprises (SMEs) in Wales using intelligent software to combat cybercrime," *European Chemical Bulletin*, vol. 12, pp. 289–302, 2023.

[35] Mirjalili, S., Mirjalili, S.M. and Lewis, A., "Grey wolf optimizer," *Advances in Engineering Software*, vol. 69, pp. 46–61, 2014.

Part III

Applications

Chapter 14

6G wireless networks for V2X communication

Challenges and opportunities

*Dhivya Priya E. L., K. Kavitha, Sharmila A.,
Suresh Chinnathampy M., and D. Nageswari*

14.1 INTRODUCTION

Recent growth of technologies has improved research on connected and autonomous vehicles, which has given the way for the design for Vehicle to everything (V2X) communication [1]. V2X is the communication between vehicles, vehicle gadgets, infrastructure, pedestrians and also with applications such as traffic management, collision avoidance and nested driving. These advantages are made possible with the design of enhanced wireless network technology. Although 5G was used in V2X communication, there are many challenges faced with the widespread network coverage and has limitations with their effectiveness. Limited coverage, achievement of high latency for applications and insufficient reliability are major challenges in the 5G V2X. Thus, a great demand for a technology that has the ability to face all these challenges is expected. And thus 6G wireless networks help to satisfy these challenges and support new possibilities. 6G V2X has more capabilities than its predecessors [2]. The network with 6G can offer ultra-high data rates, supporting for large data transfer which has been collected from different vehicles and vehicle-related gadgets. Low latency-based time-sensitive V2X applications demand 6G networks, which have the ability to satisfy the expectations. With more additional benefits than the previous technologies, 6G supports V2X communication with other vehicles, infrastructure elements, vehicle gadgets, pedestrians and other infrastructure, by facilitating cooperative traffic management, reliability and uninterrupted services without the concern for geographic regions. This chapter represents the overview of tasks, challenges and future opportunities with the utilization of 6G networks for V2X applications. It also discusses the key features and advantages of 6G V2X. The enhanced data rates with 6G enables exchange of large volumes of data, video streaming at high definition and enhancing awareness in critical situations, which helps building advanced applications such as intelligent transportation system (ITS) with decision-making at critical applications. Ultra-low latency-based system is

DOI: 10.1201/9781003522003-17

demanded for building real-time communication where vehicles can respond quickly for the changing traffic conditions and other critical situations. 6G offers seamless and simultaneous communication for massive connectivity, supporting cooperative driving scenarios and overall safety improved effective traffic scenario. Additionally, the ubiquitous coverage of 6G networks ensures reliable connectivity in both urban and remote areas, enabling continuous V2X communication and improving the overall safety and reliability of autonomous vehicles. However, the realization of the full potential of 6G networks for V2X communication requires overcoming several technical challenges. Designing efficient communication protocols that can handle the massive data traffic generated by connected vehicles, ensuring network security and privacy to protect sensitive V2X information, and effectively managing network resources to provide optimal performance are among the key challenges that need to be addressed. Furthermore, standardization and interoperability are crucial to enable seamless communication and collaboration among different V2X systems and devices. The integration of emerging technologies such as machine learning, edge computing and quantum communications into the development of 6G networks holds immense potential for further enhancing V2X communication capabilities. These technologies can facilitate intelligent data processing, enable distributed computing at the network edge for faster response times and provide secure and resilient communication channels. The challenges and opportunities with 6G V2X provides scope for researchers to build technology-enhanced networks for vehicular environments. This will create a revolution in the transportation by enabling a different era of intelligent transportation system [3–6].

14.2 OVERVIEW OF 6G NETWORKS

6G is the key for designing next generation wireless communication system that has the ability to revolutionize transportation and has the potential to overcome the limitations of previous generations. The key feature of 6G networks is ultra-high data rates which support data transmission at improved data rates. This enhanced data rate supports for transmission of large volumes of data that is collected in the vehicular and their environment. The data collected includes high-defined video streams, sensor data and other sensitive information needed for autonomous driving and building ITS. The ability to transmit and process larger data in real time enhances situational awareness, enables advanced analytics and supports intelligent decision-making.

Ultra-low latency is another important aspect of 6G network. Latency is the delay in transmission of data from one device to another and in V2X, lowest latency makes the system efficient. The lower the latency, the higher the reaction to the critical situation. 6G supports the lower latency in the order of microseconds and supports for real-time communication between vehicles, vehicle gadgets, infrastructure and pedestrians. The data transmission with such low-latency supports time critical applications with collision avoidance and emergency vehicle coordination. It also enhances the effectiveness and responsiveness of V2X communication, contributing improved safety and efficiency of transportation. Another characteristic of 6G network is massive connectivity. These networks are designed for significantly a large number of connected devices. This specific feature of 6G makes V2X communication seamless among multitude of vehicles, infrastructure elements and pedestrians. This capability supports for cooperative driving scenarios, which optimize traffic flow, enhance safety and reduce congestion. These factors support building the V2X environment in an effective manner [7–8].

In addition to the previous features mentioned, 6G aims to achieve improved coverage and has special features as represented in Figure 14.1. These special features make them provide reliable and uninterrupted

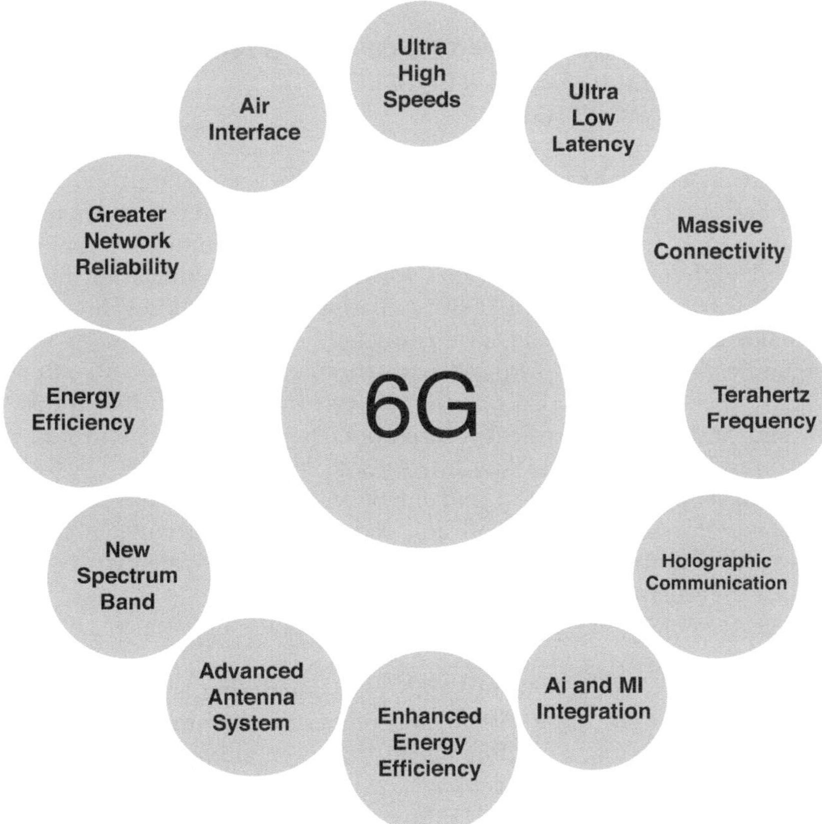

Figure 14.1 Key factors of 6G.

connectivity across various locations such as urban areas, highways and rural locations. Omni-directional coverage with the 6G V2X supports the vehicle to have continuous communication, irrespective of the location. It also supports for seamless connectivity for building efficient ITS. The intelligent system is the one that has the ability to justify for safety and reliability of autonomous vehicles, with the ability to navigate and communicate efficiently in all critical environments. 6G networks offers a host of features inclusive of ultra-high data rate, low latency, larger connectivity and Omni-directional coverage, that promise support for V2X communication, enabling advanced applications, improved safety and traffic efficiency [9]. To overcome the limitations of the already existing communication generations, 6G networks support the way by which the previous generations failed to handle things, inclusive of better communication, cooperation and interaction with their surroundings. Thus, it is analyzed 6G creates a great impact on building Intelligent Transport System with V2X, enabling realization for safety, efficiency and Omni-directional coverage [10].

14.3 NEED AND ADVANTAGES OF OF 6G NETWORKS FOR AUTONOMOUS VEHICLE COMMUNICATION

There are enormous benefits of autonomous vehicle communication, inclusive of Safety, traffic efficiency, energy consumption and others, mentioned in Figure 14.2. For enabling seamless and reliable communication, 6G is modeled with advanced capabilities for offering communication between vehicles, infrastructure and pedestrians which supports for ITS. Safety is the major concern to be considered for the design of autonomous vehicle communication which is satisfied with the incorporation of 6G communication Technology. It also has the ability to enable real-time transmission and processing of critical information, facilitating fast and accurate decision-making with ultra-low latency and high data rates. The vehicle data such as positions, speed and intentions will be exchanged between the vehicles for avoidance of collision and cooperative driving. A concrete robust and reliable communication infrastructure is provided with 6G networks, for ensuring safety of the vehicles on road.

Integration of emerging technologies into autonomous vehicle communication will enhance the 6G network-based Vehicle environment. The recent technologies such as machine learning, edge computing, quantum communications can be used along with 6G techniques for the enhancement of ITS. Based on the real-time data, Intelligent decision-making is implemented at the network edge with the machine learning algorithms and for improved security and privacy for V2V/V2I infrastructure, quantum communications are used.

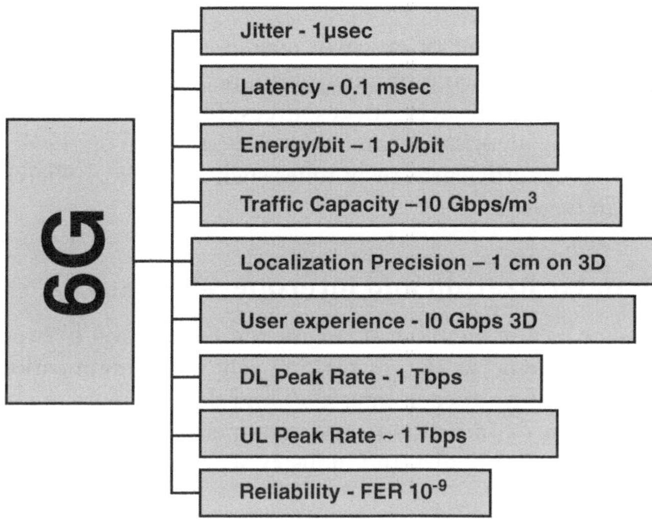

Figure 14.2 Benefits of 6G.

14.4 TECHNICAL CHALLENGES AND SOLUTIONS

Though 6G networks have several benefits, the implementation of 6G networks has many technical challenges, which are to addressed fully for the successful implementation of V2X with 6G. This section elaborates on the key challenges and provides potential solutions to overcome them.

14.4.1 Designing efficient communication protocols

A robust and efficient communication protocol is required for providing ultra-high data rates and massive connectivity in 6G V2X. Such protocols must provide low latency, high reliability and seamless connectivity for any kind of ITS-based dynamic vehicle environments. The communication techniques such as beam forming, MIMO and network slicing have to be optimized based on the requirements, to offer improved performance in the vehicular environments.

14.4.2 Ensuring network security and privacy

The wireless communication reliance increases the exposure of vehicle data and thus increases the security threat. The protection of collected data from unauthorized access is important, but finds difficult. Systems with improved secure authentication, encryption and intrusion detection are demanded to improve the integrity and privacy of V2V and V2I communication.

14.4.3 Managing network resources effectively

The effective management of network resources is very important, in the condition of rapidly growing number of connected devices. To ensure fair and efficient utilization of network resources, the dynamic resource allocation and scheduling algorithms can be developed, based on the demand. For allocating resources efficiently, techniques such as network slicing and edge computing can be used.

14.4.4 Standardization and interoperability

It plays a major role in enabling seamless communication between all connected devices in the network. For making vehicle communication efficient, the establishment of protocols, interfaces and data formats in common are demanded. To ensure compatibility and interoperability, collaborative efforts among standardization bodies and industry stakeholders are important.

14.4.5 Integration of emerging technologies

To satisfy the expectations of vehicular environment, integration of emerging technologies such as machine learning, edge computing and quantum communication is found important. This technology integration helps in the enhancement of autonomous vehicle communication. But there are many technical challenges to be addressed during the integration of 6G with V2X. The development of algorithms for real-time data processing, edge computing and security enhancement provides a wider range of research scope, which reveals the full potential of 6G network. This research scope will also support for the development of robust and efficient V2X Communication systems for building ITS.

14.5 STANDARDIZATION AND INTEROPERABILITY

This section holds the importance of assuring seamless connectivity between devices in the vehicle network. The sub-categories will be discussed in this section.

14.5.1 Common protocols and interfaces

The Transportation system being a larger interconnected network, standardization establishes common protocols and interfaces for providing better communication between the nodes of the network. Different manufacturers and service providers are able to develop compatible systems that can interoperate and exchange information reliably through this mechanism. For enabling improved safety and efficiency, establishment of common protocols is recommended.

14.5.2 Compatibility and interoperability

Irrespective of the underlying technology used by the manufactures, interoperability is the most important one, for making the devices/nodes in the network to work together. This will make the system efficient by providing better collaboration between vehicles and other traffic management systems/infrastructure. This feature will provide access and utilization to different users and vehicles for an interconnected and ITS.

14.5.3 Seamless integration and deployment

The integrated technology access with improved standardization and interoperability enhances the view of implementing ITS with V2X efficiently. This integration demands a proper framework for the smooth integration of recent technologies such as machine learning, edge computing and quantum communications. This also supports for most efficient and cost-effective deployment of new services.

14.5.4 Regulatory compliance and market adoption

The most important part for seamless V2X communication is the establishment of standard regulatory frameworks and guidelines. These standards ensure better performance of V2I infrastructure, in terms of safety, security and confidence in data sharing across vehicle network. The regulatory standards improve the industry market by providing common foundation and offer interoperable solutions.

Despite the advantages of standardization and interoperability, there are several challenges which are discussed below.

14.5.5 Complexity

A common standard of operation for a wider network such as vehicle network is very complex as it involves various technologies, applications and stakeholders. The complexities are with the emerging technologies, which leads to every time updating of the network, which demands support from industry, academia and standardization bodies. Building a network that supports all the requirements mentioned, will improve the security, privacy, and scalability of the network.

14.5.6 Evolving landscape

With the help of evolving new technologies, landscape of V2X communication is continuously emerging. These advancements must be considered during standardization, making sure the standards are relevant, up-to-date, and adaptable to developments in future.

14.5.7 Stakeholder cooperation

Cooperation and coordination between the stakeholders is very important which supports interoperability. The stakeholders include vehicle manufactures, infrastructure providers, network operators and regulatory bodies. For successful standardization and interoperability, the balance between the technical requirements, business models and competing interest is found mandatory.

14.5.8 Global alignment

To have a global alignment, standardization efforts and interoperability is considered to enable seamless cross-border communication. The efficient way of coordination with the coordinating standards offers consistent implementation and avoids fragmentation in V2X communication. The challenges to be addressed with ongoing collaboration, research and innovation. The combined work of Industry stakeholders, standardization bodies and policy makers supports for robust standards which help to promote the interoperability, compatibility and secure communication in the evolving landscape of 6G networks in building efficient ITS.

14.6 EMERGING TECHNOLOGIES IN 6G NETWORKS

For autonomous vehicle communication to have well-defined capabilities and potential of 6G network, emerging technologies play a major role. This section elaborates on the key emerging technologies and highlights the impact of 6G networks.

14.6.1 Machine learning and artificial intelligence (AI)

To enable intelligent decision-making and optimization in autonomous vehicle design, incorporation of Machine learning and AI algorithms is found more supportive. These technologies can be applied to various other areas, inclusive of resource allocation, prediction of traffic, anomaly detection and intelligent routing. 6G networks will get the ability to face various challenges with the incorporation of machine learning and AI which enhances the overall performance of V2X communication systems.

14.6.2 Edge computing

One other challenging technique among the recent technologies is Edge computing, which makes computation and data storage easier by enabling low latency and real-time processing. It also enhances the performance of

V2X communication by reducing network latency and improve response time. The incorporation of edge computing with 6G ITS supports for faster decision-making and supports for time-critical applications such as collision avoidance and traffic management.

14.6.3 Quantum communications

With the help of principles of quantum mechanics, Quantum communication offers secure and high-speed data transmission. Quantum Key Distribution (QKD) provides most relevant encryption keys, ensuring confidentiality and integrity in vehicular environment. Thus, this technique enhances security of vehicles, protecting against cyber threats and unauthorized access to sensitive information.

14.6.4 Massive internet of things

Integration of multiple devices and providence of connectivity across them is the key demand of the proposed ITS V2X system. To make the system efficient, incorporation of IoT is demanded high. With the help of IoT devices and traffic infrastructure, real-time data about the vehicular environment can be collected, accessed and processed whenever required. The data can also be involved in advanced analytics, intelligent decision making and proactive safety measures in ITS design.

14.6.5 Software-defined networking (SDN) and network function virtualization (NFV)

Softward-defined networking (SDN) and network function virtualization (NFV) make the traffic network flexible and agile for network management and services in 6G network. SDN offers centralized network management and programmability by separating the control plane from data plane. NFV enables the virtualization of network functions, allowing dynamic allocation and scaling of network resources. This integration offers efficient network performance, rapid service deployment and customization of network services for V2X communication.

14.6.6 Blockchain technology

Blockchain provides a platform for transparent transactions in a decentralized way which offers secured data transmission for 6G V2X. Thus blockchain improves the integrity and trustworthiness for V2X data communication. 6G V2X with the integration of recent technologies creates a revolution with the capabilities of the proposed network. It helps to address various challenges, enhances performance and offer new services which intend to create various applications for building safe and efficient transportation systems. However, their implementation requires careful consideration of technical, operational, and regulatory aspects to ensure compatibility, scalability, and security in the evolving landscape of 6G networks [11].

14.7 CONCLUSION

The advent of 6G wireless networks for Vehicle-to-everything (V2X) communication brings both challenges and opportunities in the realm of connected and autonomous vehicles. While 5G networks have limitations in coverage, latency, and reliability, 6G networks offer transformative capabilities for V2X communication. With ultra-high data rates, ultra-low latency, massive connectivity, and ubiquitous coverage, 6G networks enable enhanced safety, traffic efficiency, and reduced energy consumption in autonomous vehicle communication. Real-time transmission and processing of critical information support fast decision-making for collision avoidance and cooperative driving. The massive connectivity facilitates seamless communication among vehicles, infrastructure, and pedestrians, improving traffic flow and reducing congestion. Ubiquitous coverage ensures reliable V2X communication across diverse regions. However, addressing technical challenges like communication protocols, security, resource management, and standardization is crucial to fully realize 6G network potential. Integration of machine learning, edge computing, and quantum communications further enhances 6G networks for autonomous vehicle communication. 6G networks can revolutionize transportation by enabling intelligent and safe V2X communication. Overcoming challenges and leveraging opportunities presented by 6G networks will drive the development of robust V2X communication systems, contributing to safer, more efficient transportation systems.

REFERENCES

1. Bazzi et al., "Toward 6G Vehicle-to-Everything Sidelink: Nonorthogonal Multiple Access in the Autonomous Mode," in *IEEE Vehicular Technology Magazine*, vol. 18, no. 2, pp. 50–59, June 2023, doi: 10.1109/MVT.2023.3252278.
2. J. Gallego-Madrid, R. Sanchez-Iborra, J. Ortiz and J. Santa, "The Role of Vehicular Applications in the Design of Future 6G Infrastructures," in *ICT Express*, 2023, doi: 10.1016/j.icte.2023.03.011.
3. H. Ouamna, Z. Madini and Y. Zouine, "6G and V2X Communications: Applications, Features, and Challenges," in *2022 8th International Conference on Optimization and Applications (ICOA)*, Genoa, Italy, 2022, pp. 1–6, doi: 10.1109/ICOA55659.2022.9934407.

4. Z. Qadir, K.N. Le, N. Saeed and H.S. Munawar, "Towards 6G Internet of Things: Recent Advances, Use Cases, and Open Challenges," in *ICT Express*, 2022, doi: 10.1016/j.icte.2022.06.006.

5. M. Noor-A-Rahim et al., "6G for Vehicle-to-Everything (V2X) Communications: Enabling Technologies, Challenges, and Opportunities," in *Proceedings of the IEEE*, vol. 110, no. 6, pp. 712–734, June 2022, doi: 10.1109/JPROC.2022.3173031.

6. A. Bazzi, A. O. Berthet, C. Campolo, B. M. Masini, A. Molinaro and A. Zanella, "On the Design of Sidelink for Cellular V2X: A Literature Review and Outlook for Future," in *IEEE Access*, vol. 9, pp. 97953–97980, 2021, doi: 10.1109/ACCESS.2021.3094161.

7. W. Saad, M. Bennis and M. Chen, "A Vision of 6G Wireless Systems: Applications, Trends, Technologies, and Open Research Problems," in *IEEE Network*, vol. 34, no. 3, pp. 134–142, May/June 2020, doi: 10.1109/MNET.001.1900287.

8. O. Sadio, I. Ngom and C. Lishou, "Controlling WiFi Direct Group Formation for Non-critical Applications in C-V2X Network," in *IEEE Access*, vol. 8, pp. 79947–79957, 2020, doi: 10.1109/ACCESS.2020.2990671.

9. W. Saad, M. Bennis and M. Chen, "A Vision of 6G Wireless Systems: Applications, Trends, Technologies, and Open Research Problems," in *IEEE Network*, vol. 34, no. 3, pp. 134–142, 2020, doi: 10.1109/MNET.001.1900287.

10. G. Fodor et al., "Supporting Enhanced Vehicle-to-Everything Services by LTE Release 15 Systems," in *IEEE Communications Standards Magazine*, vol. 3, no. 1, pp. 26–33, March 2019, doi: 10.1109/MCOMSTD.2019.1800049.

11. P. Yang, Y. Xiao, M. Xiao et al., "6G Wireless Communications: Vision and Potential Techniques," in *IEEE Network*, vol. 33, pp. 70–75, 2019.

Chapter 15

Massive IoT access through 6G

Rucha Patel and Krupa Purohit

15.1 INTRODUCTION

15.1.1 Securing massive IoT in 6G

The creation of smart services in a range of fields, especially transportation, healthcare, and smart cities, can be made feasible by an increasingly rapidly evolving technology known as the Internet of Things (IoT). This oversupply of IoT devices has the potential to make an effect on sixth-generation (6G) systems. For sixth generation infrastructure, providing sixth generation IoT network security over hazards, mainly inventive attackers, constitutes a huge challenge. Therefore, fresh, creative designs and paradigms that are made possible by knowledge, softwarisation, and network automation must be implemented immediately. This example relies on the durability of artificial intelligence along with key networking factors like software-driven networking (SDN), network functionality virtualisation (NFV), etc. to serve as elements of promising solutions and innovative ideas for securing the IoT [1].

Experts working in various fields of engineering and computer science aim to apply the use of machine learning and deep learning methods in order to categories, anticipate, and mitigate anomalies and anomalous traffic behaviors in order to increase the ability to recognise cyber hazards. Making an intrusion detection system (IDS) inside a viable infrastructure architecture designed to support massive IoT applications is challenging. We are aware that future ideas should be based on native integrated infrastructure, despite the fact that there are many different factors to take into account for an optimum intelligent solution [2].

Practically everything must have sensors if it is to be smart. Due to the employment of various machine learning techniques, digitisation, collecting information, and continuous performance improvements can be made possible which will continue to grow and expand. With 5G, the dependability, service quality, and latency have all significantly improved [3].

DOI: 10.1201/9781003522003-18

Table 15.1 [4–6] Comparison between 5G and 6G

Standards of measurement	5G	6G
Maximum amount of data transmitted	1–10 Gb/s	1–10 Gb/s 100 Gb/s to 1 Tb/s
Lagged by	10 ms	<ms level
Number of devices	10^9	10^{12}
Mobility	350–500 km/h	>1,000 km/h
Maximum frequency	90 GHz	10 THz
Haptic interaction	Limited	Completely supported
Long-term reality	Partial	Completely supported

15.2 TECHNOLOGY COMPARISON BETWEEN 5G AND 6G

Table 15.1 shows the 5G and 6G networks show notable improvements in data transmission standards. Both can transmit 1–10 Gb/s, but 6G offers a range of 100 Gb/s to 1 Tb/s and lower latency in the millisecond range compared to the 10 ms latency in 5G. In 6G, the number of supported devices rises from 10^9 to 10^{12}, and it can handle speeds above 1000 km/h, surpassing the 350–500 km/h limit of 5G. 6G expands the frequency range to 10 THz, a substantial increase over 5G's 90 GHz. 6G brings significant improvements in haptic contact and extended reality, providing comprehensive support for both, surpassing the restricted support of 5G.

Figure 15.1 displays a detailed summary of a study that explores the connection between 6G technology and the IoT, including visions, needs, core technologies, applications, and research problems. The aim for 6G in the world of IoT involves advancing technologies like Edge Intelligence, Blockchain, Terahertz Communication, and Space-Air-Ground Underwater Communication. The basic technologies are anticipated to back many applications such as Internet of Healthcare Things, Unmanned Aerial Vehicles, Industrial IoT, Satellite IoT, and Vehicular IoT & Autonomous Driving. The study highlights key research problems that need to be tackled, including Security & Privacy, Energy Efficiency, Hardware Constraints, and the need for Standard Specifications for 6G IoT to guarantee a dependable and effective implementation of these advanced technologies [7–8].

15.3 6G NETWORKS REQUIREMENTS

The sixth generation IoT standard must be able to meet more stringent requirements than its fifth generation IoT in order to truly achieve the intelligent information society of 2030, involving complete information, large

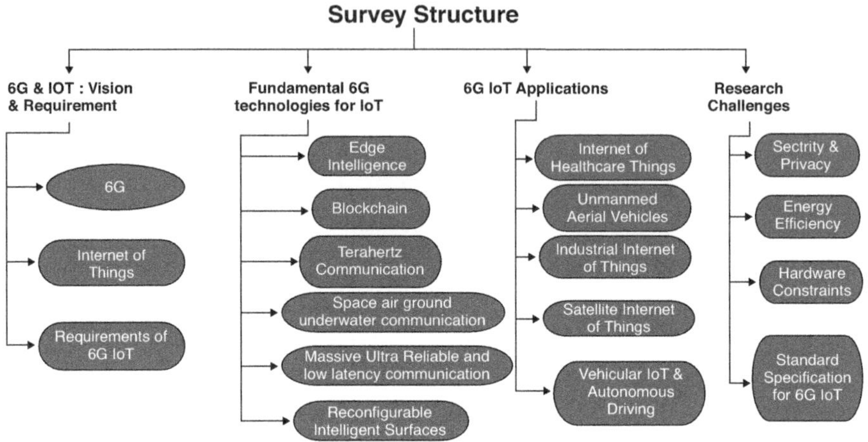

Figure 15.1 Technology comparison between 5G and 6G.

connected devices and coverage, services driven by data, and fully autonomous systems, as stated below.

15.3.1 Massive IoT connectivity

Mobile connection has been substantially increased thanks to the rise of intelligent gadgets along with the advancement of wireless communication technology. According to forecasts, global mobile traffic will increase by more than double by 2030, reaching 5,000 EB, or 80 times more than it did the year before. Large-scale communications for upcoming smart devices will be provided by broadband access platforms and low-earth orbit (LEO) satellite networks. Flying platforms like UAVs must be deployed in the future to provide smooth connection over huge-scale IoT networks since stable ground stations are insufficient to ensure constant and reliable device connections in moving IoT networks.

For example, an autonomous Unmanned Aerial Vehicles (UAV) installation concept for 6G networks is displayed, in which 33 UAVs are used to wirelessly connect 400 grounded IoT users within a 2,000 m×2,000 m region. Given the UE distribution along with the dis-continuous unmanned aerial vehicle position space, a distributed motion approach is created that enables each Unmanned Aerial Vehicles (UAV) to independently determine the ideal place in a continuous IoT area. According to simulations the use of a group of UAVs can offer a distributed deployment option with a maximum workload ratio close to 1 and a distribution time that is as much as 40% faster than the organised technique.

15.3.2 Numerous low-latency ultra-reliable IoT communications

Though Ultra Reliable and Low Latency Communication were previously created and put to use in real-world fifth generation IoT use cases this still has to be updated in order to facilitate impending 6G-IoT network applications like completely self-sufficient IoT and waving IoT devices. As an example, in future self-sufficient modes of transport, where vehicles can be controlled and guided instantaneously, the huge Ultra Reliable and Low Latency Communication is essential for sending video feeds from images to vehicle and arranging precise automotive communication on the highways in an integrated and safe manner.

In the coming intelligent networked societies will place a high value on timely information transmission. The tactile internet will predominate and provide real-time touch and actuation for mission critical IoT applications. The work uses the construction of a limited-range radio isochronous actual-time in-X subnetwork with transmission cycles smaller to a value of 0.1 ms and a failure probability lower than 106 to execute a huge URLLC-based IoT simulation for 6G IoT networks. When employing a multi-Giga Hertz spectrum to provide outstanding spatial service coverage, an extensive IoT situation with a maximum of two devices per square metre.

The times between cycles are discovered to be ten times lesser compared to the delay requirements of fifth generation wireless techniques (i.e., <0.1 ms) through semi-analytical system evaluation analysis, which may be used afterwards for extremely low delay IoT applications like self-sufficient driving, actual-time health monitoring, and industrial automation [9].

15.3.3 Enhanced communication protocols for 6G IoT

Future intelligent networks will need to make major architectural changes to enable the introduction of new vertical IoT applications to be able to cope with a number of demanding requirements (such as e-healthcare and self-driving vehicles, for example). The widespread adoption of 6G-IoT ecosystems depends on the network communication standards and protocols because of how they interact with a number of crucial computing resources like edge cloud computing and wireless networking technologies.

In fact, ETSI Multi-access Edge Computing initiative, which seeks to use an edge computing as well as communication system for IoT-connected services developed by suppliers, designers, and other parties, was just published by the Industry Specification Group of the European Telecommunications Standards Institute. The implementation of upcoming IoT applications at the outer edge of the sixth generation network, where it is anticipated that IoT processing and storage would be relocated from the network centre to the network edge, may benefit significantly from this initiative. A new Wi-Fi standard termed IEEE 802.11be Very High Throughput, which can match the peak throughput demands imposed by impending IoT applications in the sixth generation future, has recently been explored by the IEEE 802.11 group of experts.

These recommendations for communication should facilitate the efficient implementation of intelligent IoT services at the network edge by service providers.

15.3.4 Expansion of IoT network coverage

In IoT-based society of the future, it is desirable to provide complete coverage outside of the land-based networks by employing large-dimensional space-air-ground-underwater networks. Edge intelligence and unmanned aerial vehicles, which will be used to build waving ground stations in order to broaden the area of coverage currently available cellular systems in merely two-dimensional in continuing physical networks to three-dimensional in an integrated terrestrial-satellite-aerial system, will be the keys that are for accomplishing full wireless coverage.

High-altitude unmanned aerial vehicles, for instance, can be employed as adaptable flying vehicles to provide marine coverage as needed. This method of setting up terrestrial shore-based stations has promise for making the establishment of the communication system for vessels at sea simpler. For the avoidance of doubt, the distribution of vessels at an offshore site between [20 and 30] km is determined by means of a helpful automatic identification system. To provide long-term internet services and instant services of dispersed vessels, Unmanned Aerial Vehicles are set up to sail alongside vessels. Here, a 370 km stretch of coastline is flown over by an oil-powered fixed-wing unmanned aerial vehicles that has a total range of 740 km.

15.3.5 Next-generation smart IoT devices

According to predictions, upcoming sixth generation IoT networks will depend on intelligent gadgets where artificial intelligence and processing may be completely implemented on the devices, such as smartphones, automobiles, machines, and robots. Such a device-centric network addresses new problems and demands for its mobile connectivity because intelligent gadgets will not only provide or consume information but also administer and operate a network. Communication between devices and multi-hop mobile networks, both of which have been included in the 3GPP framework, could be examples of device-centric wireless solutions. By connecting to other devices at the network edge and providing services (such as intelligent control, caching, and network signaling) without the need for a centralised controller, each IoT electronic device can function as an end-consumer interface in the 6G future.

This could be further developed into demand-driven opportunistic networking, which is designed to meet certain consumer, service, or network requirements, such as lowering energy costs or raising spectrum efficiency for end-user devices. Additionally, wearable technology has rapidly advanced in recent years, gradually displacing smartphones as a key component in the creation of the 4G and 5G networks.

15.4 TAXONOMY OF 6G-IOT APPLICATIONS

Table 15.2 details numerous applications and technology suggested for 6G-based systems in various sectors. A 6G-based HIoT system in healthcare utilises ultra-reliable low-latency communications (mURLLC) and terahertz (THz) communications for remote health monitoring. It also includes virtual operation, UAVs, and blockchain for secure data transmission and mobility management. An additional application, 6G-based VIoT (Vehicle IoT), aims to facilitate connectivity for vehicle-to-everything (V2X) communication. It utilises massive machine type communications (mMTC) and deep learning (DL) for network estimation, vehicular communication modelling, and intelligence to aid autonomous driving. 6G-based UAV systems

Table 15.2 Taxonomy of 6G-IoT applications

Applications	Case study	Use of 6G technology
6G-based HIoT	System for remotely monitoring health	mURLLC and THz communications
	Virtual operation	UAVs and blockchain
	System for healthcare data communications	mURLLC
	Connected ambulance	URLLC
	Optimisation of resource allocation and health data analytics	ML
	Mobility management	ML
	Analysis of COVID-19 data	Edge intelligence
6G-based VIoT	Enabling connectivity for V2X	mMTC
	Network estimation for vehicles	ML
	Vehicular communication modelling	DL
	Vehicular intelligence	DL
	Autonomous driving	DL
	Autonomous driving	DL

(Continued)

Table 15.2 (Continued) Taxonomy of 6G-IoT applications

Applications	Case study	Use of 6G technology
6G-based UAV	Process-focused optimisation for UAV flight	UAV
	Wireless communications driven by UAV	UAV
	Optimisation of multi-UAV trajectories and resource allocation	UAV
	Intelligent UAV platforms	ML
	Federated UAV communication	FL
6G-based SIoT	IoT data collecting through satellite	UAV
	Satellite communications based on RIS	RIS
	Energy awareness mass access	Satellite communications
	Sharing of satellite spectrum	Satellite communications
6G-based IIoT	Optimal distribution of industrial resources	ML
	Transmission latency minimisation	ML
	Industrial data aggregation that is secure	Blockchain
	IIoT security improvement	Blockchain and FL

focus on improving UAV flight procedures and wireless communications. Machine learning methods enhance their intelligence and enable federated communication among UAVs. Within satellite communications (SIoT), 6G introduces IoT data gathering via satellites, satellite communications using reconfigurable intelligent surfaces (RIS), and effective spectrum sharing for widespread access. 6G prioritises efficient resource allocation and reducing transmission delays in industrial applications (IIoT) by using machine learning algorithms. It also focuses on maintaining data security during aggregation and transmission by incorporating blockchain technology and federated learning (FL) methods. The suggestions showcase the wide range of uses and possibilities of 6G technology in numerous fields, using modern communication and processing methods to tackle particular issues and improve effectiveness, safety, and intelligence in several industries.

REFERENCES

[1] Zakria Qadira, Khoa N. Lea, Nasir Saeed, and Hafiz Suliman Munawar, Towards 6G Internet of Things: Recent Advances, Use Cases, and Open Challenges, *ScienceDirect*, 9 June, 9(3), 296–312 (2022).

[2] Dinh C. Nguyen, Ming Ding, Pubudu N. Pathirana, Aruna Seneviratne, Jun Li, Dusit Niyato, Octavia Dobre, and H. Vincent Poor, 6G Internet of Things: A Comprehensive Survey, *IEEE*, 11 Aug, 9(1), 354–382 (2021).

[3] Fengxian Guo, F. Richard Yu, Heli Zhang, Xi Li, Hong Ji, and Victor C.M. Leung, Enabling Massive IoT toward 6G: A Comprehensive Survey, *IEEE*, May, 8(15), 11891–11915 (2021).

[4] Tian Feiyan, Chen Xiaoming, Zhong Caijun, and Zhang Zhaoyang, Massive Access Technology in 6G Cellular Internet of Things Network, *Chinese Journal on Internet of Things*, March, 4(1), 92–103 (2020).

[5] Hewon Choa, Sudarshan Mukherjeeb, Dongsun Kima, Taegyun Nohc, and Jemin Leed, Facing to Wireless Network Densification in 6G: Challenges and Opportunities, *ScienceDirect*, May, 9(3), 517–524 (2022).

[6] Yun Chen, Wenfeng Liu, Zhiang Niu, Zhongxiu Feng, Qiwei Hu, and Tao Jiang, Pervasive Intelligent Endogenous 6G Wireless Systems: Prospects, Theories and Key Technologies, *Digital Communications and Networks*, 6(3), 312–320 (2020).

[7] Chamith De Alwisa, Pardeep Kumar, Quoc-Viet Pham, Kapal Dev, Anshuman Kalla, Madhusanka Liyanage, and Won-Joo Hwang, Towards 6G: Key Technological Directions, *ScienceDirect*, 9(4), 525–533 (2022).

[8] Sooeon Lee, Seungheyon Lee, Yumin Choi, and Hyunbum Kim, Trustworthy Clash-Free Surveillance Using Virtual Emotion Detection in 6G-Assisted Graded Districts, *ScienceDirect*, 9(4), 754–760 (2022).

[9] Sushant Kumar Pattnaik, Soumya Ranjan Samal, Shuvabrata Bandopadhaya, Kaliprasanna Swain, Subhashree Choudhury, Jitendra Kumar Das, Albena Mihovska, and Vladimir Poulkov, Future Wireless Communication Technology towards 6G IoT: An Application-Based Analysis of IoT in Real-Time Location Monitoring of Employees Inside Underground Mines by Using BLE, *MDPI*, April, 22(9), 1–30 (2022).

Chapter 16

Dynamic test technique of analog to digital converter (ADC) for 6G communication

Radheshyam Gamad

16.1 INTRODUCTION

Section 16.2 of this chapter discusses briefly about these techniques.

ADC can be classified based on conversion speed such as slow speed ADC, medium speed ADC, high speed ADC and ultra-high speed ADC (Figure 16.1). Slow speed type ADCs are dual slope converters, and their conversion time ranges from few millisecond to few seconds. These ADCs are suitable for applications where input is DC or slowly varying. Successive approximation type ADCs are medium speed ADCs with conversion time ranging from few hundred nanoseconds to few hundred microseconds. These ADCs are more suitable for digital electronics and microprocessor-based system. Fastest 12-bit ADC in dual-channel mode, the ADC12DJ5200RF samples at 5.2 giga samples per second (GSPS) and captures instantaneous bandwidth (IBW) as high as 2.6 GHz at 12-bit resolution. In single-channel

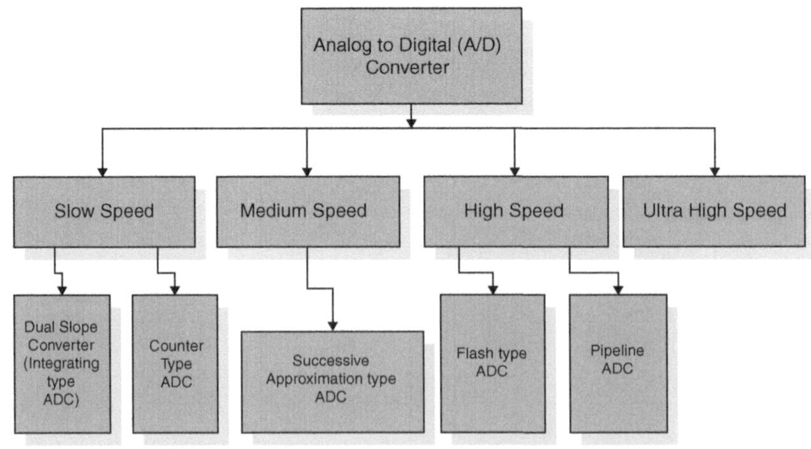

Figure 16.1 Classification of an ADC.

DOI: 10.1201/9781003522003-19

mode, the new ultra-high-speed ADC samples at 10.4 GSPS and captures IBW up to 5.2 GHz. High speed ADCs are comparator type also known as flash type with conversion time ranging from few nanosecond to few hundred nanosecond. Design of this ADC is based on comparator, so limitations are imposed due to design of comparators. With the advancement of technology, design of ADC is done to realize conversion time less than nanosecond particularly using bipolar technology.

16.2 DYNAMIC TEST METHODS OF AN ADC

For dynamic testing of an ADC, our literature survey indicates that following techniques exist like: Histogram technique, Sine wave curve fitting technique and Fast Fourier Transform Technique are mainly used for dynamic characterization of an ADC using analog as test signal [1]. Histogram technique alone can be used to determine Gain Error, Offset Error, DNL, INL and ENOB while sine wave curve fitting technique provides ENOB and FFT technique gives INL, SNR and ENOB.

16.2.1 Histogram Technique

It is employed to determine transfer characteristics of an under test ADC DNL, INL and ENOB of ADC under test. A full scale analog input is applied to ADC, and for increasing accuracy, a number of samples are taken at suitable sampling frequency and fed to the PC. Histogram is generated in frequencies vs ADC code then taken fixed number of events of every ADC code and that depends on input frequency, sampling frequency and number of samples. Here you will see if there is practical ADC, the number of events is less than ideal ADC and code bin width will be large of the practical ADC. In case of a practical ADC, missing codes are indicated when corresponding frequency of occurrence is zero [2–5]. Now by using cumulative histogram, code transition levels can be obtained, gain and offset of an ADC transfer functions are computed by fitting code transition levels in best fit straight line. DNL is determined by computing actual step size from code transition levels and its deviation from ideal step size of 1 LSB. INL is calculated as residual error after including the effects of gain and offset. ENOB is computed by estimating average quantization power of actual error at ADC.

16.2.2 Sine wave Curve Fitting Technique

It is used for obtaining of EB of ADC. Nonlinear equations are generated by equating partial derivation of sum of squared error (between actual data and best fit sine wave) with respect to amplitude, frequency, phase and DC offset of input sine wave, to zero. This algorithm of determination of best fit sine wave is very complex, and it requires initial guess of frequencies, and phase of signal for quick conversions [6, 7]. Ideal rms error is determined between data generated by digitizing best fit sine wave (determined by the sine wave curve fitting algorithm) by an ideal ADC and best fit sine wave. Large number of counts are taken at the test frequency, and best fit sine wave to the data is computed through the software. Actual rms error is determined between actual data (generated by digitizing input sine wave with actual test ADC) and best fit sine wave. The idealized best fit sine wave is then ideally digitized (in software) by a perfect N bit ADC. The rms error (actual) between actual data and the best fit sine wave is calculated. The rms error (ideal) between idealized data (generated by the perfect N bit ADC) and best fit sine wave is calculated. The ENOB value is determined from these two errors of test ADC at specified frequency [8]. Apart from practical difficulty in applying this algorithm, this method gives less accurate value of ENOB as compared to histogram method.

16.2.3 FFT Technique

FFT technique is used to estimate the linearity of an ADC. This technique is important to find frequency, amplitude and phase of analog input and unlike harmonics produced due to nonlinearity present in the test ADC and noise [9–11]. After that by taking FFT of the quantized data, the single line spectrum can be produced using digitization of a full scale mono frequency sine wave of an ideal ADC. When a real life ADC is used to sample mono frequency sine wave, FFT output spectrum contains harmonics due to non-ideal nature of test ADC, i.e. due to nonlinearity in ADC and noise present in input to ADC. It is possible to calculate SNR in dB as: $20\log(S/N)$. ENOB at test frequency, when digitizing unipolar and bipolar sine wave, can be computed by [27]:

$$\text{ENOB (unipolar)} = (\text{SNR} - 10.79)/6.02$$
$$\text{ENOB (bipolar)} = (\text{SNR} - 10.78)/6.02, \text{ respectively.}$$

16.2.4 Beat Frequency Testing Technique

The beat frequency and envelop test are qualitative tests, which provide a quick simple visual demonstration of individual local ADC dynamic failures. Errors can be seen as deviation from a perfect sine function. This chapter reports an estimation of gain and offset and their variance. Histogram technique is used for finding gain error, offset error and their variance applying sine wave as an input test signal. Mainly considered an amplitude of the signal, additive noise present in the ADC and test setup. To consider their effects on gain and offset error, first code transition levels of an ADC transfer characteristic are determined and last (highest) and first (lowest) transition levels are determined. Based on these values gain and offset error are computed as shown in Figure 16.2. Further suitable modifications are done in the proposed method for gain error and offset error computation for better accuracy. Accuracy in gain error and offset error determination are improved by modifying code transition level with zero error correction and applied for 5 to 10 bit ADC.

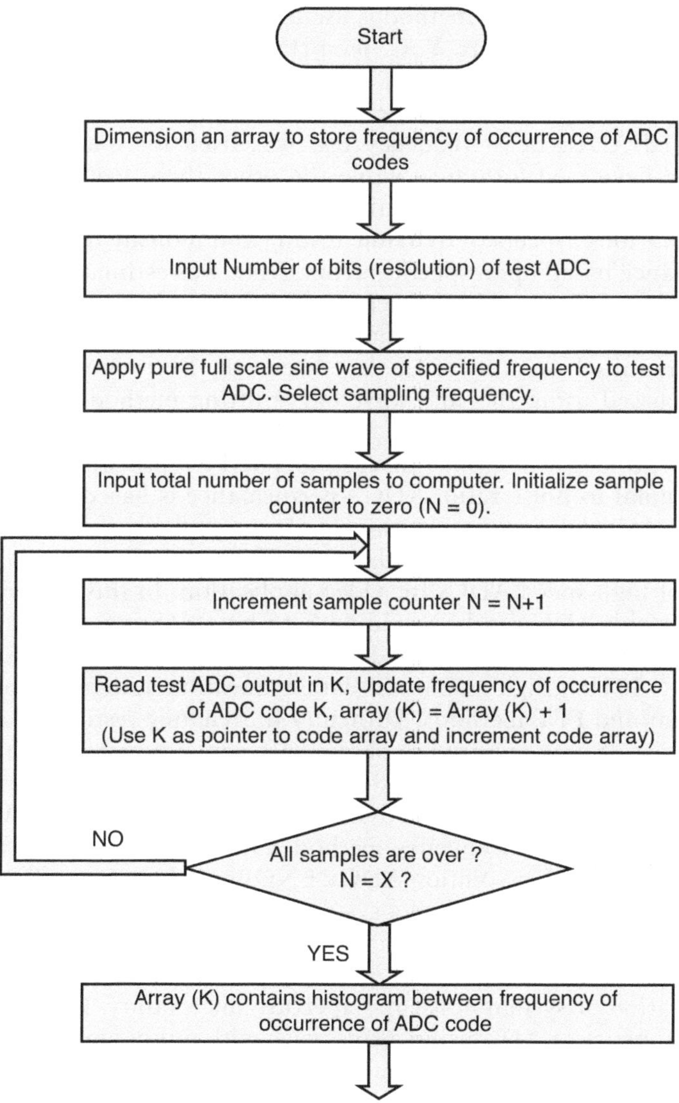

Figure 16.2 Different steps of basic histogram construction.

16.3 SOME MILESTONE WORK ON DYNAMIC TESTING OF AN ADC

M. F. Wagdy and S. S. Awad [12] have reported a new technique based on the histogram method for the computation of ENOB. ENOB was estimated based on code transition levels of ADC. The results of the proposed histogram method were compared with sine wave curve fitting and FFT methods and found histogram method was more accurate than the other two methods. This method is useful for many applications in testing and measurement systems. Y. C. Jenq [13] has proposed a method based for characterization of digitizer and determination of ENOB. A basic work presented by Peetz [14] for dynamic testing of high resolution and high-speed ADC. Beat frequency test was used for qualitative estimation of a device performance while the other three methods were used for quantitative testing using full scale sine wave as input. These tests provide various aspects of dynamic testing and indicate limitations in the performance based upon specific errors. DNL was estimated based on the measured probability of occurrence of a code and the ideal probability of occurrence of a code. Selection of input and sampling frequency to get the required accuracy was also discussed. Determination of ENOB was discussed using the sine wave curve fitting method. Limitations of sine wave curve fitting test such as initial frequency and phase guess convergence problems were also discussed. The FFT test was used for computing signal-to-noise ratio. ADC's performance is based upon the three features of the DFT spectrum namely, the noise floor, the harmonic level and spurious level. David A Hodges and his team [15] proposed dynamic testing of high-speed ADCs for video application. In this high resolution such as 16 bit ADC can be tested with 18 bit DAC to get ¼ bit precision in testing. Doerfler [16] proposed a dynamic test technique for slow sample rate (128 Hz), high resolution data acquisition system (15 bits) using histogram and FFT methods. Difficulty in dynamic testing was given as the amount of time required to obtain sufficient amount of data. Though the method was reported for testing slow-speed data acquisition system, but it is equally applicable for dynamic testing of slow-speed ADC with high resolution. Beat frequency method was not recommended due to its limitations for high-resolution ADCs. ENOB was found insufficient single result for thorough analysis of a system by sine wave curve fitting method and hence, it is also not recommended. Histogram test gives best estimate of DNL while the FFT method gives the best estimate of INL and overall indication of system noise and aperture uncertainty. Dynamic testing and diagnostics of ADC using a sine wave with unknown amplitudes and offsets were presented by M Vanden Bossche et al. [17]. In this work, Walsh transform was used to determine INL to identify the known linearity at the bit level of the ADC. Daniel Belega [18] has proposed the dynamic performance testing of the ADC algorithm. The performance of this algorithm depends upon normalized frequency accuracy. Also derived constraint depends upon the ADC resolution and the number of recorded samples. A recent work was proposed by Linus Michaeli et al. [19] for the determination of ADC nonlinearity for a successive approximation type of ADC. Nonlinearity error model was discussed and its utility for other ADC architecture was suggested. In this model, the decomposition into high- and low-frequency component was proposed with this approach for high-speed ADC test without using highly accurate instrumentation. Hsin-Wen Ting et al. [20] have proposed histogram-based test for the determination of the performance of an ADC.

16.4 ESTIMATION OF AN ADC PARAMETERS

A complete analog input is applied to test ADC and predefined large numbers of samples at a set sampling frequency are acquired. A code transition level $\hat{T}[k]$ is computed based on phase estimation.

The expressions of $\hat{T}[k]$, of an N bit ADC are [21–23]:

$$\hat{T}[k] = C - A\cos\left(\pi \frac{C_k - 1}{M}\right), K = 1, 2, ..., 2^{N-1} \tag{16.1}$$

where C_k is the cumulative histogram defined by:

$$C_k = \sum_{i=0}^{k-1} h[i] \tag{16.2}$$

where

$h[i]$ = Total number of samples received in code bin I
M = Total number of samples,
A = Amplitude of input sine wave, and
C = DC offset of input sine wave,

Between any two values in an analog signal, infinite possible voltage levels are present. When these analog values are converted into digital it has only limited possible codes based upon resolution, N, of the ADC (Possible codes = 2^N). In practical ADC the gain and offset error can be expressed as follows [24].

$$\text{Gain } G = \frac{L_{\text{ideal}} - F_{\text{ideal}}}{L - F} \tag{16.3}$$

$$\text{Offset } C = F_{\text{ideal}} - GF \tag{16.4}$$

where

L=the real last transition level, $L = T_{2^N-1}$
F=the real first transition level, $F=T_1$

Code transition level with error in ADC transfer characteristic is taken as an estimate of a transition voltage, based upon this estimate the gain and offset are:

$$\hat{G} = \frac{L_{ideal} - F_{ideal}}{\hat{L} - \hat{F}}$$ (16.5)

$$\hat{C} = F_{ideal} - \hat{G}\hat{F}$$ (16.6)

where

$\hat{L}$ is the last estimated transition voltage, $\hat{L} = \hat{T}_{2^N-1}$
$\hat{F}$ is the first estimated transition voltage, $\hat{F} = \hat{T}_1$

The top hat cover of the symbols signifies that it is an estimate value and not the actual value for the ADC under test.

After estimation of gain and offset, now we can determine gain error using Equations (16.3) and (16.5) as

$$G\_error = \hat{G} - G$$ (16.7)

where

G_error = Gain error
$\hat{G}$ = Estimated Gain
G = Actual Gain (Ref. gain nominally "1")

Using Equations (16.4) and (16.6), offset error can be determined as

$$C\_error = \hat{C} - C$$ (16.8)

where

C_error = Offset Error
$\hat{C}$ = Estimated Offset
C = Actual Offset (nominally "0")

16.4.1 Variance of Estimated Gain and Offset Error

Variance of estimated gain and offset error can be expressed as [24–27].

$$\sigma_{\hat{G}}^{\,2} \approx \left(\frac{1}{L_{\text{ideal}} - F_{\text{ideal}}}\right)^{2}\sigma_{\hat{F}}^{\,2} + \left(\frac{1}{L_{\text{ideal}} - F_{\text{ideal}}}\right)^{2}\sigma_{\hat{L}}^{\,2} \tag{16.9}$$

$$\sigma_{\hat{c}}^{\,2} \approx \left(\frac{L_{\text{ideal}}}{L_{\text{ideal}} - F_{\text{ideal}}}\right)^{2}\sigma_{\hat{F}}^{\,2} + \left(\frac{F_{\text{ideal}}}{L_{\text{ideal}} - F_{\text{ideal}}}\right)^{2}\sigma_{\hat{L}}^{\,2} \tag{16.10}$$

where

$\sigma_{\hat{G}}^{\,2}$ = Variance of the estimated gain

$\sigma_{\hat{c}}^{\,2}$ = Variance of the estimated offset

$\sigma_{\hat{F}}$ = Standard deviation of first transition

$\sigma_{\hat{L}}$ = Standard deviation of last transition

For a bipolar ADC Transfer characteristic the $T_{k_{\text{ideal}}}$ can be expressed as:

$$T_{k_{\text{ideal}}} = -\text{FS} + kQ,$$

$$L_{\text{ideal}} = -F_{\text{ideal}} = \text{FS} - Q$$

where
FS = Full Scale, k = code level and Q = Ideal code bin width
Now Equations (16.9) and (16.10) can be simplified as:

$$\sigma_{\hat{G}}^{\,2} \approx \frac{1}{4(\text{FS} - Q)^{2}}\left(\sigma_{\hat{F}}^{\,2} + \sigma_{\hat{L}}^{\,2}\right) \tag{16.11}$$

$$\sigma_{\hat{c}}^{\,2} \approx \frac{1}{4}\left(\sigma_{\hat{F}}^{\,2} + \sigma_{\hat{L}}^{\,2}\right) \tag{16.12}$$

Expressions (16.11) and (16.12) express the variance of the estimated gain and offset error as a function of the variance of the first and last estimated transition voltages [25].

Variance of the estimated transition voltage can be determined in terms of counts of cumulative histogram as [26].

$$\sigma_{\hat{T}+1}^{\,2} \approx \left(\frac{A\pi}{M}\right)^{2}\left[1 - \left(\frac{T_{K+1} - C}{A}\right)^{2}\right]\sigma_{C_{k}}^{\,2} \tag{16.13}$$

The normalized transition level for code k is

$$U_K = \frac{T_K - C}{A}, \sigma_n = \frac{\sigma}{A} \tag{16.14}$$

The DNL and INL can be estimated as follows [2, 27]:

$$\text{DNL} = \frac{G \times W[K] - Q}{Q} \tag{16.15}$$

$$\text{INL} = \frac{[K-1]Q + T[1] - T[K]}{Q} \tag{16.16}$$

where

    G = gain of an ADC,
    $W[k] = T[k+1] - T[k] = K$th code bin width of an ADC
    and Q = ideal code bin width

Estimate of DNL and INL can be expressed as below. In all formulae, this symbol ^ on top of parameter indicates the estimate value [22, 27].

Finally, estimate of DNL and INL can be expressed as:

$$\hat{DNL} = \left[\left\{ \frac{\hat{G}\left(\hat{W}[K]\right)\left(2^N - 2\right)}{\hat{L} - \hat{F}} \right\} - 1 \right] \tag{16.17}$$

$$\hat{INL} = \frac{\left\{ \left(\hat{G}\hat{T}[K] + \hat{C} \right)\left(2^N - 2\right) \right\} - T_K^{\text{ideal}}}{\hat{L} - \hat{F}} \tag{16.18}$$

where

    gain $\hat{G}$ and offset $\hat{C}$ can be estimated using Equations (16.15) and (16.16), respectively.

$$\hat{Q} = \frac{\hat{V}}{2^N - 2} = \text{Estimated code bin width}$$

$$\hat{V} = T\left[2^N - 1\right] - T[1] = \text{Estimated reduced full scale voltage}$$

$$\hat{V} = \hat{L} - \hat{F}$$

Therefore,

$$\hat{Q} = \frac{\hat{L} - \hat{F}}{2^N - 2} \text{ and}$$

$$\hat{W}[K] = \hat{T}[K+1] - \hat{T}[K]$$

16.5 PERFORMANCE ANALYSIS WITH PROPOSED TEST METHOD

Simulation of ideal 5 to n bit transfer characteristic is done and arbitrary nonlinearity error is introduced in it. Gain estimation and error in estimated gain are shown in Table 16.1 and variance computation for gain error and offset error is carried out by our method is shown in Tables 16.2–16.4 and its plots are shown in Figures 16.3–16.5. From the results it is observed that when numbers of samples are increased from 1,000 to 10,000, the variance of the gain decreases and after 10,000 samples it becomes constant.

Table 16.1 Estimate of Gain Error as a Function of the Number of Samples

No. of samples	Estimated gain	Error in estimated gain
512	1.001653	0.001653
1,024	1.001990	0.001990
2,048	1.001720	0.001720
4,096	1.001720	0.001720
8,192	1.001618	0.001618
16,384	1.001655	0.001655
20,000	1.001653	0.001653
25,000	1.001640	0.001640
30,000	1.001664	0.001664

Table 16.2 Variance of the Estimated Gain with Signal Amplitude

Signal amplitude A [V]	Variance of the estimated gain $(\sigma_{\hat{G}}^2)$ by our method
1.0	0.00225
1.1	0.0076
1.2	0.0085
1.3	0.0096
1.4	0.011
1.5	0.013
2.0	0.000393
3.0	0.000590
4.0	0.000786
5.0	0.001007

Table 16.3 Variance of the Obtained Gain with Number of Samples

Number of samples (M)	Variance of the estimated gain ($\sigma_{\hat{G}}^2$) by our method
0	0.0
200	0.017066
400	0.013934
600	0.00985
800	0.008533
1,000	0.007632
10,000	0.001007
15,000	0.001005
20,000	0.001003
30,000	0.001005

Table 16.4 Variance of the Estimated Gain as a Function of the Normalized Additive Noise Standard Deviation

Normalized additive noise standard deviation (σ_n)	Variance of the estimated gain ($\sigma_{\hat{G}}^2$) by our method
0.0	0.00086
0.02	0.004
0.04	0.0058
0.06	0.0062
0.08	0.0075
0.10	0.0088

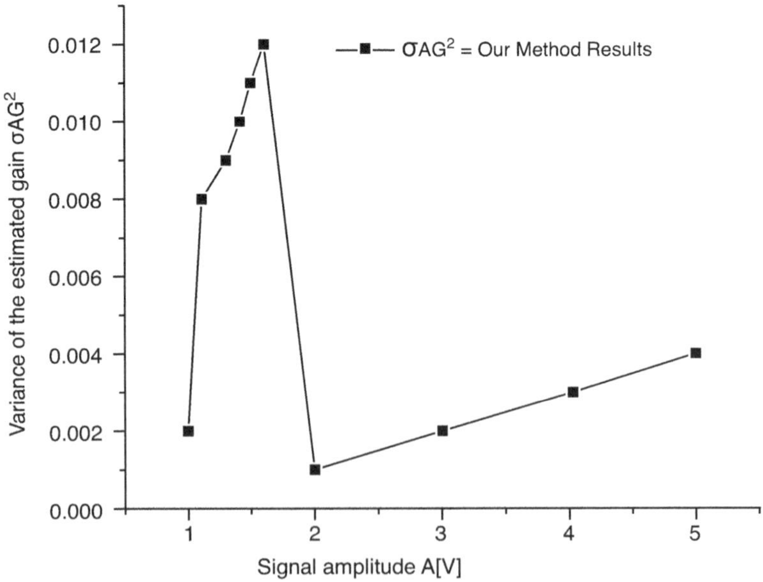

Figure 16.3 Estimated gain for 8 bit ADC as a function of the signal amplitude.

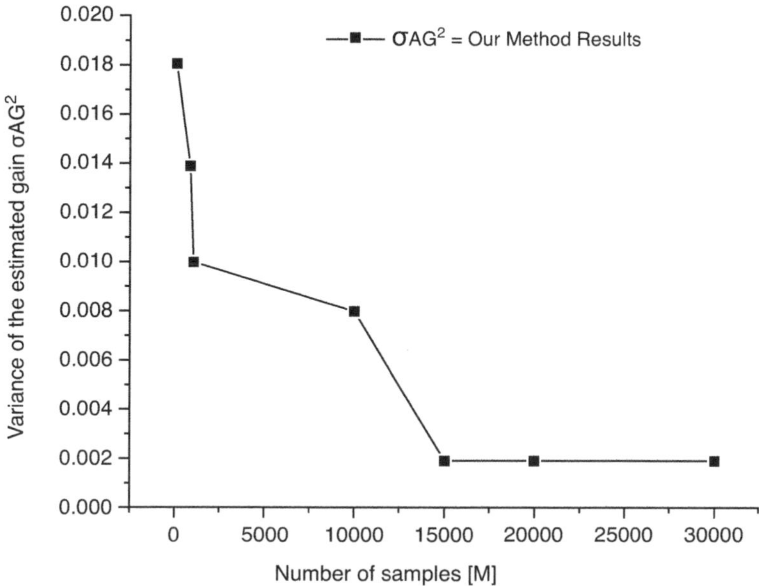

Figure *16.4* Estimated gain for 8 bit ADC as a function of number of samples.

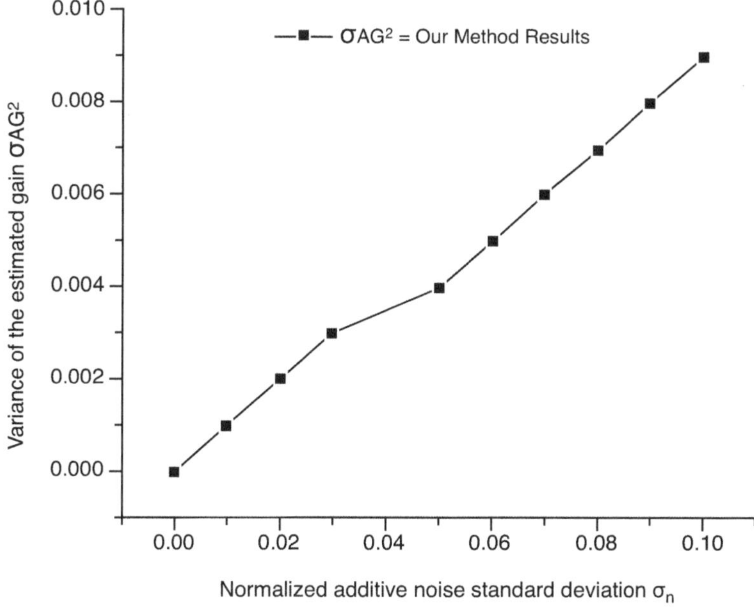

Figure *16.5* Gain estimation for 8 bit ADC as a function of the normalized additive noise standard deviation with $A=5.12\,\text{V}$.

AUTHORS BIOGRAPHY

 Dr. Radheshyam Gamad has more than 27 years of teaching and research experience. He started his career in 1996. He did his PG and Ph.D. from Rajiv Gandhi Technical University Bhopal (M.P.) in 2003 and 2010. Currently, he is a Professor in the Department of Electronics & Instrumentation Engineering, Shri G. S. Institute of Technology & Science, Indore (M.P.) and leading "Analog and mixed signal circuit design and testing, VLSI Circuit and System Design" at the discipline of Electronics & Instrumentation Engineering. He has granted four patents have been awarded. He has published two books and two book chapters. He received CV Raman award and Best paper award in 2021 for his research work. As of now, six PhD students have been awarded their PhD degree under his supervision and at present seven PhD scholars are registered apart from several M.Tech. and B.Tech. Students. He has completed three research projects funded by MHRD, MeitY, Govt. of India and one is under consideration for funding. He has Published 70 research papers in SCI, Scopus indexed International & National Journals, and International & National Conferences. He is a member of IETE, IE India and reviewer of several reputed Journals. He was the Coordinator of VDAT 2019 conference and has been associated with several reputed conferences across the globe as a member of the technical program committee.

BIBLIOGRAPHY

1. B. E. Peetz, 1983, "Dynamic Testing of Waveform Recorders", *IEEE Transactions on Instrumentation and Measurement*, Vol. Im-32, No. 1, pp. 12–17.
2. IEEE-std-1057-1994 (R-2001), "Digitizing Waveform Recorders", *IEEE Instrumentation and Measurement Transactions*.
3. F. A. Bernard, Y. Bestrand, and M. Renovell, 2000, "Toward an ADC BIST Scheme Using the Histogram Test Technique", *IEEE Transactions on Instrumentation and Measurement*. Vol. 2, pp. 1–6.
4. Jerome Blair, "Histogram measurement of ADC non-linearity using sine waves", *IEEE transaction on instrumentation and measurement*, Vol. 43, No. 3, pp. 378–383, June 1994.
5. F. C. Alegria, A. Moechitta, P. Carbone, A. Cruz Serra, and D. Petri, 2004, "Effective and Linearity Testing", Technical Report # DIT-04-041, University of Trento, Italy, Via Sommarive 14.
6. M. S. Caceci and W.P. Cacheris, 1984, "Fitting Curves to Data", *Byte*, Vol. 9, pp. 340–362.
7. HP Co-workers, 1988, "Fixed-Frequency Sine Wave Curve Fit", *Hewlett Packard Journal*, Vol. 39, p. 48.

8. A. Oppenheim and R. Schafer, 1988, "Digital Signal Processing", Englewood Cliffs, NJ, Prentice-Hall.

9. S. D. Stearus and R. A. David, 1988, "Signal Processing Algorithms", Englewood Cliffs, NJ, Prentice-Hall.

10. S. L. Marple Jr., 1987, "Digital Spectral Analysis", Englewood Cliffs, NJ, Prentice-Hall.

11. M. F. Wagdy and S. Awad, 1991, "Determining ADC Effective Number of Bits via Histogram Testing", *IEEE Transactions on Instrumentation and Measurement*, Vol. 40, No. 4, pp. 770–772.

12. A. Baccigalupi, P. Daponte, and M. D. Apuzzo, 1994, "An Improved Error Model of Data Acquisition System", *IEEE Transactions on Instrumentation and Measurement*, Vol. 43, No. 2, pp. 220–225.

13. M. Kollar and P. Michalko, 2007, "A New Approach in Differential Nonlinearity Testing of the Analog-to-Digital Converters", *International Journal on Measurement*, Vol. 40, pp. 520–526.

14. D. A. Hodges, J. Doernberg, and H.-S. Lee, 1984, "Full Speed Testing of A/D Converters", *IEEE Journal of Solid-State Circuits*, Vol. sc-19, No. 6, pp. 820–827.

15. D. W. Doerfler, 1986, "Dynamic Testing of a Slow Sample Rate, High Resolution Data Acquisition System", *IEEE Transactions on Instrumentation and Measurement*, Vol. IM-35, No. 4, pp. 477–482.

16. M. V. Bossche, J. Schoukens, and J. Renneboog, 1986, "Dynamic Testing and Diagnostics of A/D Converters", *IEEE Transaction on Circuits and Systems*, Vol. CAS-33, No. 8, pp. 775–783.

17. D. Belega and D. Dallet, 2008, "Normalized Frequency Estimation for Accurate Dynamic Characterization of A/D Converters by Means of the Three-Parameters Sine-Fit Algorithm", *International Journal on Measurement*, Vol. 41, pp. 986–993.

18. L. Michaeli, P. Michalko, and J. Saliga, 2008, "Unified ADC Nonlinearity Error Model for SAR ADC", *International Journal on Measurement*, Vol. 41, pp. 198–204.

19. H.-W. Ting, B.-D. Liu, and S.-J. Chang, 2008, "A Histogram-Based Testing Method for Estimating A/D Converter Performance", *IEEE Transaction on Instrumentation and Measurement*, Vol. 57, No. 2, pp. 420–427.

20. IEEE Standard for 2000, "Terminology and Test Methods for Analog to Digital Converter", IEEE Standard-1241.

21. D. K. Mishra, 2003, "ADC Testing Using Interpolated Fast Fourier Transform (IFFT) Technique", *International Journal of Electronics*, Vol. 90, No. 7, pp. 459–469.

22. R. S. Gamad and D. K. Mishra, 2009, "Gain Error, Offset Error and ENOB Estimation of an A/D Converter Using Histogram Technique", *International Journal of Measurement*, Vol. 42, No. 4, pp. 570–576.

23. F. Correa Algria and A. Cruz Serra, 2007, "Standared Histogram Test Precision of ADC Gain and Offset Error Estimation", *IEEE Transactions on Instrumentation and Measurement*, Vol. 56, No. 5, pp. 1527–1531.

24. D. K. Mishra, N. S. Chaudhari, and M. C. Shrivastava, 1997, "Dynamic Testing of an ADC by Least Square Error Minimization in Histogram Technique", *The Journal of Institution of Engineers (India), Electrical Engineering Division*, Vol. 78, pp. 31–35.

25. H. Daniel and D. Sheingold (Ed.), 1986, *Analog – Digital Conversion Handbook*, Third Edition, Englewood Cliffs, NJ, The Prentice – Hail.

26. P. E. Allen and D. R. Holberg, 2004, *CMOS Analog Circuit Design*, Second Edition, New York, Oxford University.

27. F. Cennamo, P. Daponte, and M. Savastano, 1992, "Dynamic Testing and Diagnostics of Digitizing Signal Analyzers", *IEEE Transactions on Instrumentation and Measurement*, Vol. 41, No. 6, pp. 840–844.

Chapter 17

Intelligent traffic management using VANET simulations with 6G for smart traffic management

Using graphing algorithm in Python

Mohd Mohsin Ali, Ayushman Pranav, and Manish Raj

17.1 INTRODUCTION: BACKGROUND OF INTELLIGENT TRAFFIC MANAGEMENT ON 6G

Intelligent traffic management plays a pivotal and multifaceted role in addressing the complex challenges confronted by contemporary transport networks. To overcome these obstacles, the utilisation of Vehicular Ad-Hoc Networks (VANETs) and sixth-generation (6G) technologies emerge as an indispensable solution. By leveraging real-time vehicle tracking, robust algorithmic frameworks, and advanced communication protocols, a transformative paradigm shift can be achieved in the domain of traffic management. This shift entails optimised traffic flow, heightened road safety, and enhanced efficiency of public transportation systems. The primary objective of this chapter is to underscore the significance of integrating VANETs and 6G communication technology within the context of intelligent traffic management. A specific emphasis is placed on exploring the practical application of Python programming, machine learning methodologies, and computer vision techniques to augment traffic flow management. Through a comprehensive and in-depth exploration of these technologies, this chapter endeavors to showcase their vast potential to revolutionise the transportation industry and significantly elevate the efficacy of traffic control measures.

17.2 SIGNIFICANCE OF PYTHON, MACHINE LEARNING, AND COMPUTER VISION IN TRAFFIC OPTIMISATION

Given its versatility, Python provides a wide range of tools and packages that are excellent for enhancing traffic conditions. It is feasible to forecast

DOI: 10.1201/9781003522003-20

Figure 17.1 Vehicle detection using Haar cascades.

traffic flow, examine traffic patterns, and make data-driven decisions for optimal traffic management using machine learning techniques. Additionally, vehicle detection, tracking, and analysis are made possible by computer vision algorithms, which improves traffic condition monitoring. We obtain a complete toolbox to address traffic issues and improve traffic control systems by fusing Python, machine learning, and computer vision.

Python uses the OpenCV package to do vehicle detection. Each frame of a video feed is loaded with the pre-trained vehicle identification model applied. To make it possible to visually identify the identified vehicles, bounding boxes are painted around them. This shows how Python may be used in conjunction with computer vision techniques to do tasks like vehicle detection, laying the groundwork for more sophisticated traffic control systems (Figure 17.1).

Certainly! Haar cascades are a well-known machine learning-based technique for object detection in both still photos and videos. They were initially developed by Viola and Jones in 2001 and have gained significant popularity since then. Haar cascades operate by training a classifier using positive and negative images. The classifier's goal is to identify specific patterns or characteristics known as Haar-like features, which are distinctive to the object being recognised. These features can include edges, lines, and diverse textures.

When applied to an image or video, the trained Haar cascade model slides a window over the content, examining each window for the presence of the target object. The classifier analyses the window and identifies it as containing Haar-like features if they are detected. This approach allows for efficient and effective object detection in various applications.

17.3 6G MAPPED VANET FOR SMOOTHER TRAFFIC FLOW

6G technology is the next iteration of cellular communication technology, succeeding 5G. Its goal is to deliver ultra-high-speed communication that is substantially quicker than present networks, allowing for data transfer rates in the terabit per second area. This high-speed connectivity will assist in developing technologies such as augmented reality, virtual reality, and holographic communication by allowing for the smooth transfer of massive volumes of data.

- Low latency is another key feature of 6G technology. The time lag between submitting a request and receiving a response is referred to as latency. 6G aims to achieve extremely low latency, in the order of microseconds, enabling near real-time communication. The importance of obtaining low latency in many applications cannot be overstated, especially in time-sensitive settings where quick reaction times are vital. Furthermore, in the field of remote robotic control, reduced latency is critical in maintaining smooth and precise robotic system manoeuvring, enabling efficient and successful teleoperation. Furthermore, immersive gaming experiences rely largely on reduced latency to give players a heightened feeling of realism and quick responsiveness, which improves overall gameplay enjoyment. As a result, reducing latency offers enormous promise for revolutionising various sectors and defining the future of technological growth.
- One of the key focal points of 6G is the establishment of extensive device connectivity, which aims to facilitate a substantially higher density of interconnected devices per unit area when compared to previous generations. This enhanced capability assumes utmost importance for applications pertaining to the Internet of Things (IoT) and smart cities, wherein seamless connectivity and communication between a multitude of devices, sensors, and vehicles are of paramount significance. The provision of such expansive connectivity stands to revolutionise these domains, empowering them with the potential to accommodate the complex interconnectivity demands of a technologically advanced future.
- VANET, or Vehicular Ad-Hoc Network, is a separate type of ad hoc network that is particularly built to permit smooth communication between automobiles and roadside infrastructure. VANET uses the intrinsic mobility of vehicles to build dynamic wireless connections, resulting in a self-organising network structure. This technology

foundation enables the deployment of a wide range of applications, such as cooperative collision warning systems, intelligent traffic management, and the provision of entertainment services within vehicles. Vehicles can successfully transmit essential information by using the capabilities of VANET, improving overall road safety, optimising traffic flow, and enriching the in-vehicle experience for occupants.

- The combination of 6G with VANET results in a powerful platform for intelligent traffic management with transformational potential. Real-time data interchange between cars and infrastructure becomes possible with 6G's high-speed connection and low-latency characteristics, enabling seamless coordination and optimisation of traffic flow. VANET helps to a more secure, efficient, and environmentally friendly transportation network with features such as cooperative collision warning, traffic signal optimisation, and dynamic route guiding. The combination of 6G and VANET has enormous promise for altering the transportation environment, ushering in an age of increased safety, increased efficiency, and sustainable mobility solutions.

17.4 REVIEW ON THE APPLICATION OF 6G MAPPED VANET FOR TRAFFIC MANAGEMENT

1. The first research paper, titled "A New Hybrid Deep Learning Algorithm for Prediction of Wide Traffic Congestion in Smart Cities" [1], introduces a cutting-edge hybrid deep learning model that combines boosted long short-term memory ensemble (BLSTME) and convolutional neural network (CNN) techniques in a synergistic manner. The proposed approach outperforms alternative deep learning algorithms in properly estimating traffic congestion inside VANETs, outperforming them by a factor of 10. Nonetheless, the study recognises numerous inherent obstacles, such as the risk of overfitting and the need to capture the complex changes in vehicle movement patterns and traffic circumstances.

2. The second scholarly publication, "6G-Based Vehicular Ad Hoc Networks: A Comprehensive Review" [2], provides an in-depth analysis of Vehicular Ad Hoc Networks (VANETs) in the context of 6G technology. Using a systematic literature review technique, the authors conduct a thorough analysis of the strengths, problems, research gaps, and limits of current VANET solutions as they correspond with the 6G paradigm. Furthermore, the report proposes research routes that are positioned to push the growth of 6G VANETs, emphasising the need of filling identified gaps and improving the efficacy of present technologies. This report throws light on the importance of additional research and development endeavours in this field of study by contributing to our collective understanding of 6G VANETs.

3. The third paper, "A Genetic Predictive Model Approach for Smart Traffic Prediction and Congestion Avoidance in Urban Transportation" [3], describes a novel hybrid deep learning algorithm that combines the boosted long short-term memory ensemble (BLSTME) and convolutional neural network (CNN) techniques. This complex model outperforms previous algorithms by a wide margin when it comes to properly estimating traffic congestion. Notably, the outcomes of the study show the possibility for precise traffic prediction and improved traffic management in real-world urban settings, with the inherent potential for minimising accidents and optimising urban transportation systems. However, the authors agree that more refining and rigorous testing are required to increase the model's performance and adaptability in a variety of traffic scenarios and geographical regions.

4. The fourth scholarly work, "A Novel Approach for Malware Detection Using Machine Learning Techniques" [4], provides a game-changing methodology for identifying and detecting harmful software using machine learning techniques. This novel technique combines many aspects such as API calls, system calls, and file system actions to assist the construction of a robust model capable of distinguishing between malicious and benign samples. Following rigorous analyses of a large malware dataset, the study achieves an amazing accuracy rate of 1,717.17%. The study employs a thorough process that includes data collecting, feature extraction, model training, and performance evaluation, resulting in a highly promising result.

5. Vehicular Ad Hoc Networks (VANETs), on the other hand, are a subset of ad hoc networks that enable seamless communication between automobiles as well as with roadside infrastructure. VANETs build dynamic wireless connections by using the intrinsic mobility of vehicles, resulting in a self-organising network architecture. VANET technology enables a wide range of applications, such as cooperative collision warning systems, intelligent traffic management, and sophisticated entertainment services for vehicle settings. VANETs, with their capacity to dynamically react to changing situations, help to improve road safety, optimise traffic flow, and provide enhanced driving experiences through informative and entertaining services [2].

6. In an assortment of ways, combining 6G with VANET can help the transport industry. 6G technology's high-speed and low-latency connection capabilities enable real-time data sharing between cars and equipment. This enables for more efficient flow of traffic synchronisation and optimising, leading to reduced congestion and higher overall effectiveness [5].

7. VANET is crucial to enhancing the transportation industry's security and effectiveness.VANET enables vehicle-to-vehicle and vehicle-to-infrastructure connectivity, enabling capabilities such as cooperative collision warning, traffic signal optimisation, and dynamic route

guiding. These features have the potential to significantly reduce the risk of accidents, improve traffic flow, and improve the overall driving experience [6].

8. The combination of 6G and VANET not only improves transportation system efficiency but also promotes environmental sustainability. Fuel consumption and resultant emissions may be considerably reduced by optimising traffic flow, reducing congestion, and using effective routing systems. This holistic approach to sustainable transportation practices reduces the negative environmental effect of the transportation system, aligning it with eco-conscious principles and contributing to a better future [7].

9. Several research papers have been published on the use of 6G-based VANETs for enhanced traffic control. One notable study presented a unique hybrid deep learning approach for reliable traffic congestion prediction inside VANET systems. The hybrid model outperformed other deep learning models in terms of accuracy, precision, and recall rates by synergistically merging boosted long short-term memory ensemble (BLSTME) and convolutional neural network (CNN) approaches. The capacity to predict traffic congestion properly has enormous promise for revolutionising traffic management practices, reducing accidents, and raising overall road safety standards [8].

10. Several research articles have undertaken in-depth studies of 6G-based VANETs, comprehensively evaluating the benefits, problems, and research gaps prevalent in this sector. These academic works give vital insights, serving as guiding lights for future study while emphasising the importance of ongoing research and development activities in 6G VANET technology [9].

11. In summary, the combination of 6G and VANET emerges as a disruptive paradigm with enormous potential to revolutionise the transportation sector environment. By combining 6G's high-speed communication, low latency, and unparalleled device scalability with VANET's vehicle-to-vehicle and vehicle-to-infrastructure communication capabilities, remarkable advances in traffic management, safety enhancement, and overall transportation system efficiency can be realised. Sustained research and development efforts in this area are critical because they lay the way for unlocking the entire range of possibilities afforded by the seamless merging of 6G and VANET, therefore crafting a visionary future for the sector of transportation [10].

17.5 IMPACT OF 6G AND VANET ON THE TRANSPORT SYSTEM

1. The advent of 6G and VANET marks a critical watershed moment with enormous revolutionary possibilities for the transport sector,

notably in terms of smarter traffic management and improved operational efficiency.

Furthermore, the low latency characteristics inherent in 6G enable near real-time communication, which is critical for time-sensitive applications, notably autonomous cars. Critical capabilities such as cooperative collision warning systems and dynamic route guiding may be implemented with surprising speed and precision using the power of 6G and VANET, resulting in better safety and efficiency within the transportation ecosystem.

2. The combination of 6G technology with Vehicle Ad-Hoc Networks (VANET) has enormous promise for revolutionising transportation. The remarkable connection capabilities of 6G enable an unprecedented number of networked devices, supporting the seamless integration of cars, sensors, and infrastructure with the aim of intelligent traffic management. By combining the capabilities of 6G with VANET, a new paradigm arises, distinguished by the harmonic interplay of cars, infrastructure, and technology, with substantial implications for optimising traffic flow, boosting safety, and improving overall transport system performance. Real-time data interchange between autos and infrastructure is now available because of 6G's high-speed and low-latency communication capabilities, resulting in decreased congestion and increased overall efficiency in the transportation system.

3. The use of 6G and VANET into transportation networks provides various advantages. It contributes to environmental sustainability by optimising traffic flow, eliminating traffic jams, and allowing optimal routing, all of which reduce fuel use and emissions. The usage of a 6G VANET for handling traffic has been investigated, with an integrated deep learning system attaining outstanding accuracy in anticipating congestion. Furthermore, detailed studies highlight the benefits, limitations, and research limitations in 6G-based VANETs, urging further study and development. Overall, the amalgamation of 6G and VANET has the potential to alter transportation by enhancing traffic management, safety, and effectiveness, and further study is required to realise their full potential.

17.6 IMPACT OF 6G AND VANET ON THE TRANSPORT SYSTEM

1. 6G and VANET (Vehicular Ad-Hoc Network) have the potential to transform transport by allowing intelligent traffic management and increasing overall efficiency. The combined use of these methods has the potential to develop a formidable framework for seamless traffic coordination and optimisation. Let us investigate the influence of 6G and VANET on the public transportation system.

2. The successor to 5G technology, 6G, provides ultra-high-speed connectivity with data transmission speeds in the terabits per second range. Massive volumes of data can be exchanged in real time because of this high-speed networking, opening the door for new technologies like virtual and augmented reality, and virtual communication. 6G's high-speed communication capabilities enable real-time data transmission between cars and infrastructure in the context of the transportation system, supporting effective traffic management.

3. Low latency is a key feature of 6G, representing an ambitious goal of reaching extremely short reaction times measured in microseconds. This remarkable achievement enables near real-time communication, which is crucial in applications requiring time-critical operations such as autonomous cars, remote robotic control, and immersive gaming experiences. The introduction of 6G technology, namely in the area of transportation, enables low latency communication, hence elevating vehicle safety and efficiency through speedy and accurate decision-making capabilities, collaborative collision warning systems, and dynamic route guiding. We create the groundwork for a future characterised by efficient and intelligent mobility solutions by adopting this cutting-edge technical paradigm.

4. Within the area of 6G, the aspect of huge device connection gains critical importance. Its primary goal is to support an exponentially increased number of networked devices per unit space, outperforming previous generations. This essential characteristic becomes vital in the realms of Internet of Things (IoT) and smart city applications. Within the context of transportation systems, enormous device connection enables the seamless integration of cars, sensors, and infrastructure, resulting in the establishment of a comprehensive network capable of administering intelligent traffic management and optimising overall efficiency. We pave the road for a future marked by increased connectivity and smarter urban settings by leveraging this transformational potential.

5. Vehicular Ad Hoc Networks (VANETs), on the other hand, are a subset of ad hoc networks that enable seamless communication between automobiles as well as with roadside infrastructure. VANETs build dynamic wireless connections by using the intrinsic mobility of vehicles, resulting in a self-organising network architecture. VANET technology enables a wide range of applications, such as cooperative collision warning systems, intelligent traffic management, and sophisticated entertainment services for vehicle settings. VANETs, with their capacity to dynamically react to changing situations, help to improve road safety, optimise traffic flow, and provide enhanced driving experiences through informative and entertaining services.

6. In an assortment of ways, combining 6G with VANET can help the transport industry. 6G technology's high-speed and low-latency connection capabilities enable real-time data sharing between cars and equipment. This enables for more efficient flow of traffic synchronisation and optimising, leading to reduced congestion and higher overall effectiveness.

7. VANET is crucial to enhancing the transportation industry's security and effectiveness.VANET enables vehicle-to-vehicle and vehicle-to-infrastructure connectivity, enabling capabilities such as cooperative collision warning, traffic signal optimisation, and dynamic route guiding. These features have the potential to significantly reduce the risk of accidents, improve traffic flow, and improve the overall driving experience.

8. The combination of 6G and VANET not only improves transportation system efficiency but also promotes environmental sustainability. Fuel consumption and resultant emissions may be considerably reduced by optimising traffic flow, reducing congestion, and using effective routing systems. This holistic approach to sustainable transportation practices reduces the negative environmental effect of the transportation system, aligning it with eco-conscious principles and contributing to a better future.

9. Several research papers have been published on the use of 6G-based VANETs for enhanced traffic control. One notable study presented a unique hybrid deep learning approach for reliable traffic congestion prediction inside VANET systems. The hybrid model outperformed other deep learning models in terms of accuracy, precision, and recall rates by synergistically merging boosted long short-term memory ensemble (BLSTME) and convolutional neural network (CNN) approaches. The capacity to predict traffic congestion properly has enormous promise for revolutionising traffic management practices, reducing accidents, and raising overall road safety standards.

10. Several research articles have undertaken in-depth studies of 6G-based VANETs, comprehensively evaluating the benefits, problems, and research gaps prevalent in this sector. These academic works give vital insights, serving as guiding lights for future study while emphasising the importance of ongoing research and development activities in 6G VANET technology.

11. In summary, the combination of 6G and VANET emerges as a disruptive paradigm with enormous potential to revolutionise the transportation sector environment. By combining 6G's high-speed communication, low latency, and unparalleled device scalability with VANET's vehicle-to-vehicle and vehicle-to-infrastructure communication capabilities, remarkable advances in traffic management, safety enhancement, and

overall transportation system efficiency can be realised. Sustained research and development efforts in this area are critical because they lay the way for unlocking the entire range of possibilities afforded by the seamless merging of 6G and VANET, therefore crafting a visionary future for the sector of transportation.

17.7 6G VANET SIMULATIONS WITH PYTHON LIBRARIES

VANETs simulate the communication between vehicles and their surroundings, allowing researchers to evaluate and optimise various aspects of vehicular communication systems. Python, with its extensive libraries and flexibility, provides a powerful platform for VANET simulations.

Overview of Python libraries for VANET simulations:

i. Networkx:

Network is a Python library used for the creation, manipulation, and study of complex networks. It provides functions to model and analyse the connectivity between vehicles in a VANET. Network visualisation, path discovery, and network statistics calculation are all important features.

ii. Viens:

Viens is a Python package designed specifically to simulate Vehicular Ad Hoc Networks (VANETs). It includes a full set of tools for simulating vehicle movement and assessing VANET communication protocols, as well as modelling realistic vehicular mobility patterns. The utilisation of real-world road maps and traffic scenarios is one of its defining features, which increases the precision and authenticity of Viens simulations. By utilising these skills, researchers and practitioners may get helpful insights into VANET dynamics and make informed decisions while developing and improving vehicular communication networks.

iii. SimuLTE:

SimuLTE is a Python library designed primarily for simulating LTE communication within automobile networks. Its primary purpose is to enable accurate modelling of LTE base stations, vehicles, and the complicated linkages between them. SimuLTE may provide researchers and practitioners with helpful insights into the performance metrics of LTE-based VANETs such as packet loss, latency, and throughput. This comprehensive knowledge allows for more informed decisions when developing and optimising vehicular communication systems, ultimately boosting the efficiency and reliability of LTE installations in VANET environments.

17.8 INTEGRATION OF 6G AND VANET IN A NON-PREEMPTIVE SMART TRAFFIC LIGHT CONTROL SYSTEM

The following steps are:

1. Vehicles outfitted with VANET (Vehicular Ad Hoc Network) technology have the ability to communicate not just with one another, but also with the traffic signal control system. This seamless interconnection allows for the exchange of essential information, such as precise vehicle position, velocity, and directional data. These vehicles significantly improve the control system's decision-making abilities by facilitating communication. The continuous flow of critical data enables the control system to make more informed and efficient choices, resulting in improved traffic management and smoother vehicle movements.

2. The integration of the traffic light control system with the cutting-edge 6G network offers a smooth and powerful communication channel, allowing the system and automobiles to exchange data in a dependable and efficient manner. This convergence provides the system with quicker gearbox speeds and more reliability, ultimately boosting the overall operational efficiency of the traffic management infrastructure. By using the expanded capabilities of the 6G network, the traffic light management system may provide faster reaction times, real-time data synchronisation, and better coordination with vehicular movements, resulting in optimum traffic flow and improved road safety.

3. The traffic light control system employs the VANET network to disseminate control signals to proximate vehicles. These signals convey precise instructions regarding actions such as stopping, proceeding, or making turns, contingent upon the prevailing status of the corresponding traffic lights.

4. The traffic light control system receives real-time traffic information from cars over the VANET network, allowing for dynamic traffic signal timing adjustments and traffic flow pattern optimisation. This vital data stream enables the system to adapt to changing traffic circumstances efficiently, guaranteeing efficient and harmonic vehicular circulation.

17.9 TRAFFIC FLOW PREDICTION WITH MACHINE LEARNING ALGORITHMS

Machine learning algorithms use previous data to estimate traffic flow. These systems make use of advanced techniques such as regression analysis, time series analysis, and neural networks. These algorithms include different

Table 17.1 Factors considered for traffic flow

Algorithm	Technique	Factors considered
Linear Regression	Regression Analysis	Time of day, day of the week, weather conditions
ARIMA	Time Series Analysis	Time of day, day of the week, weather conditions
Long Short-Term Memory (LSTM)	Neural Networks	Time of day, day of the week, weather conditions, special events
Random Forest	Ensemble Learning	Time of day, day of the week, weather conditions, special events
Support Vector Machines (SVM)	Classification	Time of day, day of the week, weather conditions

parameters like as the time of day, day of the week, weather conditions, and the occurrence of exceptional events into their computations in order to provide exact forecasts. Machine learning algorithms can give accurate forecasts about traffic patterns and aid informed decision-making processes by taking these variables into account (Table 17.1).

This part is all about comparing' the smart traffic management system with the old-school traffic management systems. We'll be looking at what each approach has going for it and what holds it back. The smart system has some great things going for it like improved efficiency, less traffic jams, and a focus on sustainability. But there are also some challenges to overcome, like the limits of technology, concerns about privacy, and making sure it can work with the existing infrastructure. Switching from the old ways to the smart traffic system is a big deal for making traffic flow better in cities. So let's dive in and check out this comparison table that highlights the key features, advantages, and limitations of both systems (Table 17.2).

Traditional Traffic Management Systems (TTMS) have served their purpose effectively over the years. However, their inherent limitations in adapting to dynamic traffic conditions render them inefficient during peak hours or in emergency situations. The reliance on manual control and human judgment within TTMS introduces the potential for errors and inconsistencies.

In contrast, Smart Traffic Management Systems (STMS) present a paradigm shift towards a more adaptive, efficient, and proactive approach to traffic management. Leveraging real-time traffic data analysis and predictive algorithms, STMS can dynamically adjust signal timings and optimise traffic flow. Moreover, STMS contributes to enhanced road safety by promptly detecting and responding to anomalies such as accidents or significant congestion.

Nonetheless, the transition to STMS is not without its challenges. Successful implementation of STMS necessitates the deployment of sophisticated technological infrastructure, including advanced sensors, data analysis tools, and high-speed communication networks. The costs associated with these

Table 17.2 Comparisons between traditional traffic management systems and smart traffic management systems

	Traditional traffic management systems	*Smart traffic management systems*
Key features	- Primarily relies on pre-set timing of traffic signals - Manual enforcement of traffic rules - Rigid, non-adaptive approach towards traffic control	- Real-time traffic data analysis and adaptive signal control - Automated enforcement through AI and IoT technologies - Proactive approach, predicting and managing traffic flow
Advantages	- Proven effectiveness over many years - Requires less technological infrastructure	- Significantly improves traffic flow efficiency - Reduces congestion and pollution - Enhances road safety through proactive monitoring and management
Limitations	- Inability to adapt to changing traffic conditions - Inefficiency during peak hours and emergencies - Relies on human control and is prone to errors	- Requires sophisticated and often expensive technology infrastructure - Concerns related to data privacy and cyber security - Complexities in integrating with existing infrastructure

technologies, coupled with their limited accessibility in certain locations, pose significant considerations. Furthermore, concerns regarding data privacy and cybersecurity arise due to the substantial reliance on collecting and processing vast volumes of data within STMS. Lastly, the integration of STMS with existing traffic infrastructure demands meticulous planning and substantial allocation of resources.

While STMS undoubtedly offers substantial advancements, addressing the challenges remains imperative for their successful adoption and widespread implementation.

17.10 CONCLUSION

Traditional traffic management systems (TTMS) have served their function for many years, but they are incapable of adapting to changing traffic circumstances, resulting in inefficiency during peak hours or crises. They rely significantly on human judgement and manual control, which are prone to mistakes and inconsistencies.

Smart Traffic Management Systems (STMS), on the other hand, use a proactive and adaptable approach to traffic management. STMS can efficiently alter signal timings and optimise traffic flow by analysing real-time traffic data and utilising predictive algorithms. They also improve road safety by quickly identifying and responding to uncommon conditions like accidents or heavy traffic.

Transitioning to STMS, on the other hand, has its own set of issues. STMS implementation necessitates a complex technological infrastructure, including modern sensors, data analysis tools, and software.

REFERENCES

[1] Kothai, G., E. Poovammal, Gaurav Dhiman, Kadiyala Ramana, Ashutosh Sharma, Mohammed A. AlZain, Gurjot Singh Gaba, and Mehedi Masud. A New Hybrid Deep Learning Algorithm for Prediction of Wide Traffic Congestion in Smart Cities. *Wireless Communications and Mobile Computing*, 2021. doi:10.1155/2021/5583874.

[2] Swain SR, Saxena D, Kumar J, Singh AK, Lee CN. An AI-Driven Intelligent Traffic Management Model for 6G Cloud Radio Access Networks, 12(6), 1056–1060, 2023. doi:10.11017/LWC.2023.325171742.

[3] Sathiyaraj R, Bharathi A, Khan S, Kiren T, Khan IU, Fayaz M. A Genetic Predictive Model Approach for Smart Traffic Prediction and Congestion Avoidance for Urban Transportation. *Wireless Communications and Mobile Computing*, 2022. doi:10.1155/2022/51738411.

[4] Ma S, Zhao M. Traffic Flow Prediction and Analysis in Smart Cities Based on the WND-LSTM Model. *Computational Intelligence and Neuroscience*, 2022. doi:10.1155/2022/70717045.

[5] Zhou Y, Liu Q, Wu J, Wang F, Huang Z, Tong W, Xiong H, Chen E, Ma J. Modeling Context-Aware Features for Cognitive Diagnosis in Student Learning. *Proceedings of the 27th ACM SIGKDD Conference on Knowledge Discovery & Data Mining*, pp. 2420–2428, Barcelona, Spain.

[6] Wang C, Xing L. An Explicit MDD-Based Method for Common-Cause Failure Analysis in Phased-Mission Systems. *2017 International Conference on Quality, Reliability, Risk, Maintenance, and Safety Engineering (QR2MSE)*, Zhangjiajie, China, 2017, pp. 655–661. doi:10.11017/QR2MSE46217.20117.17021181.

[7] Lin Y, Lin H, Chen, H. 6G: The Future of Wireless Communication. *IEEE Communications Magazine*, 58(3), 102–107, 2020. This paper discusses the future of wireless communication with a focus on 6G. The authors argue that 6G will be characterized by ultra-high-speed, ultra-low latency, and massive connectivity. They also discuss the challenges that need to be addressed in order to realize 6G.

[8] Wenxuan Long, Rui Chen, Moretti Marco, Wei Zhang, Jiandong Li. A Promising Technology for 6G Wireless Networks: Intelligent Refl ecting Surface [J]. *Journal of Communications and Information Networks*, 2021, 6(1): 1–16

[9] Khan UA, Lee SS. Multi-Layer Problems and Solutions in VANETs: A Review. *Electronics*, 8, 204, 2017. doi:10.33170/electronics8020204.

[10] Clark DE. A Cramér Rao Bound for Point Processes. *IEEE Transactions on Information Theory*, 68(4), 2147–2155, 2022, doi:10.11017/TIT.2022.3140374.

Chapter 18

Sustainable 6G communication

Design of power amplifier for wireless power transfer applications

Gaurav Bhargava and Shubhankar Majumdar

18.1 INTRODUCTION

As the world moves into the fifth-generation (5G) communication era, the demand for high-speed, dependable, and sustainable wireless networks grows. But, sixth-generation (6G) communications have received much attention from the private sector and researchers [1]. Compared to 5G, 6G will feature a more comprehensive frequency range, higher broadcast rate, spectral efficacy, increased interconnection resources, shorter delay, broader coverage, and more sophisticated interference capability to meet sustainable communication requirements. One of the most essential parts of sustainable communication is effective energy use, particularly in applications such as wireless power transfer (WPT). Wireless power transfer (WPT) has become an effective method for powering bio-implantable devices without the need for physical connections. Designing an efficient and safe PA for WPT applications presents unique challenges due to stringent size constraints, power efficiency requirements, and compatibility with the human body. This chapter investigates the design of power amplifiers (PAs) for WPT applications in the context of 6G communication, with an emphasis on sustainability and energy efficiency.

18.2 NECESSITY OF PROPOSED WORK

18.2.1 Need of WPT for 6G Communication

6G wireless technologies will use massive antenna arrays to connect with consumer devices at mmWave and sub-THz frequencies [2, 3]. With the sheer quantity of devices in 6G networks [4], traditional power sources could not be viable or practicable. By allowing wireless charging of infrastructure and devices, WPT provides a solution to the power requirements of 6G communication. This is particularly important in situations when it is not feasible or feasible to make physical connections or replace batteries. In addition to making device charging more convenient, WPT is essential to the development of self-sufficient and energy-efficient communication

DOI: 10.1201/9781003522003-21

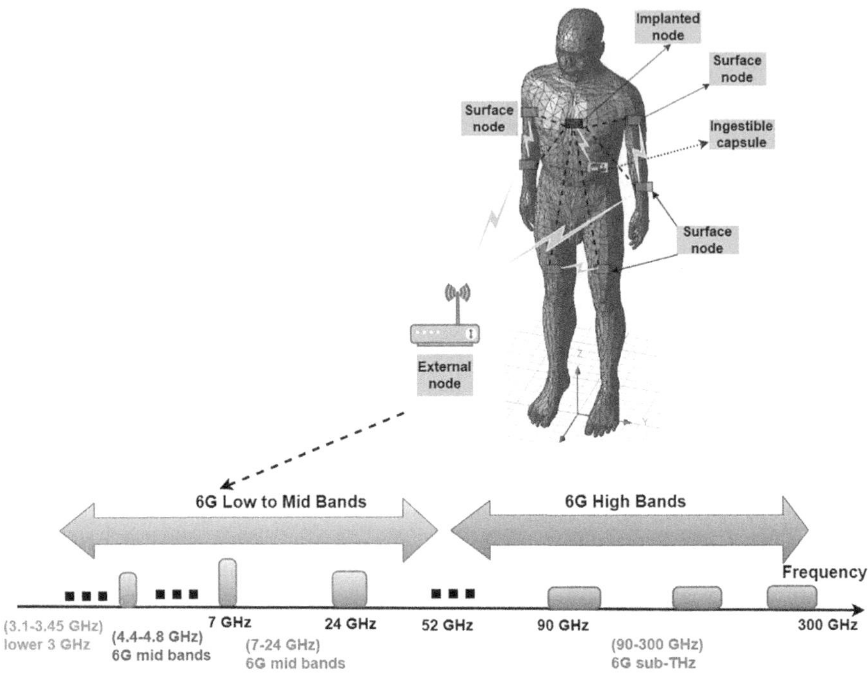

Figure 18.1 Representation of WPT to an implanted node via 6G frequency band [8].

networks. Generally, a WPT system comprises critical components such as a power source, PA, transmitter, filter [5, 6], medium of transmission, receiver, rectifier, battery, & modules for control & communication. Among these, the antenna [7] (transmitting) serves a vital role in the WPT system for transferring power to the receivers implanted for biomedical purposes. The basic structure of a WPT to implanted device inside a human phantom via a 6G frequency band [8] is shown in Figure 18.1. Numerous studies examine WPT for biomedical applications using inductive coupling [9] and capacitive coupling [10]. Inductive coupling systems have a wide variety of uses, from providing high power to railway systems and electric vehicles to providing modest power to laptop computers and cell phones. The amount of power received by a cell phone can vary depending on factors such as network technology, signal strength, distance from the cell tower, & the phone's design.

In general, mobile phones can receive power in terms of a received signal strength indicator (RSSI) ranging from −50 decibels per milliwatt (dBm) to −100 dBm. For use in applications involving the charging of mobile devices, a contactless power transfer system complete with a high-quality factor (Q) resonant tank has been created and put into operation [11]. In [12], the problem of maximizing the charging utility under the constraints of EMR (electromagnetic radiation) safety is discussed. WPT can be classified as either near-field (operating range Hz to MHz) or far-field (operating range MHz to THz) transmission. In general, near-field power transmission methods are more efficient than their far-field counterparts. Capacitive coupling, which uses electric fields; inductive coupling, which uses magnetic fields; and magnetic resonant inductive coupling, which uses a resonant circuit in both the transmitter and receiver coils, are the three most common approaches to achieving a near-field WPT [13]. In [14], A switched-capacitor-based stimulator integrated circuit that enables efficient harvesting of radio frequency (RF) power for applications involving neuro-stimulation is given. The idea of a "microbead," which is a completely integrated and wirelessly driven neural device that enables the activation of brain tissue in a spatially selective manner, is discussed in [15]. Mirbozorgi describes the idea of an inductive link for WPT to mm-sized free-floating implants (FFIs) that are spread out in a large three-dimensional space in the neural tissue without taking any care where the receiver is [16]. In [17], an efficient WPT system for real-time telemetry and wireless powering of deep-tissue implants with an antenna that works at 403, 915, 1,470, and 2,400 MHz is described. A radiating near-field (RNF) RF harvester containing a rectenna and source/transmitter with a 47.7% efficiency at 2.4 GHz was developed and produced [18]. In [19], it was proposed to use a small, battery-free device driven wirelessly by RNF WPT technology at 1,900 MHz to monitor increased intracranial pressure.

However, near-field wireless power technology has limitations regarding short range, efficiency loss, alignment and positioning requirements, and high cost. So, compared to near-field WPT, far-field WPT has some advantages, such as a longer coverage range, flexibility in device positioning, simultaneous charging of multiple devices, scalability, high power transfer efficiency, and reduced infrastructure requirements. One alternative name for far-field WPT is microwave coupling. Both far-field and close-range WPT are part of the hybrid form of wireless power transfer (HWPT). The far-field methods incorporate antennas to deliver electricity to a device. Characterizing the maximum achievable power levels in far-field RF powering-based implantable wireless myoelectric sensors while preserving safe local temperature and RF powering conditions is discussed in [20].

18.2.2 Need of PA in WPT

In advanced transmitter systems, the PA has become an increasingly important part of sending data to faraway places by amplifying the RF signal. The PA is a crucial component in the WPT system for 6G communication because it provides outstanding power transfer. The drain efficiency (DE), power added efficiency (PAE), and output power (Pout) of the PA could have a tremendous impact on the efficacy of WPT systems. A class E amplifier is developed for systems that utilize capacitively coupled wireless power transfer (CCWPT) in the MHz band [21]. Also, Casanova suggested the design and optimization of a Class E PA that utilizes inductive coupling for wireless power technology [22]. In [23], gallium nitride (GaN)-based class F PA designed at 4.6 GHz for wireless applications is discussed. A hybrid class E/F3 PA is analyzed and discussed in [24]. On reviewing some of the articles, it is suggested that there is a need for an efficient, high-power PA [25] to perform WPT from the transmitter to the receiver associated with a bio-implant device. Numeric ranges suitable for PA specifications utilized for WPT applications are:

- Power Output: The output power of a PA may vary from a few milliwatts (mW) to several watts (W), depending on the specific purpose and power requirements of the implanted device. It can, for example, range from 10 mW to 2 W.
- Efficiency: Depending on the design and operating conditions, GaN-based PAs may achieve efficiencies of 60%–70% or higher.
- Frequency Range: Take into consideration the frequency range that is suited for the wireless transfer of electricity, such as the ISM bands. The frequency of 13.56 MHz or 915 MHz is a popular option.
- Linearity: Look for a PA with good linearity characteristics. A linearity specification of better than −30 decibels relative to the carrier (dBc) is desirable for minimizing distortion.
- Thermal Management: To avoid overheating, choose a PA with acceptable thermal properties. The thermal design of the implanted device should allow for safe functioning. Thermal resistance values less than 10°C/W (degrees Celsius per watt) should be considered.
- Size and Form Factor: The PA should be as compact as feasible because implantable devices only have a minimal amount of space. A few cubic centimeters are typically a suitable size.
- Supply Voltage: The supply voltage should be compatible with the power source or battery used in the implantable device. Typical ranges can be 1.8 V to 5 V for metal oxide semiconductor (MOS) devices and 28 V to 30 V for GaN devices [26, 27], depending on the application and power requirements.
- Compliance with Medical Standards: The PA must be safe and compatible with other medical devices by adhering to applicable standards and regulations, such as those set forth by the Food and Drug Administration (FDA) or the European Medicines Agency (EMA).
- Safety Features: Build-in safeguards, such as overcurrent protection and temperature monitoring, into the power amplifier's circuitry. The normal range for overcurrent protection is 10%–150% of the operating current.

It's significant to note that the specific numeric ranges will depend on the needs of unique implantable devices; therefore, it's best to contact specialists in the area to ensure the specifications correspond with the needs of the device and comply with regulatory regulations. Please note that these are only recommendations; the amounts will depend on the needs of our implantable device and the wireless power transmission mechanism employed. It is critical to check with specialists in the field to ensure that a particular PA fits the requirements of a unique biomedical application.

18.3 MATHEMATICAL FORMULATION

In the WPT system, it is challenging to transmit the power efficiently from one end to the other end. While transmitting the power wirelessly from one antenna (at the transmitting end) to the other antenna (at the receiving end), there is a leakage of power. This leakage power generally depends on the input power fed to the transmitting antenna through PA and the received power received at the receiving antenna, as given in (18.1).

$$\text{Leakage power} = P_{\text{input}} - P_{\text{Received}} \tag{18.1}$$

The leakage power can also be formulated as given in (18.2)

$$\text{Leakage power} = \left(1 - \Gamma^2\right) \times P_{\text{input}} \tag{18.2}$$

where the reflection coefficient (Γ) is expressed as

$$\Gamma = \left(\frac{Z_{\text{in}} - Z_{\text{o}}}{Z_{\text{in}} + Z_{\text{o}}}\right) \tag{18.3}$$

by putting the values from (18.2) and (18.3) in (18.1), we get

$$\left(1 - \left(\frac{Z_{\text{in}} - Z_{\text{o}}}{Z_{\text{in}} + Z_{\text{o}}}\right)^2\right) \times P_{\text{input}} = P_{\text{input}} - P_{\text{Received}} \tag{18.4}$$

So, it is essential to find the amount of power transmitted from the transmitting side to the receiving side after the impact of leakage power. Now, to formulate the expression for power transmitted to the receiver, we need to define the equation of other factors on which the WPT depends [28].

These factors are propagation loss (P_{Loss}), absorption loss (A_{Loss}), reflection loss (R_{Loss}), transmission loss (Tr_{Loss}), mismatch loss (Mis_{Loss}), transmitter loss (T_{Loss}), medium loss (M_{Loss}), and receiver loss (Re_{Loss}), as defined in Equations (18.5)–(18.12) given below:

$$P_{\text{Loss}} = 20 \log_{10}\left(\frac{D}{D_{\text{Ref}}}\right) \tag{18.5}$$

$$A_{\text{Loss}} = 10^{(-\alpha D)} \tag{18.6}$$

$$R_{\text{Loss}} = \left(1 - \left(\frac{Z_{\text{in}} - Z_{\text{o}}}{Z_{\text{in}} + Z_{\text{o}}}\right)^2\right) \tag{18.7}$$

$$\text{Tr}_{\text{Loss}} = 10^{-(T_{\text{Loss}} + M_{\text{Loss}} + \text{Re}_{\text{Loss}})} \tag{18.8}$$

$$\text{Mis}_{\text{Loss}} = 1 - \left(\frac{Z_{\text{total}} - Z_{\text{o}}}{Z_{\text{total}} + Z_{\text{o}}}\right)^2 \tag{18.9}$$

$$T_{\text{Loss}} = 10^{-(T_{\text{coefficient}} \times D)} \tag{18.10}$$

$$M_{\text{Loss}} = 10^{-(M_{\text{attenuation}} \times D)} \tag{18.11}$$

$$\text{Re}_{\text{Loss}} = 10^{-(R_{\text{attenuation}} \times D)} \tag{18.12}$$

In the above equations, the values of the transmission coefficient ($T_{\text{coefficient}}$), medium attenuation ($M_{\text{attenuation}}$), and receiver attenuation ($R_{\text{attenuation}}$) can be preassigned by the user. Note that all the losses are measured in decibels (dB).

Now, power received by the receiving antenna [29] is expressed by (18.13)

$$P_{\text{Received}} = P_{\text{Transmitted}} \times \left[\frac{G \times \lambda}{4\pi D}\right]^2 \times \left[\frac{P_{\text{Loss}} \times A_{\text{Loss}} \times R_{\text{Loss}}}{\text{Tr}_{\text{Loss}} \times \text{Mis}_{\text{Loss}}}\right] \tag{18.13}$$

by putting the value from (18.13) in (18.4), we get

$$\left(1 - \left(\frac{Z_{\text{in}} - Z_{\text{o}}}{Z_{\text{in}} + Z_{\text{o}}}\right)^2\right) \times P_{\text{input}} = P_{\text{input}} - \left(P_{\text{Transmitted}} \left[\frac{G \times \lambda}{4\pi D}\right] \times \left[\frac{P_{\text{Loss}} \times A_{\text{Loss}} \times R_{\text{Loss}}}{\text{Tr}_{\text{Loss}} \times \text{Mis}_{\text{Loss}}}\right]\right) \tag{18.14}$$

By putting the expression of P_{Loss}, A_{Loss}, R_{Loss}, Tr_{Loss}, and Mis_{Loss} in (18.14), we get the desired expression for power transmitted to the receiver as given in (18.15).

$$P_{\text{Transmitted}} = \frac{\left(\frac{Z_{\text{in}} - Z_{\text{o}}}{Z_{\text{in}} + Z_{\text{o}}}\right)^2 \times P_{\text{input}} \times 10^{\left[(T_{\text{Loss}} + M_{\text{Loss}} + R_{\text{Loss}}) \times \left(1 - \left(\frac{Z_{\text{total}} - Z_{\text{o}}}{Z_{\text{total}} + Z_{\text{o}}}\right)^2\right)\right]}}{\left[\frac{G \times \lambda}{4\pi D}\right]^2 \times 10^{(-\alpha D)} \times \left(1 - \left(\frac{Z_{\text{in}} - Z_{\text{o}}}{Z_{\text{in}} + Z_{\text{o}}}\right)^2\right) \times 20\log_{10}\left(\frac{D}{D_{\text{Ref}}}\right)} \tag{18.15}$$

where G is the gain of the transmitter (dB), P_{input} is input power to PA (dBm), $P_{\text{Transmitted}}$ is transmitted power from the antenna of PA (dBm), P_{Received} is the received power at the receiver (dBm), the power transmitter and receiver separated by distance D (m), λ is the wavelength of electromagnetic wave, Z_{in} is input impedance (Ω), α is absorption coefficient (dB/m), Z_{total} is total impedance (Ω), D_{Ref} is the reference distance chosen to compare against the actual distance (m).

To provide a generalized mathematical formulation for the power received at implanted devices with an "i_{th}" number of surface nodes, we can consider the collective power received from all the surface nodes. Let's denote the received power as P_{Received}. The general mathematical equation for the received power of a bio-medical implant device with an "i_{th}" number of surface nodes is

$$P_{\text{Received}} = P_1 + P_2 + P_3 + \cdots + P_i \tag{18.16}$$

Here, P_1, P_2, P_3, ..., P_i represents the individual power received from each surface node. The individual power received from each surface node can be determined by considering factors such as transmission power, distance, path loss, and any other relevant parameters associated with that specific surface node. The equation for the received power from the "j_{th}" surface node can be formulated as:

$$P_{\text{Received}_j} = P_{\text{Transmitted}_j} \times G_j \times G_{\text{Receiver}_j} \times L_{\text{pathloss}_j} \tag{18.17}$$

where $P_{\text{Transmitted}_j}$ represents the transmitted power from the source to the "j_{th}" surface node, G_j represents the gain of the transmitter antenna associated with the "j_{th}" surface node; G_{Receiver_j} represents the gain of the receiver antenna on the bio-medical implant associated with the "j_{th}" surface node, and L_{pathloss_j} represents the path loss from the transmitter to the "j_{th}" surface node.

18.4 EXAMPLE OF CLASS F⁻¹ PA DESIGN

18.4.1 Design Implementation of PA

A Class F⁻¹ PA is designed by keeping load impedance with the even harmonics as open-circuited and odd harmonics as short-circuited [30]. Following the optimization of the transmission lines (TL) dimensions, a 10-W Cree CG2H40010F [31] GaN-based class F⁻¹ PA is simulated, designed, and fabricated at 4.6 GHz with the operating point ($V_{\text{gs}}=-2.7$ V and $V_{\text{ds}}=28$ V). This PA is developed on a Rogers (4350B) substrate with a dielectric constant of 3.66, a substrate height of 30 mils (0.762 mm), and a metal thickness of 35 μm. The design of the PA consists of a gate and drain bias network and an input and output matching network. TL line and radial stub-based gate and drain side bias networks are designed with resistor R1, TL lines TL1, TL2, TRstub1, and TRstub2 with $\theta=90°$, and TL3, TL4, TRstub3, and TRstub4 with $\theta=90°$, respectively. Two TL lines are linked together by connector TLs (CN1, CN2, CN3, CN4, CN5, and CN6) containing the nearby width of the TL lines. The bypass (C1, C2) and DC-blocking (C4 and C5) capacitor pairs, each having a value of 100 pF and 27 pF, respectively, are utilized at PAs input and output side circuitry.

A parallel combination of R2 of 10 Ω and C3 of 27 pF placed at the gate side of the transistor is used to make a stability network for the PA. The grouped TL, such as (TL8, TL9) and (TL5, TL6, TL7), form a network for output and input matching, respectively. TLd and TLg are placed beside the drain and gate terminals, respectively. The whole circuit of the PA is designed with the TL lines and radial stub lines, having specifications given in Table 18.1.

Table 18.1 Optimized Value of Design Parameters Utilized for PA

S. no.	Design parameters	Optimized values (mm)
1.	W1, W2	1.495
2.	L1	12
3.	W3, W4	2.5814
4.	L2	14.26
5.	L3	20
6.	L4	14.09
7.	W5, W6, W7, W8, W9	1.47416
8.	CN1 (W10, W11, W12)	1.47416
9.	CN2 (W13, W14, W15)	1.495, 1.47416, 2
10.	CN3 (W16, W17, W18, W19)	1.495
11.	CN4 (W20, W21, W22)	1.47416, 0.794, 2.5814
12.	CN5 (W23, W24, W25, W26)	2.5814
13.	CN6 (W27, W28, W29)	1.47416
14.	L5	5.8
15.	L6	12.17
16.	L7	23.51
17.	L8	3.166
18.	L9	8.22
19.	Wg	2
20.	Lg	9
21.	Wd	0.794
22.	Ld	2.693
23.	Wr1	1.495
24.	Lr1	12
25.	Wr2	2.5814
26.	Lr2	20

18.4.2 Simulation Results

Finally, we have designed the schematic of GaN-based class F^{-1} PA with the utilization of optimized values of design parameters of microstrip TLs, as shown in Figure 18.2. The harmonic balance (HB) simulation is performed to check the small and large signal parameters of the proposed PA. The whole circuit design and simulation are performed using Keysight's advanced design system (ADS) software. The transient response of current–voltage $(I–V)$ waveforms of the developed class F^{-1} PA is represented in Figure 18.3, having voltage as half sinusoidal and current as a square waveform in nature. Scattering $(S\text{-})$ parameter analysis is performed to see the performance of small signal gain (S21), input reflection coefficient (S11), reverse transmission coefficient (S12), and output reflection coefficient (S22).

The simulation results show that the value of S11 is –13 dB, S12 is –30.4 dB, S21 is 15.46 dB, and S22 is –14.46 dB. It is seen that the S-parameter results are better than –10 dB. Thereafter, we performed the HB analysis to check the effectiveness of P_{out}, DE, and PAE. The achieved simulation results of PA for DE of 64.72%, PAE of 60.54%, P_{out} of 41.37 dBm at 28 dBm P_{in}, Gain of 11.37 dB, and IMD_3 of –40.2 dBc and IMD_5 of –71.09 dBc, respectively.

18.4.3 Measurement Results

The actual area of the whole PA is 88 mm×80 mm. We have utilized a driver amplifier [32] of 17 dB gain, a signal generator (N5172B), a spectrum analyzer (N9020A), an attenuator (BW-S40W20+) from mini-circuits, a 3 dB coupler, a power sensor, and a direct current (DC) power source. Thereafter, we performed the HB to check the effectiveness of the S-parameters, P_{out}, DE, and PAE. The achieved measurement results of PA for DE of 58.4%, PAE of 55.1%, P_{out} of 38.21 dBm at 28 dBm P_{in}, and gain of 10.2 dB, IMD_3 of –32.69 dBc, IMD_5 of –64.01 dBc respectively. Also, the measured result of S11 is –11 dB, S12 is –22.8 dB, S21 is 12.5 dB, and S22 is –12.46 dB. Table 18.2 shows a comparative study of various PA designs with different devices. The measured results of the suggested PA show that both DE and P_{out} have an effective value. By observing the simulation and measurement results in Figure 18.4(a) and (b), we learned that the achieved efficiency is in the required ranges, and the output power is 38.21 dBm (7 W). Also, the linearity parameter, i.e., the IMD of PA, is better than –25 dBc. All other parameters are compatible with the requirements of such a type of PA for WPT. Also, this proposed PA follows the safety and regulatory guidelines required for performing WPT for 6G communication.

18.4.3.1 Conclusion and Future Scope

In this article, we have developed a GaN-based class F^{-1} PA utilized to perform WPT for 6G communication. Furthermore, the results are evaluated

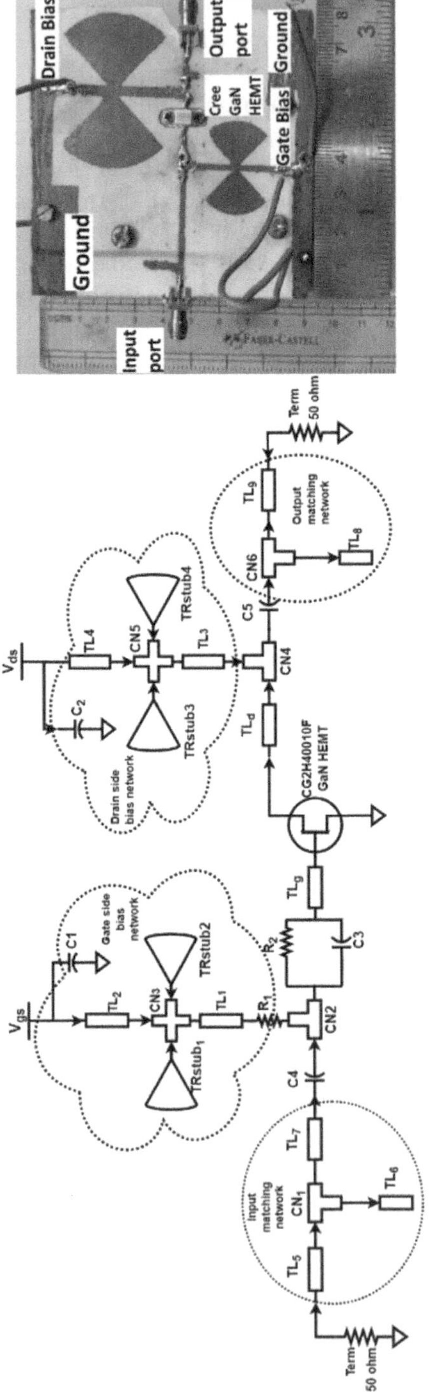

Figure 18.2 Representation of schematic with the fabricated view of the developed PA.

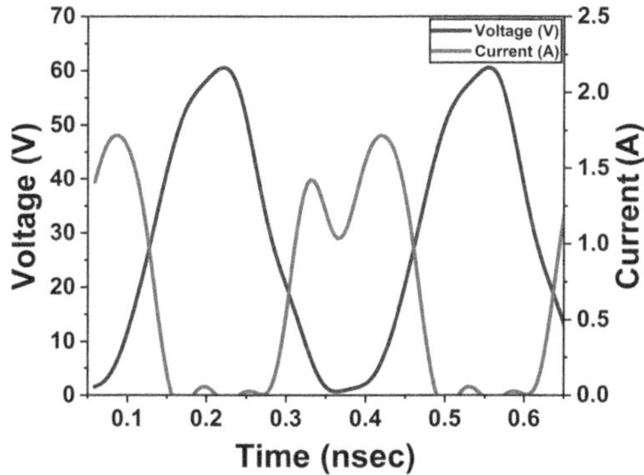

Figure 18.3 Simulated *IV* curve of 10-W PA at 28 dBm input power.

Table 18.2 State of the Art of Various Designs of PA with Different Devices

Ref.	Device	Frequency (GHz)	Biasing	Gain (dB)	DE (%)	P_{out} (dBm)	IMD_3, IMD_5 (dBc)
[33]	MACOM	1–1.2	48V, –	11–15	60–70	41–43.01	–27, –
[34]	Cree, GaN CGH40025F	0.85–5.4	30V, 680 mA	>8–9.5	>45	46.3	–
[35]	Qorvo TGF2819-FL	1–2.8	50V, 200 mA	11.1–13.1	51.6–69	50.5	–25, –
[36]	65 nm TSMC CMOS	0.3	3.48V, 72.36 mA	–	87	20.58	–
[37]	CGH40006P, CGH40025F	2.5–3.5	28V, 70 mA	14.4–17.2	45.4–67.2	46.6–47.6	–
[38]	Cree, GaN CGH40010F	0.4–2.0	28V, 40–90 mA	11	61–72	>40	–
This work	Cree, GaN CG2H40010F	4.6	28V, 175 mA	10.2	58.4	38.21	–32.69, –64.01

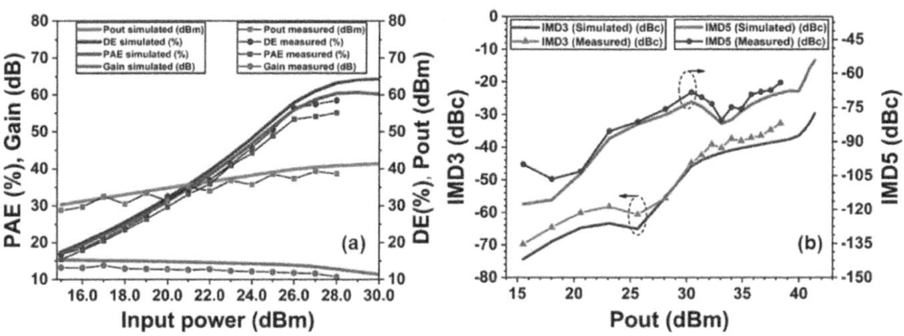

Figure 18.4 Simulated and measured results of (a) PAE, P_{out}, DE, and Gain (b) IMD_3 and IMD_5 w.r.t P_{out} of PA.

by implementing the design, simulation, fabrication, and measurement of PA. The achieved measurement results of PA for DE of 58.4%, PAE of 55.1%, P_{out} of 38.21 dBm at 28 dBm P_{in}, and gain of 10.2 dB, IMD_3 of –32.69 dBc, IMD_5 of –64.01 dBc respectively. Through WPT, the proposed PA is capable of wirelessly charging the bio-implant devices. This proposed PA represents a substantial advancement in efficiency and reliability. As we move forward into the era of 6G, the principles outlined in this chapter will play a pivotal role in shaping the future of wireless communication. With continued advances in optimization techniques, the future of PAs as transmitters promises even greater energy efficiency, enhanced dependability, and industry-changing applications. In the future, the proposed PA can be integrated with pre-distortion techniques to design linearized PA.

REFERENCES

[1] U. Gustavsson, P. Frenger, C. Fager, T. Eriksson, H. Zirath, F. Dielacher, C. Studer, A. Pärssinen, R. Correia, J.N. Matos, and D. Belo, "Implementation Challenges and Opportunities in Beyond-5G and 6G Communication," *IEEE Journal of Microwaves*, vol. 1, no. 1, pp. 86–100, Jan. 2021, doi: 10.1109/JMW.2020.3034648.

[2] W. Saad, M. Bennis, and M. Chen, "A Vision of 6G Wireless Systems: Applications, Trends, Technologies, and Open Research Problems," *IEEE Network*, vol. 34, no. 3, pp. 134–142, 2019.

[3] G. Bhargava, V. Vadalà, S. Majumdar, and G. Crupi, "Auto-Encoder Based Hybrid Machine Learning Model for Microwave Scaled GaAs pHEMT Devices," *International Journal of RF and Microwave Computer-Aided Engineering*, vol. 32, no. 11, p. e23339, 2022.

[4] M. Alsabah, M.A. Naser, B.M. Mahmmod, S.H. Abdulhussain, M.R. Eissa, A. Al-Baidhani, N.K. Noordin, S.M. Sait, K.A. Al-Utaibi, and F. Hashim, "6G Wireless Communications Networks: A Comprehensive Survey," *IEEE Access*, vol. 9, pp. 148191–148243, 2021, doi: 10.1109/ACCESS.2021.3124812.

[5] G. Bhargava and S. Majumdar, "Assessment of Filter Design for 5G Applications," *2022 IEEE VLSI Device Circuit and System (VLSI DCS)*, Kolkata, India, 2022, pp. 85–88, doi: 10.1109/VLSIDCS53788.2022.9811462.

[6] G. Bhargava and S. Majumdar, "Design of Telescopic OTA Based 6th Order Butter-worth Low Pass Filter Using 0.18 µm CMOS Technology," *2020 IEEE VLSI Device Circuit and System (VLSI DCS)*, Kolkata, pp. 489–493, 2020.

[7] G. Bhargava, S. Majumdar, G. Gugliandolo, G. Campobello, N. Donato, and G. Crupi, "Design and Validation of a Low-Cost Antenna-Based Solution for Microwave Imaging of RCC Structure," *IEEE Sensors Letters*, vol. 7, no. 4, pp. 1–4, April 2023, Art no. 3500604, doi: 10.1109/LSENS.2023.3260972.

[8] 6G Note "6G Frequency Band," 2024, https://www.nokia.com/about-us/newsroom/articles/6g-mid-band-spectrum-technology-explained/ [Online: accessed 12-March-2024].

[9] R. Shadid and S. Noghanian, "A Literature Survey on Wireless Power Transfer for Biomedical Devices," *International Journal of Antennas and Propagation*, vol. 1, p. 4382841, 2018.

[10] J. Dai and D. C. Ludois, "A Survey of Wireless Power Transfer and a Critical Comparison of Inductive and Capacitive Coupling for Small Gap Applications," *IEEE Transactions on Power Electronics*, vol. 30, no. 11, pp. 6017–6029, 2015.

[11] Q. Li and Y. C. Liang, "An Inductive Power Transfer System with a High-q Resonant Tank for Mobile Device Charging," *IEEE Transactions on Power Electronics*, vol. 30, no. 11, pp. 6203–6212, 2015.

[12] H. Dai, Y. Liu, G. Chen, X. Wu, T. He, A. X. Liu, and H. Ma, "Safe Charging for Wireless Power Transfer," *IEEE/ACM Transactions on Networking*, vol. 25, no. 6, pp. 3531–3544, 2017.

[13] A. M. Le, L. H. Truong, T. V. Quyen, C. V. Nguyen, and M. T. N. T. Nguyen, "Wireless Power Transfer Near-Field Technologies for Unmanned Aerial Vehicles (UAVS): A Review," *EAI Endorsed Transactions on Industrial Networks and Intelligent Systems*, vol. 7, no. 22, p. e5, Jan. 2020. [Online]. Available: https://publications.eai.eu/index.php/inis/article/view/248.

[14] H. Lyu, J. Wang, J.-H. La, J. M. Chung, and A. Babakhani, "An Energy-Efficient Wirelessly Powered Millimeter-Scale Neurostimulator Implant Based on Systematic Codesign of an Inductive Loop Antenna and a Custom Rectifier," *IEEE Transactions on Biomedical Circuits and Systems*, vol. 12, no. 5, pp. 1131–1143, 2018.

[15] A. Khalifa, Y. Karimi, Q. Wang, W. Montlouis, S. Garikapati, M. Stanacevic, N. Thakor, and R. Etienne-Cummings, "The Microbead: A Highly Miniaturized Wirelessly Powered Implantable Neural Stimulating System," *IEEE Transactions on Biomedical Circuits and Systems*, vol. 12, no. 3, pp. 521–531, 2018.

[16] S. A. Mirbozorgi, P. Yeon, and M. Ghovanloo, "Robust Wireless Power Transmission to mm-Sized Free-Floating Distributed Implants," *IEEE Transactions on Biomedical Circuits and Systems*, vol. 11, no. 3, pp. 692–702, 2017.

[17] A. Basir and H. Yoo, "Efficient Wireless Power Transfer System with a Miniaturized Quad-Band Implantable Antenna for Deep-Body Multitasking Implants," *IEEE Transactions on Microwave Theory and Techniques*, vol. 68, no. 5, pp. 1943–1953, 2020.

[18] B. J. DeLong, A. Kiourti, and J. L. Volakis, "A Radiating Near-Field Patch Rectenna for Wireless Power Transfer to Medical Implants at 2.4 GHz," *IEEE Journal of Electromagnetics, RF and Microwaves in Medicine and Biology*, vol. 2, no. 1, pp. 64–69, 2018.

[19] S. A. A. Shah and H. Yoo, "Radiative Near-Field Wireless Power Transfer to Scalp-Implantable Biotelemetric Device," *IEEE Transactions on Microwave Theory and Techniques*, vol. 68, no. 7, pp. 2944–2953, 2020.

[20] R. A. Bercich, D. R. Duffy, and P. P. Irazoqui, "Far-Field RF Powering of Implantable Devices: Safety Considerations," *IEEE Transactions on Biomedical Engineering*, vol. 60, no. 8, pp. 2107–2112, 2013.

[21] B. Chokkalingam, S. Padmanaban, Z. M. Leonowicz, "Class E Power Amplifier Design and Optimization for the Capacitive Coupled Wireless Power Transfer System in Biomedical Implants," *Energies*, vol. 10, no. 9, pp. 1–20, 2017.

[22] J. J. Casanova, Z. N. Low, and J. Lin, "Design and Optimization of a Class-E Amplifier for a Loosely Coupled Planar Wireless Power System," *IEEE Transactions on Circuits and Systems II: Express Briefs*, vol. 56, no. 11, pp. 830–834, 2009.

[23] G. Bhargava, P. K. Rath, and S. Majumdar, "GaN-based Class-F Power Amplifier for 5G Applications," *2022 IEEE Microwaves, Antennas, and Propagation Conference (MAPCON)*, Bangalore, 2022, pp. 1444–1449.

[24] G. Bhargava and S. Majumdar, "Highly Linearized GaN HEMT Based Class E/F3 Power Amplifier," *2021 Devices for Integrated Circuit (DevIC)*, Kalyani, India, 2021, pp. 564–568, doi: 10.1109/DevIC50843.2021.9455844.

[25] G. Bhargava, Hemant K. Dewangan, V. Vadala, S. Majumdar, and G. Crupi. "Physics-Informed Neural Network Assisted Automated Design of Power Amplifier by User Defined Specifications," *International Journal of Numerical Modelling: Electronic Networks, Devices, and Fields (Wiley)*, vol. 37, no. 3, p. e3246, 2024.

[26] S. Majumdar, A. Bag, D. Biswas, "Comparative Analysis of Parameter Extraction Techniques for AlGaN/GaN HEMT on Silicon/Sapphire Substrate," *Microelectronics Reliability*, vol. 78, pp. 389–95, 2017.

[27] S. Majumdar, D. Biswas, "Evaluating substrate's effect on RF switch performance via Verilog-A GaN HEMT model", *Microelectronics Journal*, Vol. 62, pp. no. 43-48, 2017.

[28] T. S. Rappaport, *Wireless Communications: Principles and Practice*, 2nd ed, Pearson Education India, 2010, doi: 10.1017/9781009489836.

[29] A. Goldsmith, *Wireless Communications*. New York: Cambridge University Press, 2005.

[30] S. C. Cripps, *RF Power Amplifiers for Wireless Communications*. Norwood, MA: Artech House, 2006, vol. 250, 2nd ed.

[31] Datasheet, "Cree CG2H40010F Packaged Discrete Transistor," 2022, https://assets.wolfspeed.com/uploads/2020/12/CG2H40010.pdf [Online: accessed 22-April-2022].

[32] G. Bhargava, S. Majumdar, and F. Medjdoub, "Importing Experimental Results via S2D Model in ADS Tool for Power Amplifier Design," *IETE Journal of Research*, 2024, doi: 10.1080/03772063.2024.2326588.

[33] N. Tuffy and L. Pattison, "A Compact High Efficiency GaN-Si PA Implemented in a Low Cost DFN Package with 71% Fractional Bandwidth," *2014 IEEE MTT-S International Microwave Symposium (IMS2014)*, Tampa, FL, 2014 Jun 1, pp. 1–3.

[34] H. T.-A. Nia and V. Nayyeri, "A 0.85–5.4 GHz 25-W GaN Power Amplifier," *IEEE Microwave and Wireless Components Letters*, vol. 28, no. 3, pp. 251–253, 2018.

[35] K. Tran, R. Henderson, and J. Gengler, "Design of a 1–2.8-GHz 100-W Power Amplifier with Bounded Performance Technique," *IEEE Transactions on Microwave Theory and Techniques*, vol. 67, no. 9, pp. 3707–3715, 2019.

[36] D. De Venuto, G. Mezzina, and J. Rabaey, "Automatic 3D Design for Efficiency Optimization of a Class E Power Amplifier," *IEEE Transactions on Circuits and Systems II: Express Briefs*, vol. 65, no. 2, pp. 201–205, 2018.

[37] H. Liu, C. Li, S. He, W. Shi, Y. Chen, and W. Shi, "Simulated Annealing Particle Swarm Optimization for a Dual-Input Broadband Gan Doherty Like Load-Modulated Balance Amplifier Design," *IEEE Transactions on Circuits and Systems II: Express Briefs*, vol. 69, no. 9, pp. 3734–3738, 2022.

[38] R. Krishnamoorthy, N. Kumar, A. Grebennikov, and H. Ramiah, "A High-Efficiency Ultra-Broadband Mixed-Mode Gan Hemt Power Amplifier," *IEEE Transactions on Circuits and Systems II: Express Briefs*, vol. 65, no. 12, pp. 1929–1933, 2018.

Chapter 19

6G wireless communication

Prospective technological advances

Sheeja Pon Chakravarthy

19.1 KEY TECHNOLOGIES FOR 6G-ENABLED COMMUNICATION

19.1.1 A Peak Data Rate of ≥ 1 Tb/s (100 Times of 5G)

The peak data rate targeted for sixth generation is in the order of Tb/s greater than 1 Tb/s, which is 100 times higher than that of fifth generation. It is a tremendous increase in the data at what we are going to support with sixth generation.

19.1.2 A User-Experienced Data Rate of 1 Gb/s (10× of 5G)

The user-experienced data rate considers the effects of restricted coverage and mobility, and it centers on the available data rate of numerous users. It indicates the amount of data rate that a single user will experience, which is intended to be in the range of 1 Gbps, or over ten times faster than that of the fifth generation.

19.1.3 THz Frequency Range: 0.1–10 THz

The range of frequencies in the terahertz range is very high, ranging from 100 GHz to 10 THz. This presents several challenges that we must overcome, but the benefit is that we will be able to support significant data rates at these higher frequencies, measured in terabytes per second.

19.1.4 Spectrum Efficiency of Five to Ten Times of 5G

The quantity of data transferred with the fewest transmission mistakes possible over a specific spectrum or bandwidth is referred to as spectrum efficiency. Our goal for spectrum efficiency is five to ten times higher than that of a fifth generation.

DOI: 10.1201/9781003522003-22

19.1.5 High Mobility

In wireless networks, "mobility" essentially refers to a node, also known as a Mobile Node (MN), or occasionally a subnet, shifting its point of attachment to the network while maintaining continuous network communication. The future generation of wireless networks requires very high mobility, but people anticipate a very high highly mobile environment that will be targeted in the same domain as very high-speed networks were previously handled.

19.1.6 Latency of 10–100 µs

The delay in network communication is known as network latency. It displays how long it takes for data to move across a network. High latency networks are those that have a larger delay or lag, whereas low latency networks respond quickly. There would undoubtedly be very little delay when we aim for great mobility; this should be between 10 and 100 µs.

19.1.7 Ten Times the Connectivity Density of 5G

The total number of devices providing a particular level of service per unit area is known as connection density. Our focus is on the connectivity density of nearby nodes, which is at least ten times larger than that of the fifth generation of wireless communication.

19.1.8 High Throughput, Network Capacity

The quantity of data that moves from a source to a destination during a specific length of time over a network is known as throughput. Data could be communicated utilizing a greater bandwidth transmission method to transfer information more quickly. According to some analysts, 6G networks may eventually enable internet devices to reach maximum rates of one terabit per second (Tbps). That surpasses 1 Gbps, the fastest speed currently offered by most household internet networks, by 1,000 times. It surpasses 10 Gbps, the fictitious maximum speed of 5G, by a factor of 100.

19.1.9 Enhanced Data Security

While new technology offers wonderful opportunities, there is also a higher risk of malevolent attacks. Unlocking the complete value potential of 6G will require proactively addressing and resolving the security, privacy, and trust challenges. Holographic telepresence, self-governing collaborative robot systems, widespread digital twinning, the digital revolution of our offices, and even improved personal health monitoring are all expected in the 6G world. It's obvious that privacy and security will be top priorities.

We should anticipate an increase in the number of threats due to the billions of connected devices and sensors as well as the millions of subnetworks running in untrusted domains. While machine learning and artificial intelligence will aid in the battle, they will also present new openings for hackers.

A 6G network will be able to follow a person's movements and pinpoint their exact location within a space. Biometric data breaches could be exploited for sophisticated fraud and blackmail schemes. Because more and more business models depend on regulated access to and ownership of consumer, process, and business data, the confidentiality of such data will become even more crucial. In the 6G era, risks like ransomware attacks, corporate espionage, and "deep fakes" will only get worse. Because of the unmatched processing power of quantum computing, attackers will be able to breach security protocols and cryptographic techniques that were previously assumed to be impregnable.

The prevalence of localized industrial and campus networks will increase their attractiveness to attackers and increase the risk of industrial and critical infrastructure sabotage, which might seriously harm both people and property. Trust solutions that attest to the integrity of data and systems and offer evidence of data ownership rooted in silicon and hardware include Trusted Platform Modules and Trusted Execution Environments.

Tamper-proof tracking of security claims, such as data access permissions, will be accomplished by distributed ledger technologies, such as blockchain. Distributed ledger technologies offer a safe foundation to facilitate trust building across diverse operator domains and improve use cases for the 6G era. Jamming detectors and physical layer protection will be used to protect industrial networks. They'll have countermeasures like frequency hopping and directed null steering to mitigate uplink jamming.

19.1.10 Ubiquitous Connectivity

With intelligent connection supporting a fully ubiquitous internet by the 2030s, the vision of next-generation wireless communications is centered around 6G technology. The convenience of having internet access from almost any place serves as the foundation for a lot of our daily activities. When devices can create, distribute, and process data without ever losing connectivity, this is referred to as ubiquitous connectivity.

19.2 COMPARISON BETWEEN 5G AND 6G

In addition to enhancing our smartphone experience, 5G has made new experiences possible, such as augmented reality (AR) and virtual reality (VR). Industry sectors receive more consistent data rates for these novel

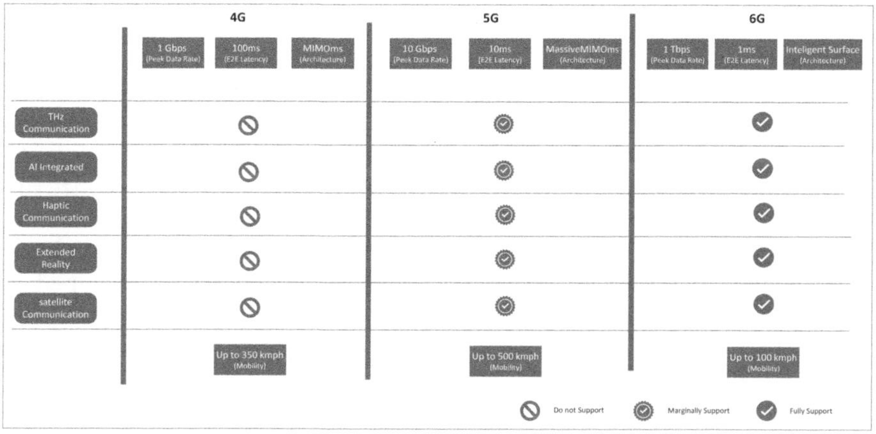

	4G			5G			6G		
	1 Gbps (Peak Data Rate)	100ms (E2E Latency)	MIMOms (Architecture)	10 Gbps (Peak Data Rate)	10ms (E2E Latency)	MassiveMIMOms (Architecture)	1 Tbps (Peak Data Rate)	1ms (E2E Latency)	Inteligent Surface (Architecture)
THz Communication	Do not Support			Marginally Support			Fully Support		
AI integrated	Do not Support			Marginally Support			Fully Support		
Haptic Communication	Do not Support			Marginally Support			Fully Support		
Extended Reality	Do not Support			Marginally Support			Fully Support		
satellite Communication	Do not Support			Marginally Support			Fully Support		
Mobility	Up to 350 kmph (Mobility)			Up to 500 kmph (Mobility)			Up to 100 kmph (Mobility)		

Legend: Do not Support · Marginally Support · Fully Support

Figure 19.1 Comparison between 4G, 5G, and 6G.

experiences due to the decreased cost-per-bit. The goal of 6G is to enhance these current services.

The future iteration of 5G, known as 6G, is a sixth-generation mobile network. Compared to 5G, the sixth-generation network will be able to operate at higher frequencies and will have enhanced speed, sophisticated data processing, seamless connectivity, and much lower latency. While mobile edge computing is an add-on feature for 5G networks, it will be a feature of 6G networks. The combined communication and computation infrastructure framework of 6G networks will incorporate edge and core computing more deeply, offering benefits including enhanced AI capabilities and enhanced support for mobile devices (Figure 19.1).

Several of the predicted 6G services cannot be realized with the current hierarchical network topologies, which stands in stark contrast to this. We distinguish between what may be theoretically feasible and what may be doable over the next ten years while assessing the advantages and disadvantages of important candidate 6G technologies.

19.3 ARCHITECTURE OF 6G

6G is a brand-new generation of network systems. Thus, the architecture design of 6G shall follow three fundamental principles: inheriting the advantages of 5G; meeting the requirement of new scenarios; and continuously enhancing the basic capabilities.

New performance metrics that represent the wider objectives of various stakeholders, such as operators, web-scalers, businesses, neutral hosts, and industry, will be introduced as we transition to the 6G era. These metrics ultimately represent societal values. The capabilities of networks and services are increasingly being measured by other metrics, such as sustainability, openness, digital inclusion, privacy, and trust. These are the kinds of new key value indicators (KVIs) that will overtake the more conventional KPIs of latency, capacity, and speed in terms of importance. We need to approach the 6G architectural blueprint with a new design methodology to construct operational networks that optimize these new KVIs. The new 6G system architecture must consider all those design objectives and criteria. All parties involved must prioritize sustainability by building networks with zero carbon footprints, in which every facet of the network's operation is planned to reduce or offset CO_2 emissions.

Compared to previous network architectures that were developed to maximize terrestrial networks, the architecture of 6G will be radically different. Three layers make up a 6G network: the *satellite layer*, the *aerial layer*, and the *terrestrial layer*. Most people agree that the vision for 6G is a unified three-dimensional network architecture. Next-generation communication networks will prioritize cross-layer optimization above forcing non-terrestrial networks (NTN) into a terrestrial framework. This will entail substantial advancements in the Radio Access Network (RAN) and the creation of a shared 6G network core, in addition to the creation and implementation of regenerative payloads, novel waveforms, and innovations in network management and orchestration.

Building a network that can serve these kinds of use cases and system objectives will necessitate several multidisciplinary architectural breakthroughs. Multiple heterogeneous and distributed public and private cloud platforms from various stakeholders can be used by a 6G network, which will make use of a cloud platform with a range of capabilities, including hardware acceleration. A new degree of programmability will be achieved by implementations to satisfy the requirements of a wide range of use cases. The architecture will have the high level of specialization and flexibility required to be implemented in highly local on-premises and personal-area networks, as well as large-scale wide area networks.

19.3.1 Space – Air – Ground – Underwater Networks

The architecture of the Space-Air-Ground Integrated Network (SAGIN) is currently in its early stages. The SAGIN ecosystem is still immature and cannot last into the actual world, even with the discovery of various important insights on the enhancement of terrestrial, aerial, and satellite systems. More robustness, scalability, agility, and flexibility are needed to meet the ideal abstraction standard. To provide ubiquitous network access, 6G will be an integrated network system that combines a regular terrestrial mobile network, space network, and undersea network. A promising network architecture that offers smooth, dependable, high-rate transmission over a very wide area is space-air-ground integrated networks. Nonetheless, in communication and network systems, the endpoints, resources, applications, and infrastructure have grown increasingly varied and complicated.

19.3.2 Three-Tier Network Descriptions

The present network protocol design, which has been incredibly successful, was in fact built on top of TCP/IP. Due to a few new issues, recent internet deficiencies guarantee the delivery of apps focused on the future. Some TCP/IP-based protocols have reduced these challenges to some extent, such as QUIC (Quick UDP Internet Connections). Unfortunately, rather than addressing the core problems with the Internet, these patch-like protocols only make things more complicated. Network layer packets now operate in a static header-payload mode that is unaffected by the needs of higher-level applications. Future IP protocols may heavily rely on meta-data and app designer instructions to facilitate a multitude of future advances and provide customized internet access. In a similar vein, further enhancements to the transport layer are necessary to meet future communication needs.

Advanced domain automation functionalities that orchestrate and automate across multiple network domains—possibly spanning multiple administrative domains, multiple stakeholders, and additional resources in the far edge and on premises beyond the traditional mobile network—will be included in the 6G architecture. For example, processing and storing the enormous volumes of data needed for services like AI/ML, XR, and the metaverse would require a major increase in computation and storage capacity.

19.4 PROSPECTIVE TECHNOLOGICAL ADVANCES

19.4.1 Terahertz Communications – 6G Test Bed

Currently, the goal is to establish a test bed for the sixth generation of communication using terahertz frequencies. Establishing a test bed for the sixth generation of communication using terahertz frequencies is a bit difficult because there is currently very little hardware support for terahertz communication because the hardware equipment needs to operate at a much higher frequency with high-speed switching capabilities, which is the first bottleneck.

19.4.2 Artificial Intelligence and Big Data Analytics

Sixth generation (6G) will enable a new wave of intelligent applications by combining AI and big data technologies. Observable, real-time activities without a device, driverless vehicles, and a host of other things.

19.4.3 Novel Radio Access Technology for 6G

For the radio access portion, hybrid-NOMA-OFDM or comparable technologies can be employed to improve connection, lower power consumption, and maximize spectrum use.

19.4.4 Super Massive Multiple Input Multiple Output (SM-MIMO)

To increase the efficiency of the spectrum and the throughput People will also be investigating SM-MIMO (Spatial Modulation – Multiple Input Multiple Output) in sixth-generation technology.

19.4.5 Laser Communications and Visible Light Communication (VLC)

Free space or visible light communication For sixth generation, optical communication with laser sources can also be included.

19.4.6 Quantum Communications and Computing

In addition, quantum communication appears to be a very appealing way to improve transmission capabilities by using qubits—a quantum bit, or qubit—as the fundamental information unit in quantum computing, which is analogous to the bit, or binary digit, in classical computing.

19.4.7 Orbital Angular Momentum Multiplexing (OAM)

Using the orbital angular momentum of the electromagnetic waves to differentiate between the several orthogonal signals, orbital angular momentum multiplexing (OAM multiplexing) is a physical layer approach for multiplexing signals carried by electromagnetic waves.

19.4.8 Energy and Spectrum Efficient Hardware and Resource Allocation

Energy and spectrum efficiency are crucial since energy is a bottleneck in this type of complicated 6G network. Advanced hardware and resource allocation strategies are required to accomplish it to increase energy and spectrum efficiency.

19.4.9 Holographic Beamforming

The 6G network's challenge of superposing radiation waves from various antenna elements is made more challenging by the requirement to build a novel beamforming system to solve the complex-domain optimization problem subject to the unusual real-domain amplitude limits.

19.4.10 Reconfigurable Intelligent Surface (RIS)

It is anticipated that 6G-and-beyond communications will enable a few cutting-edge applications, including brain-computer interfaces, high-fidelity holographic projections, digital twins, linked robotics and autonomous systems, industrial internet of things, and intelligent transportation systems. The current systems are not able to easily satisfy the high quality-of-service (QoS) needs of these applications, which include extremely high data rates, ultra-high dependability, and ultra-low latency.

19.4.11 Molecular Communications and the Internet of Nano-Things

Following the emergence of the emerging nanoscale communication paradigms of Terahertz (THz) and molecular communication (MC), the idea of a biological layer was developed. THz is thought to adopt more of a top-down strategy, whereas MC is claimed to adopt a bottom-up approach, drawing inspiration from natural signaling mechanisms.

19.4.12 Semantic Communication

Semantic communications are now seen as a critical component that will enable 6G networks in the future. Referring to Shannon's information theory, communication has always sought to ensure that messages are received correctly, regardless of their intended meaning. Sensing, location, localization, and orientation are all included in the larger framework of semantic communication. While positioning may be thought of as both a data point in and of itself and as a sort of metadata to communication, sensing can be thought of as a communication in which we are not in control of the information source. Semantic and goal-oriented communication toward 6G can be understood as a fusion of diverse communication modalities, much like the brain orchestrates multi-channel inputs.

19.4.13 Unmanned Aerial Vehicle

With improved communication capabilities, unmanned aerial vehicles (UAVs) are seen as crucial components of 6G networks because of their affordable and adaptable deployment options. The goal of 6G is to create an all-encompassing network that can connect people everywhere—in space, on land, and underwater. In the media, public, defense, and civic spheres, among others, unmanned aerial vehicles (UAVs) hold great promise. They have novel and practical uses in situations when human lives could otherwise be in jeopardy. Furthermore, when compared to single UAV systems, many UAV systems can perform the same tasks with greater economy, accuracy, and efficiency when working together on a mission. But before unmanned aerial vehicles can be effectively employed to provide safe, dependable, and context-focused networks, a few issues need to be addressed.

19.4.14 Integration of Access-Backhaul Networks

Using the same technology and specifications as access lines, integrated access and backhaul, or IAB, enables wireless backhaul links, making it a cost-effective and versatile solution for deploying ultra-dense fifth- and sixth-generation (5G and 6G) networks.

Chapter 20

6G applications and use cases

*Sreejith L. Das, Om Prakash, V. Jayachandra Naidu,
Saumya Das, Subrata Chowdhury, and
T. Somassoundaram*

20.1 INTRODUCTION

As technological advancements persist, the advent of 6G networks [1] presents an array of fascinating uses and scenarios. 6G networks, which build on the principles set by earlier wireless communication generations, promise unmatched connectivity, breakneck speeds, and revolutionary capabilities [2]. These networks are poised to revolutionise various sectors, including healthcare, transportation, education, entertainment, and beyond. With 6G, we can anticipate advancements such as ultra-high-definition streaming, immersive virtual and augmented reality experiences, seamless connectivity for Internet of Things (IoT) devices, and real-time collaboration on a global scale. Smart cities, driverless cars, remote surgery, environmental monitoring, and customised healthcare are some areas where 6G has applications. 6G networks can completely transform how people work, live, and interact by pushing the envelope of what is now feasible and opening the door to a new era of innovation and connectedness (Figure 20.1).

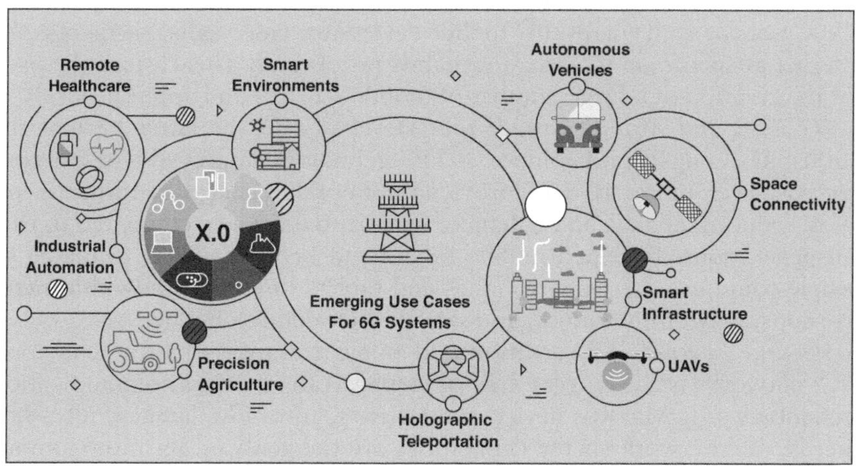

Figure 20.1 Evolution of wireless communication technologies.

DOI: 10.1201/9781003522003-23

Through a remarkable progression throughout time, wireless communication technologies [3] have revolutionised how we connect and communicate with one another. The development of wireless networks, from the earliest days of simple radio transmissions to today's complex and pervasive networks, has completely changed many facets of our lives, including international connectivity, commercial operations, and interpersonal contact. An outline of the significant turning points in the development of wireless communication technology is given in this article [4].

Radio Invention: The development of radio in the late 1800s laid the groundwork for wireless communication. In 1895, wireless telegraphy pioneer Guglielmo Marconi successfully sent the first wireless signal 1.6 km away. This discovery established the foundation for subsequent developments in wireless communication.

Analog Cellular Systems: The advent of analog cellular systems in the 1980s marked the next significant advancement in wireless communication. These systems separated geographical areas into cells, each serviced by a base station, using frequency-division multiplexing. Advanced Mobile Phone Systems (AMPS) and other analog cellular networks made it possible to move about more and have better voice quality.

Digital Cellular Systems: A critical turning point in wireless communication was the switch from analog to digital cellular systems. The 1980s saw the introduction of first-generation (1G) systems, which relied on methods for converting analog to digital data. But thanks to their enhanced voice quality, increased capacity, and introduction of text messaging, second-generation (2G) technologies like the Global System for Mobile Communications (GSM) and Code Division Multiple Access (CDMA) were the ones that revolutionised wireless communication.

Third-generation (3G) wireless technologies and mobile internet: These developments ushered in a new era of wireless communication. 3G networks allow users to send emails, use mobile devices for video calls, and access the internet at speeds up to several megabits per second. This cleared the path for the creation of a large number of mobile services and applications.

4G LTE and Mobile Broadband: Developed in the latter part of the 2000s, 4G Long-Term Evolution (LTE) technology improved wireless connectivity even more. 4G LTE provided noticeably faster data transmission rates, reduced latency, and enhanced spectrum efficiency compared to earlier generations. It helped mobile broadband become widely available so people could use their smartphones and tablets to enjoy bandwidth-intensive apps, play online games, and stream high-definition videos.

5G and Beyond: The creation and implementation of fifth-generation (5G) networks is the current area of emphasis for wireless communication technology [5]. Massive device connectivity, ultra-low latency, ultra-fast speeds, and network slicing capabilities are the goals of 5G. Autonomous vehicles, smart cities, the Internet of Things (IoT), augmented reality (AR), and virtual reality (VR) are just a few of the industries it promises to transform. Research and development initiatives are underway to investigate increasingly faster and more sophisticated wireless technologies, such as 6G. These initiatives go beyond 5G.

Since the development of radio, wireless communication technologies have advanced significantly. Every iteration of cellular technology—from analog to digital, mobile internet, 4G LTE, and now, ongoing 5G deployments—brings notable improvements in speed, capacity, and functionality. These developments have completely changed the way we interact and communicate with one another, creating new opportunities and influencing how wireless communication will develop in the future.

20.2 THE EMERGENCE OF 6G

Although 5G technology is still widely used, 6G, the next generation of wireless communication, is fast approaching. Although 5G is still being implemented and incorporated into many industries, experts and academics are working hard to investigate how 6G might unlock previously unheard-of levels of innovation, speed, and connectivity. A summary of the emergence of 6G, as well as its expected features and uses, is given in this article [6].

The Need for 6G [7]: The increasing need for even faster and more dependable wireless connectivity is driving the development of 6G. Networks that can handle large data transfers, extremely low latency, and a substantially higher number of connected devices are required as technologies like virtual reality, driverless vehicles, smart cities, and the Internet of Things continue to advance. These criteria are what 6G seeks to address, opening up hitherto unthinkable uses.

Expected qualities: Although research and development on 6G are still in their infancy, experts have suggested several essential qualities and traits. Among them are:

a. Unprecedented Speeds: 6G promises to deliver data transfer speeds never before possible, possibly up to terabits per second (Tbps). Instantaneous downloads, smooth streaming of 8K or higher definition video, and data-intensive applications like holographic communication and real-time virtual reality would be possible.

b. Ultra-Low Latency: 6G is anticipated to achieve a sub-millisecond response time by drastically cutting latency compared to earlier generations. Applications that need real-time communication, such as industrial automation, remote surgery, and driverless cars, would need this.

c. Hyper-Connectivity: 6G plans to enable up to a million connected devices per square kilometer, an enormous amount of connectivity. This would allow the Internet of Things to be integrated smoothly, allowing a wide range of sensors and gadgets to interact and communicate to create a brilliant and connected environment.

d. Energy Efficiency: 6G seeks to be more energy-efficient than its predecessors in light of the growing concern for sustainability. This entails lowering the energy needs of connected devices, streamlining the power usage in network infrastructure, and putting cutting-edge energy-saving strategies into practice.

e. Integration of AI: 6G networks are anticipated to rely heavily on artificial intelligence (AI). Network performance can be improved using AI algorithms and machine learning approaches to control traffic, allocate resources more efficiently, and optimise network performance. Advanced applications like context-aware services and intelligent edge computing may also be made possible by AI-powered technology.

Potential Uses: It is anticipated that 6G will significantly influence several different sectors and industries. Among the possible uses are:

a. Immersive Extended Reality [8] (XR): 6G can lay the groundwork for incredibly immersive VR, AR, and MR experiences by fusing ultra-high resolutions, real-time interaction, and seamless communication with these three virtual reality technologies.

b. Infrastructure and Smart Cities: 6G can create more intelligent and productive cities by improving connectivity across diverse elements like utilities, public services, and transportation networks. This could result in enhanced public safety systems, intelligent energy networks, and optimised traffic management.

c. Healthcare and Remote Surgery [9]: Thanks to 6G's extremely low latency and excellent dependability, doctors can operate on patients in real-time from a distance. The delivery of healthcare could be revolutionised by this, especially in emergencies or rural places.

d. driverless Vehicles and Intelligent Transportation [10]: By enabling dependable and very responsive communication between automobiles, infrastructure, and pedestrians, 6G networks may facilitate the mainstream use of driverless vehicles. This would improve

20.3 OBJECTIVES OF THE CHAPTER

The chapter on "The Emergence of 6G" aims to achieve the following objectives:

Please give a Brief Overview: The idea of 6G will be covered in this chapter, along with its importance in wireless communication technology. It will draw attention to the necessity of 6G networks to satisfy the expanding needs for connectivity, speed, and capacity.

Examine Expected aspects: This chapter will go into the elements that people are looking forward to from 6G, like ultra-low latency, hyper-connectivity, energy efficiency, and the integration of artificial intelligence. It will describe how these characteristics may transform wireless communication and make new services and applications possible.

Examine Possible Uses: This chapter will review the possible uses of 6G in several fields and businesses. It will examine how 6G networks can support intelligent transportation systems, driverless cars, remote surgery, smart cities and infrastructure, and immersive extended reality experiences. Readers will better understand the wide-ranging effects of 6G on several facets of society by reading this topic.

Emphasise Research and Development [11]: This chapter will give readers an overview of the current 6G research and development initiatives. It will go over the steps research institutions, businesses, and standards organisations took to investigate and mold the direction of 6G technology. This section will emphasise the value of interdisciplinary methods and the collaborative aspect of 6G development.

Examine Challenges and Considerations [12]: This chapter will discuss the issues and worries surrounding the development of 6G. It will cover spectrum availability, infrastructure deployment, security, privacy, and regulatory considerations. By exploring these problems, readers will comprehend the intricacies associated with the shift to 6G networks.

Give Prospective Views: Discuss prospective advancements beyond 6G, including research on 7G and beyond, which will round up the chapter. It will show how quickly technology develops and how wireless communication innovation is constantly needed.

The chapter's overall goal is to present a thorough overview of the advent of 6G, including its expected properties, possible uses, existing research projects, difficulties, and future outlooks. It will be an invaluable tool for those who want to know how wireless communication technologies are developing and how 6G might change the game.

20.4 ULTRA-FAST AND RELIABLE CONNECTIVITY

Developing an ultra-fast and dependable network is a significant goal for creating 6G connectivity. 6G promises to provide previously unheard-of speeds and reliability to satisfy the growing demand for seamless connectivity and high-speed data transfer. As data-intensive applications like virtual reality, cloud gaming, and high-definition video streaming proliferate, ultra-fast bandwidth is essential to provide flawless real-time experiences and rapid downloads. Furthermore, maintaining wireless communication reliability is critical for consistent user experiences and unbroken connectivity. 6G networks aim to deliver high availability, enhanced resilience, and effective error correction mechanisms by utilising cutting-edge technology such as massive

MIMO [13], beamforming, millimeter-wave communication, and optimised network algorithms. 6G networks' ultra-fast and dependable connectivity can completely change a number of industries and make ground-breaking applications possible, such as immersive extended reality experiences, real-time remote surgeries, and intelligent transportation systems.

20.4.1 Unleashing Unprecedented Speeds

Unlocking previously unheard-of speeds is a primary goal for 6G network development. Networks carrying data at lightning-fast speeds are desperately needed as technology advances and data consumption soars. 6G promises to introduce groundbreaking developments in wireless communication to push the speed envelope. 6G networks can offer near-instantaneous downloads, smooth streaming of high-resolution material, and real-time interactive applications. Expected speeds are as high as terabits per second [14] (Tbps). These previously unheard-of speeds will revolutionise how we access and consume media and spur the creation of cutting-edge services and applications in a wide range of industries. The potential to unleash previously unheard-of speeds in 6G networks opens up possibilities for the future of connectivity and communication, ranging from immersive extended reality experiences to sophisticated data analytics and artificial intelligence.

20.4.2 Enabling Seamless Streaming and Gaming

Providing smooth gaming and streaming experiences is essential to developing a 6G network [15]. Networks that can serve bandwidth-intensive applications without interruptions or buffering are in greater demand as high-definition video streaming platforms proliferate and online gaming gains popularity. With ultra-fast speeds, ultra-low latency, and enhanced network dependability, 6G seeks to satisfy this need. 6G networks can guarantee seamless and continuous streaming of high-resolution material by utilising cutting-edge technologies and optimisation strategies, enabling users to participate in immersive entertainment experiences in real-time. Furthermore, 6G networks can provide a smooth gaming experience by reducing latency, facilitating real-time multiplayer games, and supplying robust network connectivity. The next generation of interactive applications and entertainment will be made possible by 6G networks' seamless gaming and streaming capabilities, improving user experiences and opening up new opportunities for developers and content creators.

20.4.3 Advancements in Data Transfer and Sharing

As 6G networks grow, data sharing and transfer improvements will be critical components. Networks that can manage the enormous volume of data securely and efficiently are required due to exponential data generation and consumption development. 6G promises to introduce novel technologies and approaches to transform data sharing and transfer.6G networks can offer almost instantaneous data transfers with ultra-fast speeds and ultra-low latency, making it easier to share big files, multimedia material, and real-time information quickly and effectively. Furthermore, enhancing data security and privacy will be a primary priority of 6G networks, guaranteeing the protection of confidential data while in transit. To protect data integrity and confidentiality, robust data management systems, secure authentication methods, and advanced encryption techniques will be used. These improvements in 6G networks' data transfer and sharing capabilities will enable people, organisations, and sectors to work together in real-time, share information quickly, and seize new chances for productivity and innovation.

20.5 INTERNET OF THINGS (IOT) REVOLUTION

One of the main goals of developing 6G networks is to broaden the scope of the Internet of Things (IoT) [16]. Networks that can support an increasingly growing number of linked devices and facilitate smooth communication between them are required as the IoT ecosystem develops. 6G promises hyper-connectivity, ultra-low latency, and increased energy efficiency to solve this problem. With the capacity to accommodate a colossal quantity of devices per square kilometer—roughly one million devices—6G networks will lay the groundwork for a brilliant and interconnected Internet of Things environment. Numerous applications, including smart cities, industrial automation, agriculture, healthcare, and more, will be made possible by this broadening of the Internet of Things. 6G will enable real-time data collection, analysis, and decision-making using cutting-edge technologies and protocols. This will enable businesses and individuals to fully utilise the potential of IoT and spur innovation across various industries. A new era of connectivity will be ushered in by the seamless integration of IoT devices and the proliferation of intelligent systems enabled by 6G. This will also open the door for revolutionary improvements in automation, efficiency, and quality of life.

20.5.1 Expanding the Horizons of IoT

One of the main goals of developing 6G networks is to broaden the scope of the Internet of Things (IoT). Networks that can support an increasingly growing number of linked devices and facilitate smooth communication between them are required as the IoT ecosystem develops. 6G promises hyper-connectivity, ultra-low latency, and increased energy efficiency to solve this problem. With the capacity to accommodate a colossal quantity of devices per square kilometer—roughly one million devices—6G networks will lay the groundwork for a brilliant and interconnected Internet of Things environment. This broadening of the Internet of Things' horizons [17] will make various applications possible, including smart cities, industrial automation, healthcare, agriculture, and more. 6G will enable real-time data collection, analysis, and decision-making using cutting-edge technologies and protocols. This will allow businesses and individuals to fully utilise the potential of IoT and spur innovation across a range of industries. A new era of connectivity will be ushered in by the seamless integration of IoT devices and the proliferation of intelligent systems enabled by 6G. This will also open the door for revolutionary improvements in automation, efficiency, and quality of life.

20.5.2 Smart Cities and Infrastructure

One important use of cutting-edge technologies, such as the Internet of Things (IoT) and wireless connectivity, is in smart cities and infrastructure [18]. Smart cities seek to improve sustainability, maximise resource use, and improve the quality of life for their citizens by utilising automation, data analytics, and networking. The potential for smart cities and infrastructure becomes even further with the introduction of 6G networks. Transportation systems, public services, utilities, and infrastructure are just a few elements of a smart city that can communicate and coordinate easily, thanks to 6G networks' hyper-connectivity, ultra-fast speeds, and ultra-low latency. This enables enhanced public safety systems, intelligent energy grids, real-time environmental monitoring, and optimised traffic management. Smart cities may efficiently manage resources, lessen traffic, and offer their residents individualised and effective services by utilising 6G's potential. A significant advancement in creating resilient, sustainable, and livable urban environments that put efficiency, convenience, and environmental conscience first is the incorporation of 6G in smart city infrastructure.

20.5.3 Industrial Automation and Manufacturing

With the introduction of cutting-edge technologies like the Internet of Things (IoT) and wireless connectivity, industrial automation and manufacturing [19] have experienced a revolutionary impact. These technologies have transformed Traditional manufacturing methods, allowing for greater flexibility, efficiency, and output. The possibilities of manufacturing and industrial automation are expected to soar with the advent of 6G networks. With ultra-fast speeds, ultra-low latency, and vast device connectivity offered by 6G networks, machines, sensors, and control systems can easily communicate and coordinate. In manufacturing operations, this makes intelligent decision-making, predictive maintenance, and real-time monitoring easier. The seamless integration of robots, artificial intelligence, and data analytics is made possible by 6G's hyper-connectivity and high data transfer speeds, which enable autonomous production lines and adaptable manufacturing processes. The potential for greater productivity, decreased downtime, enhanced quality control, and more customisation capabilities is unlocked by the integration of 6G in industrial automation and production. Industries may welcome the era of smart factories and usher in a new era of manufacturing that is highly automated, intelligent, and sensitive to market demands by utilising 6G's potential.

20.5.4 Healthcare and Remote Monitoring

Developments in wireless communication technology, particularly the Internet of Things (IoT) and telemedicine, have drastically changed healthcare and remote monitoring [20]. The possibilities for remote monitoring and healthcare are significantly expanded with the introduction of 6G networks. With 6G networks' rapid speeds, incredibly low latency, and dependable connectivity, medical data can be transmitted seamlessly, and patients can be remotely monitored in real-time. This makes it possible for medical professionals to conduct telemedicine consultations, monitor patients' vital signs from a distance, and evaluate the efficacy of therapies in real-time. Furthermore, using cutting-edge technologies like augmented reality (AR) and virtual reality (VR) for remote surgeries, immersive medical education, and surgical training is made possible by 6G's high data transfer rates and low latency. By facilitating access to specialised medical services and promoting healthcare results, the integration of 6G in healthcare and remote monitoring holds the potential to completely transform patient care, especially in underprivileged and remote areas. It makes continuous monitoring, early identification of health problems, and prompt management possible, eventually improving patient outcomes, lowering healthcare costs, and improving general well-being.

20.5.5 Agriculture and Environmental Monitoring

Thanks to the integration of wireless communication technologies, especially the Internet of Things (IoT) and data analytics, agriculture, and environmental monitoring have advanced significantly [21]. The potential for revolutionising ecological monitoring and agriculture is enhanced dramatically with the advent of 6G networks. 6G networks facilitate smooth communication and data sharing between agricultural equipment, sensors, and devices by offering hyper-connectivity, ultra-fast speeds, and ultra-low latency. This makes it possible to monitor various characteristics in real-time, including temperature, humidity, soil moisture, and crop growth. This gives farmers the information to allocate resources and manage crops efficiently. Furthermore, 6G networks make integrating cutting-edge technologies like autonomous robots and drones easier, making farming more accurate and productive. 6G networks make real-time data collecting and analysis possible in environmental monitoring, which helps with efficient environmental management and conservation initiatives. By providing farmers and environmental agencies with timely information, integrating 6G technology in agriculture and environmental monitoring facilitates sustainable practices, effective resource utilisation, and improved resilience to climate change. Agriculture can increase its productivity, environmental friendliness, and resilience by utilising 6G networks, which will help ensure food security and protect natural resources.

20.6 HOLOGRAPHIC COMMUNICATION AND TELEPRESENCE

While holographic communication and telepresence [22] have long been the stuff of science fiction, they are starting to come true with the development of advanced technologies like 6G networks. These networks provide the infrastructure needed for immersive holographic communication and telepresence experiences, offering ultra-fast speeds, ultra-low latency, and high bandwidth. Holographic communication breaks down the barriers of traditional video conferencing by allowing people to interact with each other as though they were physically present in the same place. This technology significantly impacts healthcare, education, distance collaboration, and other fields. Holographic communication and telepresence, which break down barriers and increase the sense of presence and engagement, have the potential to completely transform how people connect and cooperate across distances with 6G networks. The way we connect, collaborate, and see the world will change due to the combination of 6G networks with holographic communication and telepresence, ushering in a new era of immersive and interactive communication experiences.

20.6.1 The Rise of Holographic Communication

The emergence of holographic communication is a fascinating trend changing how we communicate and engage with one another. Holographic communication, made possible by cutting-edge technology like 6G networks, virtual reality (VR), and augmented reality (AR), is completely changing how people communicate today. Holographic communication adds three-dimensional depictions of people instead of two-dimensional screens, making the experience more lifelike and immersive. Holographic communication is made even more frictionless by 6G networks, which offer breakneck speeds and minimal latency, enabling high-quality, real-time interactions. This technology significantly affects business, education, healthcare, and entertainment, among other fields. It makes it possible to collaborate remotely, hold virtual meetings, have lifelike teleconferences, and have better telepresence experiences. With the growth of holographic communication, there are more opportunities for global connectivity, as distances no longer separate people, and a sense of presence is fostered even when they are physically away. Holographic communication can revolutionise communication, collaboration, and worldview as technology develops further, providing an insight into a day when immersive and lifelike interactions will be standard.

20.6.2 Real-Time Collaboration and Teleconferencing

In today's linked world, real-time collaboration and teleconferencing are becoming more and more critical, and the emergence of cutting-edge technology like 6G networks is expected to improve these capabilities further. Sixth-generation (6G) networks offer ultra-fast speeds, ultra-low latency, and large bandwidth, making real-time teleconferencing and collaboration easy. Professionals in various places can work together in real-time, instantly exchanging ideas, papers, and information. With its immersive high-definition pictures and crisp audio, video conferencing lets people communicate more effectively by removing the obstacles of physical distance. Furthermore, 6G networks enable cutting-edge functionalities like virtual reality (VR) and augmented reality (AR), which provide attendees with an immersive and captivating teleconference experience. This technology has significant business ramifications because it allows teams worldwide to collaborate easily, increases productivity, and minimises the need for travel. A future where remote teams may work together without difficulty, overcoming geographical barriers and promoting a sense of connection and collaboration regardless of physical location is made possible by integrating 6G networks in real-time collaboration and teleconferencing.

20.6.3 Education and Remote Learning

Incorporating new technology has brought about a dramatic revolution in education and remote learning [23]. The introduction of 6G networks has the potential to improve these educational experiences further. 6G networks offer ultra-low latency, ultra-fast speeds, and dependable connectivity—all of which lay the groundwork for flawless remote learning. Pupils from all over the world can access instructional resources, take part in virtual debates, and participate in real-time interactive classes. Students can explore virtual simulations, perform virtual experiments, and participate in hands-on learning experiences by integrating augmented reality (AR), virtual reality (VR), and immersive technologies made possible by 6G networks. Moreover, 6G networks allow excellent instructional content, such as interactive presentations, multimedia files, and video lectures, to be transmitted quickly, guaranteeing that students can access a wide variety of learning tools. Geographical obstacles could be removed, access to high-quality education could be improved, and lifelong learning could be encouraged by incorporating 6G in education and remote learning. It makes it possible for learners to acquire knowledge and skills flexibly and dynamically, promotes worldwide collaborations, and offers personalised and adaptable learning experiences.

20.6.4 Healthcare and Virtual Surgeries

With the introduction of cutting-edge technology like 6G networks, healthcare and virtual surgeries [24] have the potential to transform the medical industry completely. With 6G networks' ultra-fast speeds, ultra-low latency, large bandwidth, real-time data, precise instructions, and high-definition images, they may all be easily transmitted. By utilising these skills, virtual surgeries allow highly qualified doctors to perform remote surgical procedures from a distance. Surgeons can do complex surgeries with accuracy and three-dimensional visualisation of patient anatomy thanks to augmented reality (AR) and virtual reality (VR) technologies. A more realistic surgical experience is offered by incorporating haptic feedback devices, which further improve the tactile sense. Virtual operations may eliminate geographical limitations, reducing patient travel and related dangers while providing specialised treatment to underserved areas and permitting collaborative surgeries with professionals globally. The ultra-low latency of 6G networks guarantees the slightest delay, essential for making real-time decisions and coordinating actions during surgical procedures. Virtual operations and healthcare hold great potential for advancing surgical procedures and education, increasing patient access to high-quality treatment, and improving patient outcomes as 6G networks develop.

20.7 AUTONOMOUS VEHICLES AND INTELLIGENT TRANSPORTATION SYSTEMS

The way we commute and navigate our cities is being drastically changed by autonomous cars and intelligent transportation systems (ITS) [25], and the addition of cutting-edge technologies like 6G networks presents many opportunities for their continued advancement. 6G networks allow infrastructure, other ITS components, and autonomous cars to communicate and coordinate seamlessly thanks to their ultra-fast speeds, ultra-low latency, and high connection. This makes it easier for real-time data to be exchanged, enabling cars to judge based on current knowledge of traffic patterns, potential hazards on the road, and maritime routes. The low latency of 6G networks guarantees quick reaction times, which is essential for ensuring autonomous cars' effectiveness and safety. Moreover, colossal device connectivity supported by 6G networks allows for the smooth integration of vehicle-to-vehicle (V2V) and vehicle-to-infrastructure (V2I) communication, promoting coordinated and cooperative driving behavior. Road safety, traffic flow, and congestion are all benefited by these connectivity and communication capabilities developments. The future of transportation is expected to be entirely transformed by the incorporation of 6G technology into intelligent transportation systems and driverless cars. This will open the door to more effective, sustainable, and connected mobility solutions.

20.7.1 Accelerating the Era of Autonomous Vehicles

We are getting closer to a time when self-driving cars are a regular sight on the roads, thanks to the incorporation of cutting-edge technologies like 6G networks. With 6G networks, autonomous vehicles, and their environment can communicate seamlessly and reliably thanks to their high connectivity, ultra-low latency, and rapid speeds. This makes transmitting data in real-time easier, enabling cars to see their surroundings, decide what to do, and drive safely and effectively. The low latency of 6G networks guarantees quick reaction times, essential for making snap decisions in challenging driving situations. Massive device connectivity is another benefit of 6G networks. This allows for smooth communication between vehicles and infrastructure and between cars, encouraging coordinated and cooperative driving practices. Consequently, this improves autonomous cars' general safety, efficacy, and efficiency. With the ongoing development of 6G networks, autonomous vehicles can utilise new functions, including expanded situational awareness, sophisticated artificial intelligence algorithms, and better sensor fusion. The development and implementation of autonomous cars are accelerated by integrating 6G networks, improving accessibility, efficiency, and safety for all users.

20.7.2 Enhancing Vehicle-to-Vehicle Communication

Improving vehicle-to-vehicle (V2V) communication [26] is crucial in developing autonomous vehicles' capabilities. Incorporating cutting-edge technology like 6G networks is considered essential in this field. 6G networks facilitate smooth and dependable communication between cars on the road by providing breakneck speeds, extremely low latency, and high connection. This makes it possible for cars to communicate in real time about their position, speed, and trajectory, which promotes coordinated and cooperative driving practices. V2V communication improves with 6G networks, allowing for quicker reaction times and a lower chance of mishaps. The low latency of 6G networks guarantees almost immediate communication, making it possible to send crucial safety-related data accurately and on time. Moreover, more vehicles may be integrated into the vehicle-to-vehicle (V2V) communication ecosystem thanks to the high connectivity capabilities of 6G networks, which promotes a more extensive and dependable network. 6G networks' enhanced V2V communication enhances traffic flow, increases road safety, and fosters a more productive and peaceful driving environment. V2V communication, which uses 6G networks, can revolutionise vehicle-to-vehicle interactions and move us closer to a time when linked and autonomous cars can safely and efficiently navigate our roadways.

20.7.3 Traffic Management and Road Safety

Road safety and traffic management are two of the most critical factors in contemporary transport systems, and integrating cutting-edge technologies like 6G networks has a lot of promise to improve these areas. Real-time data interchange and communication between automobiles, infrastructure, and traffic management systems are made possible by 6G networks' breakneck speeds, extremely low latency, and high connection. This makes real-time monitoring and analysis of traffic flow, congestion, and road conditions possible, as well as dynamic and adaptive traffic management. Critical information may be quickly shared over 6G networks due to their low latency, which enables prompt interventions and incident response while driving. Furthermore, 6G networks facilitate the integration of cutting-edge technology such as data analytics, predictive modeling, and artificial intelligence, allowing traffic management systems to make data-driven decisions and optimise traffic flow. Incorporating 6G networks also contributes to increased road safety because vehicles can connect with infrastructure and each other, exchanging information about possible hazards, the state of the roads, and other vital safety data. This promotes coordinated and cooperative driving behaviors, lowering the chance of collisions and raising the road safety standard. Traffic management systems may become more effective, responsive, and intelligent by utilising the capabilities of 6G networks. This will improve traffic flow, lessen congestion, and increase road safety for all road users.

20.7.4 Real-Time Updates and Predictive Analytics

Predictive analytics and real-time updates are revolutionising several industries, and the addition of cutting-edge technologies like 6G networks is expected to improve these capacities even more. Massive volumes of data can be processed and transmitted seamlessly in real time because of 6G networks' strong connection, ultra-low latency, and rapid speeds. This allows companies and organisations to get quick updates on essential data, like consumer behavior, market trends, and operational insights. Companies can move quickly and more efficiently with access to real-time updates, which also helps them stay competitive. Moreover, enterprises can use real-time and historical data to predict future trends, patterns, and results by integrating predictive analytics powered by 6G networks. This makes proactive decision-making, risk reduction, and business process optimisation possible. In manufacturing, supply chain management, finance, and customer service, 6G networks' real-time updates and predictive analytics transform company operations. It allows them to stay on top of trends, adapt quickly to changes, and seize fresh chances for development and achievement.

20.8 ADVANCED HEALTHCARE SERVICES

Using state-of-the-art technologies, such as 6G networks, transforms advanced healthcare services and opens new avenues for improved patient outcomes. The lightning-fast speeds, low latency, and strong connection of 6G networks enable seamless communication and data sharing between medical device manufacturers, patients, and healthcare providers. This facilitates remote healthcare services, telemedicine consultations, and real-time patient vitals monitoring. Healthcare professionals can collaborate with experts, view patient records from a distance, and make choices based on the best available information. Modern technologies like virtual reality (VR) and augmented reality (AR), which enhance medical education by providing lifelike simulations and remote surgical consultations, are easier to integrate with 6G networks. Moreover, the extensive device connectivity of 6G networks enables the integration of wearable technologies, remote monitoring systems, and intelligent healthcare infrastructure, enabling the Internet of Medical Things (IoMT). Better illness management, early detection of health issues, and customised treatment options are the outcomes. Advanced healthcare services are made possible by 6G networks, which lower healthcare costs, increase patient accessibility to specialised care, and improve the efficacy and efficiency of healthcare delivery.

20.8.1 Remote Surgeries and Telemedicine

Telemedicine and remote surgery are rapidly transforming the healthcare sector, and when cutting-edge technologies like 6G networks are added to the mix, these techniques will only become more innovative. 6G networks are perfect for seamless remote surgeries and telemedicine consultations because they provide robust connectivity, extremely low latency, and fast speeds. With haptic feedback devices, high-definition graphics, and real-time video feeds, surgeons may operate on patients in various locations with accuracy and immersion.

This improves patient outcomes by removing the need for patients to travel and enabling everyone, regardless of physical location, to have access to specialised surgical expertise. 6G networks enable digital communication between medical personnel and patients via telemedicine, enabling remote consultations, diagnosis, and treatment. Patients can get medical advice, prescriptions, and follow-up care online. 6G networks provide immediate data interchange, secure and dependable communication, and enhanced interactions for telemedicine and remote procedures, thereby increasing the effectiveness, efficiency, and accessibility of medical services. It changes how healthcare is experienced and viewed by enabling medical professionals to reach underserved regions, provide timely treatments, and improve patient outcomes.

20.8.2 Real-Time Monitoring and Personalised Healthcare

Incorporating state-of-the-art technologies such as 6G networks ushers in a new era of proactive and customised patient care by revolutionising real-time monitoring and personalised healthcare. Sixth-generation (6G) networks facilitate the seamless exchange of real-time health data from medical sensors, wearables, and remote monitoring systems by combining high connection, ultra-low latency, and quick speeds. Healthcare practitioners are able to monitor patients' physiological traits, vital signs, and health trends in real time due to this continuous data flow. 6G networks make it easier to gather and analyse data by combining artificial intelligence (AI) and machine learning algorithms, providing personalised healthcare recommendations and useful insights. People are empowered to actively manage their health thanks to early detection of health issues, timely interventions, and customised treatment plans made possible by real-time monitoring and personalised healthcare. Thanks to this technology, which also enhances the remote delivery of healthcare services, healthcare providers may now monitor and manage patients' symptoms remotely, hold virtual consultations, and provide individualised care regardless of their physical location. The integration of 6G networks in real-time monitoring and personalised healthcare is revolutionising the healthcare paradigm. Better patient outcomes, empowering people to live healthier lifestyles, and a move towards proactive and preventative healthcare practices are the results of this.

20.8.3 Improved Access to Medical Services

Improved access to medical services is one significant advantage made feasible by implementing cutting-edge technologies like 6G networks. 6G networks' extremely fast speeds, extremely low latency, and high connectivity make it possible to expand healthcare services to underdeveloped areas, remote locations, and marginalised populations. Patients can now acquire medical consultations, diagnosis, and treatment remotely because of the increasing availability of telemedicine and remote healthcare services, which also reduce access barriers and eliminate the need for travel.

Furthermore, the integration of wearable technology, remote monitoring systems, and intelligent healthcare infrastructure is made possible by 6G networks, enabling real-time data gathering and transfer. Allowing medical personnel to track chronic diseases, monitor patients' health conditions remotely, and provide timely interventions improves the quality and timeliness of care. The increased accessibility of medical services made possible by 6G networks has great promise for reducing healthcare disparities, establishing connections between individuals residing in remote or isolated areas, and ensuring that everyone has access to essential medical information and resources. It is crucial for enhancing healthcare fairness and developing inclusive healthcare systems.

20.8.4 Applications in Rural and Underserved Areas

Allowing medical personnel to track chronic diseases, monitor patients' health conditions remotely, and provide timely interventions improves the quality and timeliness of care. The increased accessibility of medical services made possible by 6G networks has great promise for reducing healthcare disparities, establishing connections between individuals residing in remote or isolated areas, and ensuring that everyone has access to essential medical information and resources. It is crucial for enhancing healthcare fairness and developing inclusive healthcare systems. Healthcare specialists, medical advice, consultations, and remote monitoring are all available to healthcare practitioners. Children who reside in rural areas can also participate in online learning environments, access educational resources online, and receive top-notch training. Additionally, 6G networks can support intelligent agriculture solutions, enabling farmers to monitor crops and optimise irrigation, increasing agricultural output, and providing them with access to real-time meteorological information. By bringing cutting-edge applications to underserved and rural areas, 6G networks can boost general development and well-being, empower communities, and close the digital divide.

20.9 ENVIRONMENTAL MONITORING AND SUSTAINABILITY

To handle global environmental concerns, environmental sustainability and monitoring are essential [21], and the incorporation of cutting-edge technology like 6G networks is necessary in this field. For environmental monitoring, 6G networks' ultra-fast speeds, ultra-low latency, and high connectivity allow real-time data collection, processing, and distribution. Connected to 6G networks, sensors, and IoT devices can collect data on biodiversity, air and water quality, climate conditions, and biodiversity, offering critical new perspectives on the state of the environment. Using this information, environmental hazards may be evaluated, pollution sources can be located, and well-informed decisions on sustainable resource management can be supported. Furthermore, 6G networks make deploying intelligent energy and smart grids easier, allowing for real-time energy usage optimisation and monitoring, which increases energy efficiency and lowers carbon emissions. Enhancing environmental monitoring and sustainability programs using 6G networks can result in more efficient conservation efforts, improved resource management, and the preservation of ecosystems for future generations.

20.9.1 Smart Grids and Energy Optimisation

The way we produce, distribute, and use energy is being entirely transformed by smart grids and energy optimisation, and this change is being accelerated by the incorporation of cutting-edge technology like 6G networks. Thanks to their ultra-fast speeds, ultra-low latency, and high connectivity, 6G networks facilitate smooth communication and data exchange between different parts of the energy ecosystem, such as power plants, renewable energy sources, smart meters, and consumer appliances. This makes optimising, regulating, and monitoring energy use easier in real-time. 6G networks-powered smart grids allow for two-way communication, giving utility companies access to real-time data on system conditions, voltage levels, and energy consumption. Using this data, demand response systems, dynamic pricing, and intelligent load balancing may optimise energy distribution and reduce waste. Consumers can make educated decisions, track their energy usage in real time, and modify their usage habits to conserve energy and money. 6G networks facilitate the integration of cutting-edge technologies like artificial intelligence (AI) and machine learning, which improve energy optimisation algorithms and allow predictive modeling and intelligent energy management systems. Using 6G networks, smart grids, and energy optimisation may integrate renewable energy sources, improve grid stability, and save energy consumption, all of which contribute to developing a more resilient and sustainable energy infrastructure.

20.9.2 Real-Time Environmental Sensors

Real-time environmental sensors are revolutionising environmental monitoring and management techniques, which are made possible by integrating cutting-edge technology like 6G networks. 6G networks offer the infrastructure required for seamless data collection and transmission from sensors deployed in various environmental situations. These networks boast ultra-fast speeds, ultra-low latency, and high connectivity. These sensors can make real-time measurements of variables like radiation, noise, soil conditions, water and air quality, and radiation quality. The instantaneous transmission of the data gathered by these sensors via 6G networks enables prompt analysis and interpretation. Environmental sensors that operate in real-time offer insightful information on the condition of the environment, facilitating immediate interventions and well-informed decision-making. They make it easier to identify environmental concerns, detect pollution episodes early, and track the effects of pollution on the environment. Policymakers, scientists, and environmentalists can take preemptive steps to safeguard and preserve the environment with the help of this real-time information. Real-time environmental sensors are essential for improving ecological monitoring programs, supporting sustainability programs, and bringing about beneficial environmental changes by utilising the power of 6G networks.

20.9.3 Climate Monitoring and Conservation

Climate conservation and monitoring are crucial to solving the issues posed by climate change, and the incorporation of cutting-edge technologies like 6G networks is revolutionising these efforts. Real-time climate data collection, analysis, and distribution from various sources, such as weather stations, satellites, and environmental sensors, are made possible by 6G networks' breakneck speeds, extremely low latency, and high connection. Scientists, decision-makers, and environmentalists may track climatic trends, evaluate the effects of climate change, and create successful conservation plans with the help of this real-time data. Artificial intelligence (AI) and machine learning algorithms can be integrated with 6G networks to analyse vast volumes of climate data, spot trends, and produce precise predictions. This information makes it possible to take preventative action to lessen the effects of climate change, save delicate ecosystems, and preserve biodiversity. Moreover, 6G networks facilitate communication and cooperation between governments, research institutes, and international organisations, creating a worldwide network for exchanging information and data about climate change. Using 6G networks can facilitate improving climate monitoring and conservation initiatives, resulting in better strategies, knowledgeable decision-making, and a sustainable future for the Earth.

20.9.4 Promoting Sustainable Development

Encouraging sustainable development is essential to solving world problems and guaranteeing future generations' prosperity. The advancement of sustainable development methods in various sectors is greatly aided by integrating cutting-edge technologies like 6G networks. 6G networks facilitate the smooth transmission of information across stakeholders by offering ultra-fast speeds, ultra-low latency, and high connection. This encourages collaboration and knowledge-sharing. This makes applying sustainable techniques in industries like waste management, transportation, energy, and agriculture easier. 6G networks facilitate the integration of intelligent transportation systems, precision agriculture, smart grids, and effective waste management systems. This integration reduces environmental impact, improves operational efficiency, and optimises resource utilisation. Furthermore, by offering chances for participation, personalised recommendations, and real-time information, 6G networks enable people and communities to participate actively in sustainable practices. We can create a more resilient, fair, and ecologically conscious society that satisfies present demands without jeopardising the capacity of future generations to satiate their wants by fostering sustainable growth through the integration of 6G networks.

20.10 ENHANCED PUBLIC SAFETY AND EMERGENCY RESPONSE

Implementing cutting-edge technologies such as 6G networks significantly improves emergency response and public safety systems. 6G networks facilitate real-time communication, data interchange, and emergency services coordination because of their ultra-fast speeds, ultra-low latency, and high connection. This enhances public safety by enabling prompt and efficient emergency response. Furthermore, 6G networks facilitate the integration of diverse sensors, security cameras, and Internet of Things devices, enabling all-encompassing situational awareness and prompt identification of possible hazards. Real-time streaming video and sensor data allow responders to assess circumstances more precisely and make well-informed decisions. Furthermore, 6G networks enable predictive analytics and machine learning algorithms to analyse massive volumes of data to spot trends and anticipate possible emergencies, allowing for proactive risk mitigation. Using cutting-edge technologies and 6G networks, emergency responders can save lives, limit damage, and guarantee community safety by equipping themselves with essential tools and resources.

20.10.1 Transforming Public Safety Systems

Enhancing emergency response, disaster management, and public safety through cutting-edge technologies like 6G networks requires transforming public safety systems. 6G networks allow emergency responders, law enforcement agencies, and other public safety organisations to collaborate in real-time through data interchange, ultra-fast speeds, and high connection. This enables prompt and well-coordinated response actions by facilitating the quick distribution of vital information, such as event notifications, position monitoring, and situational updates. Moreover, 6G networks facilitate the integration of cutting-edge technology, such as artificial intelligence (AI) algorithms, surveillance systems, and Internet of Things (IoT) devices, improving risk assessment, threat identification, and surveillance capabilities. With 6G networks, real-time video feeds, biometric data, and sensor information can all be quickly delivered. This allows for more effective and efficient decision-making as well as actionable insights. Communities can benefit from more excellent emergency response, improved disaster readiness, and increased general safety and security by utilising 6G networks to modernise public safety systems.

20.10.2 Real-Time Video Streaming and Sensor Data

In many areas, real-time video streaming and sensor data are essential, and their capabilities are being revolutionised by the incorporation of cutting-edge technology like 6G networks. 6G networks allow continuous and uninterrupted real-time video streaming, making viewing live video feeds from surveillance cameras, drones, and other sources possible. These networks also have extremely low latency and fast speeds. Applications for this are extensive and include traffic control, public safety, and remote monitoring. Instantaneous situational awareness made possible by real-time video streaming facilitates quicker reaction times and better decision-making. Furthermore, 6G networks facilitate the transfer of sensor data from various sources, such as smart devices, Internet of Things (IoT) devices, and environmental sensors. The real-time sensor data offers insights into several aspects, including air quality, temperature, humidity, and motion detection. It makes real-time data monitoring and analysis possible for individuals and organisations for smart home systems, industrial automation, and environmental monitoring. Combining sensor data with real-time video streaming via 6G networks creates new opportunities for improved situational awareness, real-time tracking, and data-driven decision-making in various industries.

20.10.3 AI and Machine Learning Integration

Significant breakthroughs in several industries are being made possible by the fusion of cutting-edge technology like 6G networks with AI (Artificial Intelligence) and machine learning [27]. 6G networks, with their ultra-fast speeds, ultra-low latency, and high connectivity, offer the infrastructure required for adequate data processing and transmission, which fosters the development of AI and machine learning algorithms. Massive volumes of data may be analysed and interpreted in real-time using 6G networks with AI and machine learning, providing insightful information and enabling well-informed decision-making. This integration can be used by healthcare, finance, manufacturing, and transportation sectors to boost productivity, encourage innovation, and optimise procedures. Algorithms for artificial intelligence (AI) and machine learning can be used to automate complex operations, find patterns, identify anomalies, and forecast results. This can increase productivity and save costs. Additionally, personalised experiences like recommendation engines, virtual assistants, and targeted advertising are made possible by integrating AI and machine learning with 6G networks. These services are provided to individuals based on their interests and behavior. The future of a technology-driven society can be significantly influenced by the integration of AI and machine learning with 6G networks, which has the potential to revolutionise industries and accelerate technological progress.

20.10.4 Predictive Analytics and Emergency Preparedness

The power of cutting-edge technologies like 6G networks combined with predictive analytics transforms disaster response and preparedness. 6G networks allow massive volumes of data from many sources to be collected, analysed, and transmitted in real time because of their breakneck speeds, low latency, and high connection. Predictive analytics algorithms applied to this data can yield insightful analysis and forecasts about future emergencies and their effects. Predictive analytics can more accurately identify patterns, spot anomalies, and foresee possible emergencies by examining past data, weather patterns, social media feeds, and sensor information. This facilitates the proactive allocation of resources, evacuation plans, and mitigation actions by authorities and emergency responders before the occurrence of an incident. Moreover, emergency responders can communicate and work together more easily using 6G networks, which enables real-time information sharing, resource coordination, and decision-making. Predictive analytics combined with 6G networks allows law enforcement to anticipate events better, act quickly, and lessen their effects, eventually saving lives and protecting communities.

20.11 ETHICAL CONSIDERATIONS AND CHALLENGES

As advanced technologies like 6G networks continue to evolve, it is essential to consider their ethical implications and address the associated challenges. Ethical considerations [28] arise from various aspects, including data privacy, security, algorithmic bias, and social impact. Protecting individuals' privacy rights while harnessing the power of data is a critical challenge. Striking the right balance between data collection and usage for societal benefit while respecting privacy is essential. Moreover, ensuring the security of 6G networks is vital to prevent data breaches and cyber-attacks. Algorithmic bias and fairness need to be addressed to avoid perpetuating social inequalities and discrimination. Transparent and accountable AI systems are necessary to ensure that decision-making processes are fair and unbiased. Additionally, the ethical use of emerging technologies requires addressing the impact on labor markets, employment, and social norms.

Collaborative efforts between policymakers, industry leaders, researchers, and the public are crucial to establishing robust ethical frameworks, regulations, and guidelines. By actively addressing ethical considerations and challenges, we can shape the development and deployment of 6G networks in a way that respects human rights, promotes social good, and fosters trust and confidence in these transformative technologies.

20.11.1 Privacy and Data Security

The emergence of advanced technologies like 6G networks brings with it a range of ethical considerations and challenges. As these technologies become more integrated into our daily lives, it is essential to address issues such as data privacy [12], security, and fairness. The massive amount of data generated and transmitted through 6G networks raises concerns about the collection, storage, and usage of personal information. Striking a balance between utilising data for societal benefit while respecting individuals' privacy rights becomes a significant challenge. Additionally, ensuring the security of 6G networks becomes crucial to prevent unauthorised access, data breaches, and cyber-attacks. The fairness and transparency of AI algorithms are also ethically significant since they influence decision-making procedures and strengthen prejudice or discrimination. It is imperative to create ethical frameworks, rules, and guidelines to address these ethical issues and regulate the responsible use of 6G technology. Policymakers, industry stakeholders, and the general public must work together to resolve these moral quandaries and guarantee that 6G networks are implemented in a way that upholds people's rights and values, encourages inclusivity, and is transparent.

20.11.2 Digital Divide and Inclusivity

Equality and the digital gap are urgent issues that need to be addressed in light of cutting-edge technology development like 6G networks. 6G networks have great promise for innovation and connection, but it's essential to ensure everyone can benefit from them. The term "digital divide" describes the differences in opportunities, educational attainment, and socioeconomic advancement between people with access to digital technologies and those without access. Contrarily, inclusivity highlights the necessity of giving everyone equal access to opportunities, irrespective of location, socioeconomic background, or other demographic characteristics. It will take coordinated measures to close the digital divide and advance inclusivity, including building broadband infrastructure, lowering the cost of internet access, and offering training and digital literacy. It is imperative to mitigate socio-economic obstacles, language difficulties, and cultural sensitivities to guarantee that a diverse range of people may reap the advantages of 6G networks. By prioritising inclusivity and tackling the digital divide, it is possible to fully realise the potential of 6G networks, promote equitable development, and empower people worldwide.

20.11.3 Environmental Impact and Sustainability

It is crucial to consider how cutting-edge technologies, like 6G networks, will affect the environment and improve sustainability as they develop. Significant energy resources are needed to install and maintain 6G networks, which may increase carbon emissions and worsen environmental conditions. As a result, when developing and running 6G networks, prioritising sustainable practices, renewable energy sources, and energy-efficient infrastructure is essential. It is also essential to consider the life cycle of infrastructure and electronic devices to reduce electronic waste, including responsible manufacturing, recycling, and disposal methods. Additionally, 6G networks can improve conservation, resource management, and environmental monitoring. 6G networks can enable real-time monitoring of biodiversity, water, and air quality, resulting in educated decision-making and proactive environmental protection actions. This can be achieved by integrating environmental sensors, IoT devices, and data analytics. 6G networks can assist in solving global environmental concerns and contribute to a greener future by using a holistic approach that balances technology growth and environmental sustainability.

20.11.4 Regulation and Policy Development

To fully realise the potential of cutting-edge technologies like 6G networks and ensure their responsible and ethical use, regulation and policy formulation [29] are essential. Given how quickly technology is developing, legislators must continue proactively adjusting laws to take advantage of new opportunities and difficulties. Several factors, including data privacy, security, competitiveness, net neutrality, and ethical considerations, should be the main emphasis of effective regulation and policy creation. Achieving a balance between advancing innovation and protecting the rights and interests of individuals and the community at large is imperative. Government agencies, business partners, and academic institutions must work together to create robust frameworks that uphold consumer rights, encourage fair competition, and handle new moral and social issues. Furthermore, international collaboration and standardisation initiatives are required to guarantee interoperability, harmonisation, and consistency across several locations. Policymakers may create an environment that maximises the benefits of 6G networks while protecting public interests and advancing societal well-being by enacting comprehensive and progressive regulations and policies.

20.11.5 Recap of 6G Applications and Use Cases

The emergence of 6G networks promises a wide range of applications and use cases that have the potential to revolutionise various sectors. Ultra-fast and reliable connectivity will unleash unprecedented speeds, enabling seamless streaming, gaming, and immersive experiences. Advancements in data transfer and sharing will facilitate efficient communication and collaboration while expanding the horizons of the Internet of Things (IoT) and enabling smart cities, industrial automation, and personalised healthcare. Holographic communication and telepresence will enable realistic virtual interactions, transforming the way we communicate and collaborate. Real-time collaboration and teleconferencing will enhance remote work and global connectivity. Education and remote learning will be enriched with interactive and personalised experiences. Healthcare will benefit from virtual surgeries, remote monitoring, and improved access to medical services. Autonomous vehicles and intelligent transportation systems will transform transportation, while traffic management and road safety will be enhanced through real-time updates and predictive analytics. Moreover, advanced healthcare services, including real-time monitoring and personalised healthcare, will become more accessible. Applications in rural and underserved areas will bridge the digital divide and promote inclusivity. Environmental monitoring, sustainability, and smart grids will contribute to a greener future. The potential of 6G networks is vast, and as they continue to develop, we can expect to witness significant advancements and transformative changes across various domains.

20.11.6 Anticipating the Societal Impact

Anticipating the societal impact [30] of advanced technologies like 6G networks is crucial as we embark on this technological revolution. While 6G networks hold immense potential for connectivity, innovation, and economic growth, it is essential to consider their wider implications on society. The deployment and integration of 6G networks will bring about transformative changes in various aspects of our lives, including communication, healthcare, transportation, education, and more. Recognising potential advantages and difficulties and taking proactive measures to overcome them are part of anticipating the societal impact. Inclusion must be promoted to provide fair access to 6G networks and close the digital divide. Protecting people's rights and upholding public confidence also means prioritising ethics, security, and data privacy. It also calls for rethinking the nature of labor in the era of automation and artificial intelligence and addressing the possibility of job displacement. By foreseeing the social implications of 6G networks and formulating approaches and regulations that foster favorable consequences, we can harness the advantages of this innovation to construct a more enduring, equitable, and thriving community.

20.11.7 Looking Ahead: Opportunities and Challenges

As we look to the future, many potential problems are associated with the emergence of new technology, such as 6G networks. Regarding prospects, 6G networks have the potential to unleash hitherto unheard-of speeds, connections, and data capacities, resulting in groundbreaking developments across a range of industries, including environmental monitoring, healthcare, transportation, and education. These networks have the power to transform entire industries, boost productivity, and raise standards of living. But these prospects also provide essential obstacles. Ethical issues such as algorithmic fairness and data privacy must be considered to guarantee responsible use. It is imperative to foster diversity and bridge the digital divide to prevent societal disparities from worsening. To reduce carbon emissions and electronic waste, 6G networks' sustainability and environmental impact must also be monitored appropriately. Policymakers, industry leaders, and society as a whole must collaborate to navigate these challenges, seize the opportunities, and shape the future of 6G networks in a way that benefits humanity and preserves the well-being of the planet. By proactively addressing these challenges, we can maximise the potential of 6G networks and create a more equitable, sustainable, and technologically advanced future.

REFERENCES

[1] Tomkos, Ioannis, et al. "Toward the 6G network era: Opportunities and challenges." *IT Professional* 22.1 (2020): 34–38.

[2] Shahraki, Amin, et al. "A comprehensive survey on 6G networks: Applications, core services, enabling technologies, and future challenges." arXiv preprint arXiv:2101.12475 (2021).

[3] Chowdhury, Mostafa Zaman, et al. "The role of optical wireless communication technologies in 5G/6G and IoT solutions: Prospects, directions, and challenges." *Applied Sciences* 9.20 (2019): 4367.

[4] Osseiran, Afif, Jose F. Monserrat, and Patrick Marsch, eds. *5G Mobile and Wireless Communications Technology*. Cambridge University Press, 2016.

[5] Hu, Fei, ed. *Opportunities in 5G Networks: A Research and Development Perspective*. CRC Press, 2016.

[6] Elmeadawy, Samar, and Raed M. Shubair. "6G wireless communications: Future technologies and research challenges." *2019 International Conference on Electrical and Computing Technologies and Applications (ICECTA)*. IEEE, 2019.

[7] Jiang, Wei, et al. "The road towards 6G: A comprehensive survey." *IEEE Open Journal of the Communications Society* 2 (2021): 334–366.

[8] Zhang, Miao, et al. "Toward 6G-enabled mobile vision analytics for immersive extended reality." *IEEE Wireless Communications* 30.3 (2023): 132–138.

[9] Kaiser, M. Shamim, et al. "6G access network for intelligent internet of healthcare things: Opportunity, challenges, and research directions." *Proceedings of International Conference on Trends in Computational and Cognitive Engineering: Proceedings of TCCE 2020.* Springer Singapore, 2020.

[10] Deng, Xiaoheng, et al. "A review of 6G autonomous intelligent transportation systems: Mechanisms, applications and challenges." *Journal of Systems Architecture* 142 (2023): 102929.

[11] De Alwis, Chamitha, et al. "Survey on 6G frontiers: Trends, applications, requirements, technologies and future research." *IEEE Open Journal of the Communications Society* 2 (2021): 836–886.

[12] Siriwardhana, Yushan, et al. "AI and 6G security: Opportunities and challenges." *2021 Joint European Conference on Networks and Communications & 6G Summit (EuCNC/6G Summit).* IEEE, 2021.

[13] Dala Pegorara Souto, Victoria, et al. "Emerging MIMO technologies for 6G networks." *Sensors* 23.4 (2023): 1921.

[14] Sarkar, Pia, and Arijit Saha. "Evolution of 6G and terahertz communication." In: Saha, A., Biswas, A., Ghosh, K., Mukhopadhyay, N. (eds) *Optical to Terahertz Engineering.* Singapore: Springer Nature Singapore, 2023, 45–58.

[15] Bui, Van-Phuc, et al. "Game networking and its evolution towards supporting metaverse through the 6G wireless systems." arXiv preprint arXiv:2302.01672 (2023).

[16] Nguyen, Dinh C., et al. "6G Internet of Things: A comprehensive survey." *IEEE Internet of Things Journal* 9.1 (2021): 359–383.

[17] Lv, Zhihan, et al. "Big data analytics for 6G-enabled massive internet of things." *IEEE Internet of Things Journal* 8.7 (2021): 5350–5359.

[18] Kulkarni, Rachana, and Spoorthi Kulkarni. "How 6G has an influence on smart cities: An overview." *International Journal of Engineering Research & Technology* 10.5 (2021): 832–838.

[19] Tanwar, Sudeep, et al. "Blockchain-assisted industrial automation beyond 5G networks." *Computers & Industrial Engineering* 169 (2022): 108209.

[20] Nayak, Sabuzima, and Ripon Patgiri. "6G communication technology: A vision on intelligent healthcare." *Health Informatics: A Computational Perspective in Healthcare* 932 (2021): 1–18.

[21] Zhang, Fan, et al. "6G-enabled smart agriculture: A review and prospect." *Electronics* 11.18 (2022): 2845.

[22] De Alwis, Chamitha, et al. "Towards 6G: Key technological directions." *ICT Express* 9 (2022).

[23] Ali, Ashraf, and Abdullah Alourani. "An investigation of cloud computing and E-learning for educational advancement." *International Journal of Computer Science & Network Security* 21.11 (2021): 216–222.

[24] Kharche, Shubhangi, and Jayshree Kharche. "6G intelligent healthcare framework: A review on role of technologies, challenges and future directions." *Journal of Mobile Multimedia* 19.3 (2023): 603–644.

[25] Song, W., Rajak, S., Chinnadurai, S., Dang, S., Liu, R., & Li, J. (2022). Deep Learning Enabled IRS for 6G Intelligent Transportation Systems: A Comprehensive Study. *IEEE Transactions on Intelligent Transportation Systems*, 24(11), 12973–12990.

[26] Morandi, Filippo, et al. "A probabilistic codebook technique for fast initial access in 6G vehicle-to-vehicle communications." *2021 IEEE International Conference on Communications Workshops (ICC Workshops)*. IEEE, 2021.

[27] Mahmood, M. Rezwanul, et al. "A comprehensive review on artificial intelligence/machine learning algorithms for empowering the future IoT toward 6G era." *IEEE Access* 10 (2022): 87535–87562.

[28] Srivastava, Vandana, Tripti Mahara, and Pooja Yadav. "An analysis of the ethical challenges of blockchain-enabled E-healthcare applications in 6G networks." *International Journal of Cognitive Computing in Engineering* 2 (2021): 171–179.

[29] Matinmikko-Blue, Marja, Seppo Yrjölä, and Petri Ahokangas. "Spectrum management in the 6G era: The role of regulation and spectrum sharing." *2020 2nd 6G Wireless Summit (6G SUMMIT)*. IEEE, 2020.

[30] Yrjölä, Seppo, Petri Ahokangas, and Marja Matinmikko-Blue. "Visions for 6G futures: A causal layered analysis." *2022 Joint European Conference on Networks and Communications & 6G Summit (EuCNC/6G Summit)*. IEEE, 2022.

Index